PROPERTY OF
OKLAHOMA CITY COMMUNITY COLLEGE
KEITH LEFTWICH MEMORIAL LIBRARY

WIRELESS SENSOR NETWORKS

WIRELESS SENSOR NETWORKS

KAP ARCHIEF

Edited by

C. S. RAGHAVENDRA
University of Southern California

KRISHNA M. SIVALINGAM
University of Maryland, Baltimore County

TAIEB ZNATI
National Science Foundation / University of Pittsburgh

PROPERTY OF
**OKLAHOMA CITY COMMUNITY COLLEGE
KEITH LEFTWICH MEMORIAL LIBRARY**

Kluwer Academic Publishers
Boston/Dordrecht/London

Distributors for North, Central and South America:
Kluwer Academic Publishers
101 Philip Drive
Assinippi Park
Norwell, Massachusetts 02061 USA
Telephone (781) 871-6600
Fax (781) 871-6528
E-Mail: <kluwer@wkap.com>

Distributors for all other countries:
Kluwer Academic Publishers Group
Post Office Box 322
3300 AH Dordrecht, THE NETHERLANDS
Telephone 31 78 6576 000
Fax 31 78 6576 254
E-Mail: <services@wkap.nl>

 Electronic Services <http://www.wkap.nl>

Library of Congress Cataloging-in-Publication Data

Wireless Sensor Networks
C. S. Raghavendra, Krishna M. Sivalingam, Taieb Znati (Eds.)
ISBN 1-4020-7883-8
eBook ISBN 1-4020-7884-6

Copyright © 2004 by Kluwer Academic Publishers

All rights reserved. No part of this work may be reproduced, stored in a retrieval system, or transmitted in any form or by any means, electronic, mechanical, photocopying, microfilming, recording, or otherwise, without prior written permission from the Publisher, with the exception of any material supplied specifically for the purpose of being entered and executed on a computer system, for exclusive use by the purchaser of the work.

Permission for books published in Europe: permissions@wkap.nl
Permissions for books published in the United States of America: permissions@wkap.com

Printed on acid-free paper.

Printed in the United States of America

This book is dedicated to our families.

Contents

Dedication v
Contributing Authors xi
Preface xiii

Part I BASICS

1
Sensor Networks: A Bridge to the Physical World 3
Jeremy Elson and Deborah Estrin

2
Communication Protocols for Sensor Networks 21
Weilian Su, Özgür B. Akan, and Erdal Cayirci

3
Energy Efficient Design of Wireless Sensor Nodes 51
Vijay Raghunathan, Curt Schurgers, Sung Park and Mani Srivastava

Part II NETWORK PROTOCOLS

4
Medium Access Control in Wireless Sensor Networks 73
Wei Ye and *John Heidemann*

5
A Survey of MAC Protocols for Sensor Networks 93
Piyush Naik and Krishna M. Sivalingam

6
Dissemination Protocols for Large Sensor Networks 109
Fan Ye, Haiyun Luo, Songwu Lu and Lixia Zhang

7
Routing On A Curve 129
Dragos Niculescu and Badri Nath

8
Reliable Transport For Sensor Networks ... 153
Chieh-Yih Wan, Andrew T. Campbell and Lakshman Krishnamurthy

Part III DATA STORAGE AND MANIPULATION

9
Data-centric Routing and Storage in Sensor Networks ... 185
Ramesh Govindan

10
Compression Techniques for Wireless Sensor Networks ... 207
Caimu Tang and Cauligi S. Raghavendra

11
Fundamental Limits of Networked Sensing ... 233
Bhaskar Krishnamachari and Fernando Ordóñez

Part IV SECURITY

12
Security for Wireless Sensor Networks ... 253
Sasikanth Avancha, Jeffrey Undercoffer, Anupam Joshi and John Pinkston

13
Key Distribution Techniques for Sensor Networks ... 277
Haowen Chan, Adrian Perrig, and Dawn Song

14
Security in Sensor Networks: Watermarking Techniques ... 305
Jennifer L. Wong, Jessica Feng, Darko Kirovski and Miodrag Potkonjak

Part V LOCALIZATION AND MANAGEMENT

15
Localization in Sensor Networks ... 327
Andreas Savvides, Mani Srivastava, Lewis Girod and Deborah Estrin

16
Sensor Management ... 351
Mark Perillo and Wendi Heinzelman

Part VI APPLICATIONS

17
Detecting Unauthorized Activities Using A Sensor Network ... 375
Thomas Clouqueur, Parameswaran Ramanathan and Kewal K. Saluja

18
Analysis of Wireless Sensor Networks for Habitat Monitoring 399
Joseph Polastre, Robert Szewczyk, Alan Mainwaring, David Culler and John Anderson

Index 425

Contributing Authors

Ozgur Akan, *Georgia Institute of Technology*
John Anderson, *College of the Atlantic, Bar Harbor*
Sasikanth Avancha, *University of Maryland, Baltimore County*
Andrew Campbell, *Columbia University*
Erdal Cayirci, *Istanbul Technical University*
Haowen Chan, *Carnegie Mellon University*
Thomas Clouqueur, *University of Wisconsin, Madison*
David Culler, *University of California, Berkeley*
Jeremy Elson, *University of California, Los Angeles*
Deborah Estrin, *University of California, Los Angeles*
Jessica Feng, *University of Southern California, Los Angeles*
Lewis Girod, *University of California, Los Angeles*
Ramesh Govindan, *University of Southern California, Los Angeles*
John Heidemann, *University of Southern California, Los Angeles*
Wendi Heinzelman, *University of Rochester*
Anupam Joshi, *University of Maryland, Baltimore County*
Darko Kirovski, *Microsoft Research*
Bhaskar Krishnamachari, *University of Southern California, Los Angeles*
Lakshman Krishnamurthy, *Intel Labs*
Songwu Lu, *University of California, Los Angeles*
Haiyun Luo, *University of California, Los Angeles*
Alan Mainwaring, *Intel Research Lab, Berkeley*
Piyush Naik, *University of Maryland, Baltimore County*
Badri Nath, *Rutgers, The State University of New Jersey*
Dragos Niculescu, *Rutgers, The State University of New Jersey*

Fernando Ordonez, *University of Southern California, Los Angeles*
Sung Park, *University of California, Los Angeles*
Mark Perillo, *University of Rochester*
Adrian Perrig, *Carnegie Mellon University*
John Pinkston, *University of Maryland, Baltimore County*
Joseph Polastre, *University of California, Berkeley*
Miodrag Potkonjak, *University of California, Los Angeles*
Cauligi Raghavendra, *University of Southern California, Los Angeles*
Vijay Raghunathan, *University of California, Los Angeles*
Parameswaran Ramanathan, *University of Wisconsin, Madison*
Kewal Saluja, *University of Wisconsin, Madison*
Andreas Savvides, *Yale University*
Curt Schurgers, *University of California, Los Angeles*
Krishna Sivalingam, *University of Maryland, Baltimore County*
Mani Srivastava, *University of California, Los Angeles*
Weilian Su, *Georgia Institute of Technology*
Robert Szewczyk, *University of California, Berkeley*
Caimu Tang, *University of Southern California, Los Angeles*
Jeffrey Undercoffer, *University of Maryland, Baltimore County*
Chieh-Yih Wan, *Columbia University*
Jennifer Wong, *University of Southern California, Los Angeles*
Fan Ye, *University of California, Los Angeles*
Wei Ye, *University of Southern California, Los Angeles*
Lixia Zhang, *University of California, Los Angeles*

Preface

Wireless Sensor Networks is a fast growing and exciting research area that has attracted considerable research attention in the recent past. This has been fueled by the recent tremendous technological advances in the development of low-cost sensor devices equipped with wireless network interfaces. The creation of large-scale sensor networks interconnecting several hundred to a few thousand sensor nodes opens up several technical challenges and immense application possibilities. Sensor networks find applications spanning several domains including military, medical, industrial, and home networks. Wireless sensor networks have moved from the research domain into the real world with the commercial availability of sensors with networking capabilities. Companies such as Crossbow (www.xbow.com) and Sensoria (www.sensoria.com) have emerged as suppliers of the necessary hardware and software building blocks.

Some of the key challenges deal with the scalability of network protocols to large number of nodes, design of simple and efficient protocols for different network operations, design of power-conserving protocols, design of data handling techniques including data querying, data mining, data fusion and data dissemination, localization techniques, time synchronization and development of exciting new applications that exploit the potential of wireless sensor networks.

The purpose of this book is to present a collection of excellent chapters from leading researchers on various aspects of wireless sensor networks. The submitted articles were closely reviewed by the editors and their graduate students. The published articles are the revised versions based on these comments. We realize that technology is so rapidly changing that it is nearly impossible to keep up with the changes. However, we have endeavored to capture a substantial subset of the key problems and known solutions to these problems. Some of the chapters present a survey of past research on important problems while others deal with specific problems and solutions.

The potential audience for this book includes practitioners who are interested in recent research work that has not appeared in textbooks; those who are interested in survey articles on specific topics without having to pore over larger textbooks; and graduate students and others who are starting research in this field. Our hope is that the reader gains valuable insight into the mainstream ideas behind the technology and is inspired to go forth and innovate new ideas and technologies.

ORGANIZATION OF THE BOOK

The book is organized into six parts. Part I presents the motivation, basic concepts and introductory material on wireless sensor networks. The first chapter, by Elson and Estrin, presents on overview of sensor networks including the basic facilitating technologies and components and the challenges faced. Chapter 2, by Su. et. al. presents a comprehensive survey of the necessary sensor networking protocols. Raghunathan et. al. present the hardware aspects of sensor networks with particular emphasis on energy efficient hardware design.

Part II delves more into networking protocols for sensor networks. A comprehensive survey of medium access protocols developed for sensor networks is presented in Chapter 4 (by Ye and Heidemann) and Chapter 5 (by Naik and Sivalingam). Routing protocols that are necessary for data dissemination are discussed in Chapter 5 by Ye et. al. Niculescu and Badrinath present a novel approach to routing in sensor networks based on a technique called trajectory-based forwarding. The issue of reliable transport protocol design is addressed by Wan, Campbell and Krishnamurthy in Chapter 8.

In addition to networking, data management is an important challenge given the high volumes of data that are generated by sensor nodes. Part III contains three chapters related to data storage and manipulation in sensor networks. Chapter 9 by Govindan discusses recent work on data-centric routing and storage. Tang and Raghavendra then present data compression techniques based on spatial and temporal correlation. Chapter 10, by Krishnamachari and Ordonez, presents a theoretical approach to identify fundamental performance limits on information routing in sensor networks.

Part IV deals with security protocols and mechanisms for wireless sensor networks. Chapter 12 by Avancha et. al. presents on overview of recent security related research and a framework for implementing security in wireless sensor networks. Chapter 13, by Chan, Perrig and Song, reviews several key distribution and key establishment techniques for sensor networks. Security based on watermarking techniques is presented in Chapter 14 by Wong et. al.

Part V deals with localization and management techniques. A survey of localization techniques is first presented in the chapter by Savvides et. al. Chapter 16, by Perillo and Heinzelman, presents a discussion of sensor management techniques that includes assigning different roles to the sensor nodes (forwarding or sensing) and topology control.

Part VI presents chapters related to applications of sensor networks. The first chapter by Clouqueur et. al. deals with target detection techniques based on sensor networks. The final chapter presents the details of a habitat monitoring project in Great Duck Island off the coast of Maine.

ACKNOWLEDGMENTS

We would like to acknowledge the help of Uttara Korad whose help was monumental in the compilation of the book chapters. She spent countless hours working with the word processing software to get everything just right. We also like to acknowledge the valuable assistance of Minal Mishra, Piyush Naik, Shantanu Prasade, Aniruddha Rangnekar, Rama Shenai, Manoj Sivakumar, Mahesh Sivakumar and Sundar Subramani, graduate students at University of Maryland, Baltimore County. We also gratefully acknowledge our research sponsors – Cisco Systems, National Science Foundation, Air Force Office of Scientific Research, Intel Corporation and UMBC. We are also grateful to Alex Greene and Melissa Sullivan at Kluwer Academic Publishers for their help and patience, without which this book would not have been possible.

Prof. C. S. Raghavendra, *Professor*
University of Southern California
Email: raghu@usc.edu

Prof. Krishna Sivalingam, *Associate Professor*
University of Maryland, Baltimore County
Email: krishna@umbc.edu, krishna_siva@ieee.org

Prof. Taieb Znati, *Program Director/Professor*
NSF/University of Pittsburgh
Email: tznati@nsf.gov, znati@cs.pitt.edu

I

BASICS

Chapter 1

SENSOR NETWORKS: A BRIDGE TO THE PHYSICAL WORLD

Jeremy Elson and Deborah Estrin
Center for Embedded Networked Sensing
University of California, Los Angeles
Los Angeles, CA 90095
{jelson,destrin}@cs.ucla.edu

Abstract

Wireless sensor networks are a new class of distributed systems that are an integral part of the physical space they inhabit. Unlike most computers, which work primarily with data created by humans, sensor networks reason about the state of the world that embodies them. This bridge to the physical world has captured the attention and imagination of many researchers, encompassing a broad spectrum of ideas, from environmental protection to military applications. In this chapter, we will explore some of this new technology's potential and innovations that are making it a reality.

Keywords: Wireless Sensor Networks, Sensor Network Applications, In-network processing, automatic localization, time synchronization, actuation, environmental monitoring, distributed signal processing, security and privacy.

1.1 THE QUAKE

It was in the early afternoon of an otherwise unremarkable Thursday that the Great Quake of 2053 hit Southern California.

The earth began to rupture several miles under the surface of an uninhabited part of the Mohave desert. Decades of pent-up energy was violently released, sending huge shear waves speeding toward greater Los Angeles. Home to some 38 million people, the potential for epic disaster might be only seconds away. The quake was enormous, even by California standards, as its magnitude surpassed 8 on the Richter scale.

Residents had long ago understood such an event was possible. This area was well known for its seismic activity, and had been heavily instrumented by scientists for more than a century. The earliest data collection had been primitive, of course. In the 1960's, seismometers were isolated devices, each simply recording observations to tape for months at a time. Once or twice a year, seismologists of that era would spend weeks traveling to each site, collecting the full tapes and replacing them with fresh blanks. If they were lucky, each tape would contain data from the entire six months since their last visit. Sometimes, they would instead discover only a few hours of data had been recorded before the device had malfunctioned. But, despite the process being so impractical, the data gathered were invaluable—revealing more about the Earth's internal structure than had ever been known before.

By the turn of the century, the situation had improved considerably. Many seismometers were connected to the Internet and could deliver a continuous stream of data to scientists, nearly in real-time. Experts could analyze earthquakes soon after they occurred, rather than many months later. Unfortunately, instrumenting undeveloped areas this way remained a problem, as networked seismometers could only be deployed where infrastructure was available to provide power and communication.

Some researchers at that time worked on early-warning systems that would sound an alarm in a city moments before an earthquake's arrival. If a sensor was close enough to the epicenter, and the epicenter was far enough from a population center, the alarm could be raised 20 or 30 seconds before the city started to shake. The idea was promising. But the lack of a pervasive sensor-support infrastructure was a serious impediment to deploying large-scale networks in an area like the Mohave desert.

But, in the half-century leading up to the Great Quake of 2053, technological advances changed everything. Pervasive infrastructure was no longer required for pervasive sensing. And, perhaps equally important, the job of responding to an alarm was no longer solely the domain of *people*. With the help of its new sensory skin, the city itself could protect its inhabitants.

By the mid 2040's, the vast, desolate expanse of the desert floor was home to nearly a million tiny, self-contained sensors. Each had a processor, memory, and radio that could be used to communicate and co-operate with others that were nearby. Most were almost invisibly small. There were dozens of varieties, each with a different observational specialty. In the desert, the atmospheric, chemical, organic, and seismic sensors were most popular.

It was just a few dozen of those seismometers—closest to the epicenter—that first sensed unusual acceleration in the ground, nearly the instant that shock from the Great Quake reached the desert's surface. Individual sensors could not be trusted, of course, but as the number of confirmed observations grew, so did the likelihood that this event was not simply random noise, or a malfunction. It was real. But what was it?

In those first few milliseconds, information was sketchy. To conserve energy, many sensors had been off. Those that happened to be active were only monitoring the ground with low fidelity. The network did not yet have enough detailed data to distinguish an earthquake from the demolition of a far-away building, or a boulder rolling down a hill.

More data were needed quickly, from a much larger area. Electronic word of the anomaly spread. In a few tenths of a second, the earth's movement had the full attention of thousands of seismometers within a few miles of the epicenter. In seconds, most saw the same shock waves as they raced past with a frightening magnitude. The network soon reached consensus: this was an earthquake. It was a dangerous one.

Small earthquakes are commonplace, and typically only of academic interest. For efficiency's sake, information of such quakes might not be transmitted outside of the desert for a minute or two. But a quake as big as that one in 2053 was a matter of public safety and time was of the essence. Seismometers immediately recruited the neighboring sensors that had the fastest, longest-range, highest-power radios. Several summarized the important details and sent urgent messages on their way. The information hopped from one node to the next, streaking wirelessly across the desert floor. After 41 miles, it finally reached the first sign of civilization: a wired communication access point. Four seconds had passed since the quake began. Once on the wired grid, the alarm spread almost instantly to every city in the area.

The new generation of smart structures in Los Angeles learned of the quake nearly thirty seconds before it arrived. Thousands of bridges, freeways, high-rise buildings, underground pipelines, and even some private homes were informed of the important details, such as the impending shock's magnitude and frequency composition. The alarm reached a dizzying array of embedded computers, from millions of microactuators attached to internal beams to the enormous adjustable shock-absorbers in building foundations. Structures throughout the city quickly de-tuned themselves so as to prevent resonance and collapse. These electronic earthquake countermeasures had been mandated by building codes for more than twenty years.

Tense moments passed. Sirens blared as every traffic light turned red and every elevator stopped and opened at the nearest floor. The city seemed to stand still, holding its breath. Finally, the silence was broken, at first by a low rumble, then a deafening roar. The earth rolled and shook violently. Sensors on the ground and within each structure monitored the dynamics of the motion. Each continued to make defensive adjustments. Swaying and groaning, buildings and bridges were strained, but most survived, keeping their occupants unharmed.

Even with the countermeasures, the city could not completely escape damage. Older buildings were particularly susceptible, and many older homes collapsed. Rescue crews arrived quickly with Portable Emergency Survivor Locators. Each was a nylon package the size of a wine bottle, containing thousands of tiny, self-propelled sensors that could disperse themselves as the package was thrown over or inside of the target area. Sensors were automatically activated when landing, and began to cooperatively explore their environment. Back at the rescue truck, a map of the structure began to appear. The structure itself was mapped using radar and acoustic reflections. People were visible as heat sources. As sensors penetrated deeper into the the structure, the map grew and gained additional detail.

Sensors throughout the city's water system went onto high alert, ready to divert contaminants to a safe disposal area if any toxins were detected above the threshold safe for human exposure. An hour after the quake, chemical sensors in several older residential areas began to detect abnormal traces of natural gas—the result of ruptures in the labyrinth of pipes that snaked beneath the city. The newest pipes had sensors embedded every few inches along their outer surface, allowing instant and unambiguous location of leaks. But, many of the older pipes had not yet been replaced, so the far coarser job of detecting leaks with in-ground sensors began. After several minutes, the arrays were able to compute three-dimensional concentration gradients. They hypothesized the most likely set of points where pipes were broken. The upstream valves nearest to each point were commanded to close. On one city block, the shutdown did not come in time; the leak contaminated the area, risking an explosion. The sensor array warned residents and dispatched a crew to clean up the spill.

Meanwhile, in a forgotten corner of Los Angeles, a small fire started in an abandoned lumber yard. In another era, under the cover of the chaos, the flame might have gone unnoticed for hours, blazing wildly out of control. This particular fire, however, could not escape attention. Some years before, local high school students had scattered sensors that measured temperature and airborne particulates. They'd gathered data for a one-week class project on air pollution, but even now, unmaintained and long forgotten, some sensors still functioned. The combination of smoke and heat provoked their fire-warning reflex. A few miles down the road, in a sleepy volunteer fire department, a map to the yard popped up on a screen.

By Monday, Southern California had returned to normal. The 2053 quake came and went, thanks largely due to the pervasive sensors that had been woven into both our technological fabric and the natural environment. Even 50 years earlier, a similar quake would have caused widespread destruction. To many people in 2003, using technology to prevent such a catastrophe must have seemed fanciful and improbable. It seemed as improbable as 2003's globally interconnected and instantly searchable network of nearly a billion commodity computers must have seemed to those in 1953. Indeed, perhaps even as improbable as 1953's "electronic brain" must have seemed to those in 1903, who would soon experience their own Great Earthquake in San Francisco.

1.2   OBSERVATION OF THE PHYSICAL WORLD

Automatic detection, prevention, and recovery from urban disasters is but one of many potential uses for an emerging technology: *wireless sensor networks*. Sensor networks have captured the attention and imagination of many researchers, encompassing a broad spectrum of ideas. Despite their variety, all sensor networks have certain fundamental features in common. Perhaps most essential is that they are embedded in the real world. Sensors *detect* the world's physical nature, such as light intensity, temperature, sound, or proximity to objects. Similarly, actuators *affect* the world in some way, such as toggling a switch, making a noise, or exerting a force. Such a close relationship with the physical world is a dramatic contrast to much of traditional computing, which often exists in a virtual world. Virtual-world computers deal exclusively in the currency of information invented by humans, such as e-mail, bank balances, books and digital music.

Sensor networks are also large collections of nodes. Individually, each node is autonomous and has short range; collectively, they are cooperative and effective over a large area. A system composed of many short-range sensors lends itself to a very different set of applications than one that uses a small number of powerful, long-range sensors. The difference can be illustrated by considering an age-old question: is the sky clear or cloudy today? Answering this question is easy, even on a large scale, by using only a few long-range sensors such as satellite imagers. A satellite works because clouds have two important properties: they can be seen from far away, and a strategically placed sensor has an unobstructed view of a large area. A single satellite can detect the cloud cover of an entire hemisphere.

Such a centralized, long-range approach fail in complex, cluttered environments where line-of-sight paths are typically very short. For example, satellites are not very good at detecting individual animals in a forest, objects in a building, or chemicals in the soil. Moving to a distributed collection of shorter-range sensors can dramatically reduce the effect of clutter. By increasing the number of vantage points, it is more likely that an area will be viewable, even when line-of-sight paths are short.

Many interesting phenomena can not be effectively sensed from a long distance. For example, temperature and humidity are both very localized. Unlike clouds, they are not easily observable from afar, even by a sensor with a line of sight. Distributed sensing improves the signal-to-noise ratio because sensors are closer to the phenomena they are observing. This allows greater fidelity, and in the extreme, can reveal phenomena that are invisible when viewed from far away.

Distributed sensing has been successful for weather observations. For example, in the United States, temperature and humidity information is provided by over 600 automatic weather observation systems (AWOS) installed near airports. This distributed (but wired) sensor network produces local observations using short-range sensors. While the system is temporally continuous, it is spatially sparse, and therefore can not be used to sense phenomena at a greater resolution than several miles. This is an important and fundamental limitation: each sensor site requires a support infrastructure, including connections to the power and communications grids. The high overhead cost makes spatially dense sensing prohibitively impractical.

Wireless sensor networks can address the issue of observing the environment at close range, densely in space, and frequently in time. This has the potential to *reveal previously unobservable phenomena in the physical world*, making sensor networks attractive to a broad spectrum of scientists [7]. Historically, new tools for observing the world around and within us have heralded new eras of understanding in science, as Galileo's telescope did for astronomy, or the microscope did for biology.

1.3 TECHNOLOGICAL TRENDS

Automated sensing, embedded computing, and wireless networking are not new ideas. However, it has only been in the past few years that computation, communication, and sensing have matured sufficiently to enable their integration, inexpensively, at low power, and at large scale.

Microprocessors have undergone a slow but inexorable transformation in the decades since their introduction. In 1965, Gordon Moore famously predicted that there would be an exponential growth in the number of transistors per integrated circuit. His rule of thumb still remains true. Year after year, manufacturers vie for customers' attention with products that boast the highest speed, biggest memory, or most features.

While such high-end technology has taken center stage, a dramatic but much quieter revolution has also taken place at the low end. Today's smallest processors can provide the same computational power as high-end systems from two or three decades ago, but at a tiny fraction of the cost, size and power.

In 1969, a system called the Apollo Guidance Computer was on board the manned spacecraft that made the journey from the Earth to the Moon's surface. The AGC was custom-designed over the course of ten years, and for some time afterwards was likely the most advanced computer in existence. Its core had a 2 megahertz clock and could access 74,000 bytes of memory. Today, in 2003, a commercial off-the-shelf micro-controller with the same capabilities costs about $1, uses one ten-thousandth the power, and one hundred-thousandth the volume and mass.

That AGC-equivalent computational power has become so readily available is only part of the story. Small micro-controllers have been commonplace for so-called "embedded" applications for many years—computers, for example, that run digital radios, engine valve timing, and thermostats. Indeed, some 97% of processors manufactured are destined for such applications, rather than for personal computers on desks. But embedded processors have typically lived in isolation, each working independently. Without communication, their sole, tenuous link to one another is the humans who use them.

Wireless communication facilitates the integration of individual processors into an interconnected collective. The mass consumption of wireless devices has driven the technology through an optimization similar to microprocessors, with modern radios consuming less power, space, and money than their predecessors.

Sensor networks differ significantly in their communication model from typical consumer wireless devices. The primary focus of communication is nodes' interaction with *each other*—not delivery of data to the user. In sensor networks, a user does not micro-manage the flow of information, in contrast to browsing the Internet or dialing a cellular phone. Users are not aware of every datum and computation, instead being informed only of the highest-level conclusions or results.

This matches our experience in the natural world. For example, a person can feel a mosquito bite without being conscious of the individual sensations—a characteristic buzz, a flying insect seen peripherally, the tingling of hairs on the skin—that were synthesized to create the illusion of a mosquito bite "sensor." Similarly, in sensor networks, the goal is not necessarily to provide a complete record of every sensor reading as raw *data*, but rather to perform synthesis that provides high-level *information*.

The goal of delivering synthesized, high-level sensing results is not simply convenience. It is, in fact, a first-order design principle, because such designs conserve the network's most precious resource: energy.

1.4 CORE CHALLENGES

In sensor networks, energy is valuable because it is scarce. Deployment of nodes without a support infrastructure requires that most of them be *untethered*, having only finite energy reserves from a battery. Unlike laptops or other hand-held devices that enjoy constant attention and maintenance by humans, the scale of a sensor network will make manual energy replenishment impossible. Though certain types of energy harvesting are conceivable, *energy effi-*

ciency will be a key goal for the foreseeable future. This requirement pervades all aspects of the system's design, and drives most of the other requirements.

Fundamental physical limits dictate that, as electronics become ever more efficient, *communication* will dominate a node's energy consumption [17]. The disproportionate energy cost of long-range vs. short-range transmission (r^2 to r^4), as well as the need for spatial frequency re-use, precludes communication beyond a short distance. The use of local processing, hierarchical collaboration, and domain knowledge to convert raw data into increasingly distilled and high-level representations—or, *data reduction*—is key to the energy efficiency of the system. In general, a perfect system will reduce as much data as possible as early as possible, rather than incur the energy expense of transmitting raw sensor values further along the path to the user.

Another fundamental challenge in sensor networks is their *dynamics*. Over time, nodes will fail—they may run out of energy, overheat in the sun, be carried away by wind, crash due to software bugs, or be eaten by a wild boar. Even in fixed positions, quality of RF communication links (and, thus, nodes' topologies) can change dramatically due to the vagaries of RF propagation. These changes are a result of propagation's strong environmental dependence, and are difficult to predict in advance. Traditional large-scale networks such as the Internet work in the face of changing configurations and brittle software partly because the number of people maintaining and using the network has grown along with the size of the network itself. In contrast, there may be a single human responsible for thousands of nodes in a single sensor network. Any design in which each device requires individual attention is infeasible. This leads to another important requirement: sensor networks must be *self-configuring*, working without fine-grained control from users. They must also be *adaptive* to changes in their environment—not make a single configuration choice, but continually change in response to dynamics.

1.5 RESEARCH DIRECTIONS

These basic challenges in wireless sensor networks have sparked a number of different research themes. In this section, we give a broad overview of many of these areas. It is meant as a general guide to the types of work currently underway, not a comprehensive review.

1.5.1 TIERED ARCHITECTURES

Although Moore's law predicts that hardware for sensor networks will inexorably become smaller, cheaper, and more powerful, technological advances will never prevent the need to make tradeoffs. Even as our notions of metrics such as "fast" and "small" evolve, there will always be compromises:

nodes will need to be faster *or* more energy-efficient, smaller *or* more capable, cheaper *or* more durable.

The choice of any *single* hardware platform will make compromises. The diverse needs of a sensor network can be satisfied only through a *tiered architecture*, a design that is a composition of platforms selected from a continuum of design points along axes of capability, energy requirements, size, and price. Small, numerous, cheap, disposable nodes can be used more effectively by also populating the network with some larger, faster, and more expensive hardware. Smaller devices will trade functionality and flexibility for smaller form factor, lower price, and better energy efficiency.

An analogy can be made to the memory hierarchy commonly found in desktop computer systems. Modern microprocessors typically have expensive and fast on-chip cache, backed by slower but larger L2 cache, main memory, and ultimately on-disk swap space. This organization, combined with a tendency in computation for locality of reference, results in a memory system that appears to be as large and as cheap per byte as the swap space, but as fast as the on-chip cache memory. In sensor networks, where localized algorithms are a primary design goal, similar benefits are possible by using using a heterogeneous spectrum of hardware. A tiered network may seem to be as cheap, portable, disposable, embeddable and energy-efficient as the tiniest nodes, while appearing to have larger nodes' larger storage capacity, higher-speed computation, and higher-bandwidth communication.

To date, much of the research and system construction has been based on a two- or three-tiered architecture. The highest tier is typically a connection to the Internet, where sensor networks can merge with traditional wired, server-based computing. At the smallest end, currently, are platforms such as TinyOS [11] on the Berkeley Mote [12]. Motes have minimal storage and computation capacity, but have integrated sensors, are small enough to embed in the environment unobtrusively, and use energy slowly enough to run from small batteries. In between these two tiers are "micro-servers"—computers with power on the order of a PDA, running from large batteries or a solar panel, and running a traditional operating system such as Linux.

1.5.2 ROUTING AND IN-NETWORK PROCESSING

Routing is a topic that arises almost immediately in any network as soon as it is large enough to require multiple hops—that is, if there is a pair of nodes that are not directly interconnected. In sensor networks, as in the Internet, this is of course the case. However, there is an important difference in the routing used by sensor networks. The Internet, and much of the earlier research in ad-hoc wireless networks, was focused on building the network as a *transport mechanism*—that is, a way to route packets to a particular endpoint. In a sensor network, efficiency demands that we do as much in-network processing (e.g., data reduction) as possible. Instead of blindly routing packets to a far-away endpoint, many applications do processing *at each hop* inside the network— aggregating similar data, filtering redundant information, and so forth.

For example, emerging designs allow users to task the network with a high-level query such as "notify me when a large region experiences a temperature over 100 degrees" or "report the location where the following bird call is heard" [10, 15]. If a node can correlate an incoming audio stream to the desired pattern *locally*, and report only the time and location of a match, the system will be many orders of magnitude more efficient than one that transmits the complete time-series of sampled audio.

The routing necessary to facilitate these queries is not simply end-to-end routing by node address. Routing must often be integrated with and influenced by the application, in stark contrast to Internet-style routing where the two are nearly always separated.

One early example of such a routing system is *Directed Diffusion* [10]. Unlike Internet routing, which uses node end-points, Directed Diffusion is data-centric. Data generated by sensor nodes are identified by attribute-value pairs. "Sinks", or nodes that request data, send "interests" into the network. Data generated by "source" nodes that match these interests "flow" toward the sinks. At each intermediate node, user software is given the opportunity to inspect and manipulate the data before it is transmitted to the next hop. This allows application-specific processing inside the network, but also places an additional burden on application developers. Unlike the Internet, where routing is an abstraction that can be essentially ignored, the hop-by-hop routing in Directed Diffusion requires software that is aware of, and perhaps even influences, the routing topology.

1.5.3 AUTOMATIC LOCALIZATION AND TIME SYNCHRONIZATION

Some of the most powerful benefits of a distributed network are due to the integration of information gleaned from multiple sensors into a larger world-view not detectable by any single sensor alone. For example, consider a sensor network whose goal is to detect a stationary phenomenon P, such as that depicted in Figure 1.1. P might be a region of the water table that has been polluted, within a field of chemical sensors. Each individual sensor might be very simple, capable only of measuring chemical concentration and thereby detecting whether or not it is within P. However, by fusing the data from all sensors, *combined with knowledge about the sensors' positions*, the complete network can describe more than just a set of locations covered by P: it can also compute P's size, shape, speed, and so forth. The whole of information

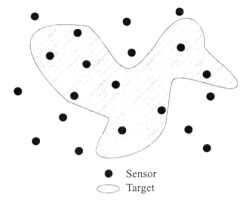

Figure 1.1. Individually, sensors may only be capable of a binary decision: they are within the sensed phenomenon, or they are not. If the sensor positions are known, integration of information from the entire field allows the network to deduce the *size* and *shape* of the target, even though it has no "size" or "shape" sensors.

has become greater than the sum of the parts: the network can deduce the size and shape of P even though it does not have a "size" or "shape" sensor.

Nearly every sensor network does this type of data fusion, but it is only possible if the sensors have known positions. Positions may be absolute (latitude and longitude), relative (20 meters away from the other node), or logical (inside the barn vs. on the pasture). But, in any of these cases, sensor networks need *automatic localization*. That is, nodes require the capability of localizing themselves after they have been deployed. This is necessary both due to the large scale of deployments, making manual surveys impractical, and due to dynamics that can change a node's position after its initial deployment.

For any phenomenon that is time-varying or mobile, time synchronization is also a crucial service necessary to combine the observations of multiple sensors with each other. For example, synchronized time is needed to integrate multiple position estimates into a velocity estimate.

In some cases, the Global Positioning System (GPS) can and does provide nodes with both their position and a global clock. However, it requires line of sight to several satellites, which is not available inside of buildings, beneath dense foliage, underwater, when jammed by an enemy, or during Mars exploration. In addition, in many contexts, GPS receivers are too expensive in terms of their energy consumption, size, or cost. GPS may not be practical, for example, in a network made up entirely of nodes on the scale of a dust-sized mote, or even the Berkeley COTS mote.

A number of researchers have developed schemes for automatic, ad-hoc localization. Localization schemes often use acoustic time-of-flight ranging [8, 21, 23] or RF connectivity to beacons of known position [3, 19]. Many time-of-flight measurement schemes incorporate various forms of sensor network time synchronization [6, 9].

1.5.4 DISTRIBUTED SIGNAL PROCESSING

For decades, the signal processing community has devoted much research attention to seamless integration of signals from multiple sources, and sources with heterogeneous sensing modalities. The signal processing literature sometimes refers to this as *array processing*; with heterogeneous sensors, it is often called *data fusion*. There are many applications, such as signal enhancement (noise reduction), source localization, process control, and source coding. It would seem to be a natural match to implement such algorithms in distributed sensor networks, and there has been great interest in doing so. However, much of the extensive prior art in the field assumes *centralized sensor fusion*. That is, even if the sensors gathering data are physically distributed, they are often assumed to be wired into a single processor.

Centralized processing makes a number of assumptions that are violated in sensor networks. For example, in a centralized processor, data can be shared among sensors at (effectively) infinite bandwidth. In sensor networks, communication is expensive in terms of energy, limited in bandwidth, nontrivial in its routing, and unreliable on both short and long timescales. A centralized fusion point also assumes implicit time synchronization—sensor channels sampled by the same processor also share a common timebase. In sensor networks, distributed sensors are on independent nodes with independent clocks; time synchronization must be made explicit.

In sensor networks, signal processing is a crucial building block. It is, for example, responsible for target tracking, signal identification and classification, localization, and beam-forming. All of these are processes that resolve low-level sensor signals (e.g., acoustic or seismic) into higher level sensors (e.g., "car" or "earthquake"). In sensor networks, the challenge is get the best signal processing results given the bandwidth and computational constraints of the sensing platform.

1.5.5 STORAGE, SEARCH AND RETRIEVAL

Sensor networks can produce a large volume of raw data—a continuous time-series of observations over all points in space covered by the network. In wired sensor networks, that mass of data is typically aggregated in a large, centralized database, where it later processed, queried and searched. The easiest path toward the adoption of new wireless sensor networks might be to provide users with a familiar model. For example, we might conceive a sensor network with a well-known declarative query interface such as SQL, allowing them to exploit traditional data mining techniques to extract interesting features or event information from the data. However, standard database assumptions about resource constraints, characteristics of data sources, reliability and availability no longer hold in a sensor network context, requiring significant modifications to existing techniques.

Resource constraints introduce possibly the most fundamental difference compared to a traditional database approach. Data mining over a massively distributed database which is under energy, bandwidth and storage constraints is daunting. The data cannot be transmitted to and stored at a central repository without adversely impacting lifetime of the network since limited energy resources are available at each node. An alternative approach could be to store data locally at sensor nodes, and query such data on-demand from users. However, the need to operate unattended for many years, coupled with cost and form factor constraints on sensor nodes, limits the amount of storage available on nodes.

In addition to storage constraints, processing and memory constraints on sensor nodes, especially on the low-end such as the Berkeley Motes, adds a dimension to optimize in such systems. The need for in-network processing necessitates that each sensor node act as a data processing engine, an implementation of which is challenging on nodes that function under tight CPU and memory constraints.

Traditional databases are not suitable when the data source is continuous and needs to be processed in real time, which might be required in some sensor network applications. Typical query processing deals with stored data on disk, rather than streaming sensor data. New techniques will be required for online processing of data streams that do not assume availability of significant secondary storage or processing power.

Reliability and availability of sensor nodes need to be considered in the construction of data processing engines. Since sensor data is transmitted over wireless connections with varying channel characteristics, the database should be able to handle variable delay and different delivery rates from different nodes. Further, these delays should be handled with low energy overhead— that is, minimal listening duration over the radio.

1.5.6 ACTUATION

In many cases, a sensor network is an entirely passive system, capable of detecting the state of the environment, but unable to change it or the network's relationship to it. *Actuation* can dramatically extend the capabilities of a network in two ways. First, actuation can enhance the sensing task, by pointing cameras, aiming antennae, or repositioning sensors. Second, actuation can *affect* the environment—opening valves, emitting sounds, or strengthening beams [2].

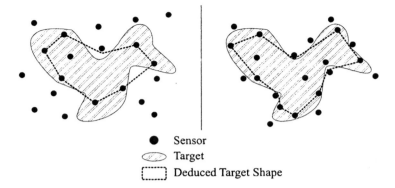

Figure 1.2. left) A sensor field might deduce the shape of a target area (e.g., contaminated soil) by drawing a convex hull around the sensors that are within the contamination, with concave deviations around uncontaminated sensors. *right)* If the sensors are *mobile*, they can collaboratively adjust their positions, estimating the shape of the target with greater accuracy.

One of the most commonly described forms of actuation is sensor mobility. For example, Sibley *et al.* developed a prototype "RoboMote," which brings autonomous mobility to the Berkeley Mote sensor platform [22]. They propose a number of ways to leverage such mobile sensors, including *actuated boundary detection*. The size and shape of a target region can be determined with more precision if the sensors can move, as shown in Figure 1.2. Others have proposed using mobile nodes to harvest energy from the environment, delivering it to stationary sensors [20, 16].

Mobility is also central to the DARPA SHM program [1]. In that system, nodes collaboratively construct a map of their relative positions using acoustic localization. Each node is equipped with a mobility subsystem consisting of small rocket motors. The goal is to keep the target area covered by sensors, even as nodes fail or are moved. If the sensors detect that a gap in the field has opened, nearby nodes fire their rockets and move in to fill the open space.

1.5.7 SIMULATION, MONITORING, AND DEBUGGING

Simulation and debugging environments are important in any large-scale software development project. In sensor networks, a number of factors make the use of innovative development models particularly important.

Large-scale sensor networks will not be able to simply send all the raw sensor data back to a centralized recording point. Energy and channel capacity constraints will require as much localized processing as possible, delivering only a (small) high-level sensing result. As we described earlier, a perfect system will reduce as much data as possible as early as possible, rather than incur the energy expense of transmitting raw sensor values further along the path to the user.

For a system designer, there is an unfortunate paradox intrinsic to this ideal: the data that must be discarded to meet the energy and channel capacity constraints are necessary for the evaluation and debugging of the data reduction process itself. How can a designer evaluate a system where, by definition, the information necessary for the evaluation is not available? That is, how can we be sure that the final, high-level sensing result delivered by the system is an accurate reflection of the state of the environment—when sensor networks are, by definition, deployed where we have insufficient energy and channel capacity to record all the raw data?

This fundamental problem makes simulation crucial in sensor networks. A simulator may be the only environment in which a sensor network can both run at a large scale and record each raw datum. This allows the "environment" as observed by the sensors to be reconciled with the high-level, distilled result delivered by the software being tested. Several example of these systems exist, including TOSSIM [14], EmStar [4], and sensor network extensions to GloMoSim and ns-2.

1.5.8 SECURITY AND PRIVACY

In sensor networks, many unique challenges arise in ensuring the security of sensor nodes and the data they generate [18, 5, 13]. For example, the fact that sensors are embedded in the environment presents a problem: the physical security of the nodes making up the network can not be assured. This can make security significantly different than in Internet servers. In sensor networks, attackers may modify node hardware, replace it with malicious counterparts, or fool sensors into making observations that do not accurately reflect the environment. To a single temperature sensor, a small match may look no different than a forest fire. Algorithms for ensuring network-wide agreement are crucial to detecting attacks because we can no longer assume the security of individual nodes, or the data they generate.

The limited resources on the smallest sensor nodes also can pose challenges. Many encryption schemes are impractically resource-intensive, consuming far more energy, memory, and computational time than would be required to send a raw, unprotected data value. In addition, protection of data from eavsedropping en-route—particularly important in wireless networks—traditionally implies end-to-end encryption. That is, data is encrypted as soon as it is created, transmitted through the network, and finally received by a secured server where decryption keys can be stored without danger of exposure. Unfortunately, finite energy drives the need for in-network processing in sensor networks, which confounds the traditional security schemes. Nodes inside the network can not perform application-specific processing on encrypted data. Decrypting the data at each node implies decryption keys are stored at each node; unfortunately, the node hardware itself is exposed, and can not be assumed to be out of reach of attackers.

As with many types of information technology throughout history, sensor networks also raise important questions about the privacy of individuals. Certain aspects of privacy have gradually eroded due to various forces—for example, the tracks we leave behind by using credit, the ubiquity of surveillance cameras, and the seeming omniscience of Internet search engines. Sensor networking, similarly, is a technology that can be used to enrich and improve our lives, or turned into an invasive tool. As sensor networks become more widespread, they will become an important point to consider in the continuing debate between public information and private lives.

1.6 CONCLUSIONS

Recent advances in miniaturization and low-cost, low-power electronics have led to active research in large-scale networks of small, wireless, low-power sensors and actuators. Pervasive sensing that is freed from the burden of infrastructure will revolutionize the way we observe the world around us. Sensors networks will automatically warn us of invisible hazards—contaminants in the air we breathe or the water we drink, or far-away earthquakes soon to arrive. Sensor networks can open the eyes of a new generation of scientists to phenomena never before observable, paving the way to new eras of understanding in the natural sciences.

Sensor networks will eventually be integral to our homes and everyday lives in ways that are difficult to imagine today. Computers themselves were once specialized tools limited to the esoteric domain of rote mathematical computation. It was inconceivable at the time that people would want computers in their homes, yet in the span of a single generation they have become household fixtures. Perhaps, someday, electronic sensing will be as natural to us as our own innate senses. Only time will tell.

ACKNOWLEDGEMENTS

This chapter was supported by the National Science Foundation, via the Center for Embedded Networked Sensing (CENS), NSF Cooperative Agreement #CCR-0120778. Additional support was provided by Intel Corporation. The authors are grateful for contributions by Deepak Ganesan, Lewis Girod, and Michael Klein.

REFERENCES

[1] DARPA Advanced Technology Office (ATO). Self-healing minefield. http://www.darpa.mil/ato/programs/SHM/.

[2] A.A. Berlin, J.G. Chase, M.H. Yim, B.J. Maclean, M. Oliver, and S.C. Jacobsen. MEMS-based control of structural dynamic instability. *Journal of Intelligent Material Systems and Structures*, 9(7):574–586, 1998.

[3] Nirupama Bulusu, John Heidemann, Deborah Estrin, and Tommy Tran. Self-configuring localization systems: Design and experimental evaluation. *ACM Transactions on Embedded Computing Systems*, page To appear., May 2003.

[4] Jeremy Elson, Solomon Bien, Naim Busek, Vladimir Bychkovskiy, Alberto Cerpa, Deepak Ganesan, Lewis Girod, Ben Greenstein, Tom Schoellhammer, Thanos Stathopoulos, and Deborah Estrin. EmStar: An environment for developing wireless embedded systems software. Technical Report 0009, Center for Embedded Networked Sensing, University of California, Los Angeles, March 2003. http://lecs.cs.ucla.edu/publications.

[5] Laurent Eschenauer and Virgil D. Gligor. A key-management scheme for distributed sensor networks. In Vijay Atlury, editor, *Proceedings of the 9th ACM Conference on Computer and Communication Security (CCS-02)*, pages 41–47, New York, November 18–22 2002. ACM Press.

[6] Jeremy Elson, Lewis Girod, and Deborah Estrin. Fine-grained network time synchronization using reference broadcasts. In *Proceedings of the 5th ACM Symposium on Operating System Design and Implementation (OSDI-02)*, Operating Systems Review, pages 147–164, New York, December 9–11 2002. ACM Press.

[7] National Center for Atmospheric Research. Cyberinfrastructure for environmental research and education. Technical report, Sponsored by the National Science Foundation, September 2003. http://www.ncar.ucar.edu/cyber/.

[8] Lewis Girod and Deborah Estrin. Robust range estimation using acoustic and multimodal sensing. In *Proceedings of the IEEE/RSJ International Conference on Intelligent Robots and Systems (IROS 2001)*, March 2001.

[9] S. Ganeriwal, R. Kumar, and M. B. Srivastava. Timing-sync protocol for sensor networks. In *Proceedings of the First ACM Conference on Embedded Networked Sensor Systems (SenSys 2003)*, Los Angeles, November 2003.

[10] John Heidemann, Fabio Silva, Chalermek Intanagonwiwat, Ramesh Govindan, Deborah Estrin, and Deepak Ganesan. Building efficient wireless sensor networks with low-level naming. In *Proceedings of the Symposium on Operating Systems Principles*, pages 146–159, Chateau Lake Louise, Banff, Alberta, Canada, October 2001. ACM.

[11] Jason Hill, Robert Szewczyk, Alec Woo, Seth Hollar, David Culler, and Kristofer Pister. System architecture directions for networked sensors. In *Proceedings of the Ninth International Conference on Architectural Support for Programming Languages and Operating Systems (ASPLOS-IX)*, pages 93–104, Cambridge, MA, USA, November 2000. ACM.

[12] J.M. Kahn, R.H. Katz, and K.S.J. Pister. Next century challenges: mobile networking for Smart Dust. In *Proceedings of the fifth annual ACM/IEEE international conference on Mobile computing and networking*, pages 271–278, 1999.

[13] Y.W. Law, S. Etalle, and P.H. Hartel. Assessing security-critical energy-efficient sensor networks. In *Proceedings of IFIP WG 11.2 Small Systems Security Conference*, Athens, Greece, May 2003.

[14] Philip Levis, Nelson Lee, Matt Welsh, and David Culler. Tossim: Accurate and scalable simulation of entire tinyos applications. In *Proceedings of the First ACM Conference on Embedded Networked Sensor Systems (SenSys 2003)*, Los Angeles, November 2003.

[15] Samuel R. Madden, Michael J. Franklin, Joseph M. Hellerstein, and Wei Hong. Tag: a tiny aggregation service for ad-hoc sensor networks. In *Proceedings of the Fifth Symposium on Operating Systems Design and Implementation (OSDI)*, pages 131–146, Boston, MA, USA, December 2002.

[16] Gregory J. Pottie and Rodney Brooks. Towards a robotic ecology. DARPA ISAT Study, 1999.

[17] Gregory J. Pottie and William J. Kaiser. Wireless integrated network sensors. *Communications of the ACM*, 43(5):51–58, May 2000.

[18] Adrian Perrig, Robert Szewczyk, Victor Wen, David Culler, and J. D. Tygar. SPINS: Security suite for sensor networks. In *Proceedings of the Seventh Annual International Conference on Mobile Computing and Networking (MOBICOM-01)*, pages 189–199, New York, July 16–21 2001. ACM Press.

[19] Maurizio Piaggio, Antonio Sgorbissa, and Renato Zaccaria. Autonomous navigation and localization in service mobile robotics. In *Proceedings of the IEEE/RSJ International Conference on Intelligent Robots and Systems (IROS 2001)*, March 2001.

[20] Mohammad Rahimi, Hardik Shah, Gaurav Sukhatme, John Heidemann, and Deborah Estrin. Energy harvesting in mobile sensor networks. Technical report, Center for Embedded Networked Sensing, 2003. http://www.cens.ucla.edu/Project-Descriptions/energyharvesting/EnergyHarvesting.pdf.

[21] Andreas Savvides, Chih-Chien Han, and Mani Srivastava. Dynamic Fine-Grained localization in Ad-Hoc networks of sensors. In *Proceedings of the Seventh Annual International Conference on Mobile Computing and Networking (MOBICOM-01)*, pages 166–179, New York, July 16–21 2001. ACM Press.

[22] Gabriel T. Sibley, Mohammad H. Rahimi, and Gaurav S. Sukhatme. Robomote: A tiny mobile robot platform for large-scale sensor networks. In *Proceedings of the IEEE International Conference on Robotics and Automation (ICRA2002)*, 2002.

[23] Kamin Whitehouse and David Culler. Calibration as parameter estimation in sensor networks. In *Proceedings of the first ACM International Workshop on Wireless Sensor Networks and Applications*, pages 59–67, 2002.

Chapter 2

COMMUNICATION PROTOCOLS FOR SENSOR NETWORKS

Weilian Su,[1] Özgür B. Akan,[2] and Erdal Cayirci[3]

[1] *Broadband and Wireless Networking Laboratory*
School of Electrical and Computer Engineering
Georgia Institute of Technology
Atlanta, GA 30332-0250 U.S.A
weilian@ece.gatech.edu

[2] *Broadband and Wireless Networking Laboratory*
School of Electrical and Computer Engineering
Georgia Institute of Technology
Atlanta, GA 30332-0250 U.S.A
akan@ece.gatech.edu

[3] *Istanbul Technical University*
80626 Istanbul Turkey
cayirci@cs.itu.edu.tr

Abstract   This chapter describes about the challenges and essence of designing communication protocols for wireless sensor networks. The sensor nodes are densely deployed and collaboratively work together to provide higher quality sensing in time and space as compared to traditional stationary sensors. The applications of these sensor nodes as well as the issues in the transport, network, datalink, and physical layers are discussed. For applications that require precise timing, different types of timing techniques are explored.

Keywords: Wireless Sensor Networks, Sensor Network Applications, Transport Layer, Networking Layer, Data Link Layer, Medium Access Control, Error Control, Physical Layer, and Time Synchronization Schemes.

INTRODUCTION

Recent advances in wireless communications, digital electronics, and analog devices have enabled sensor nodes that are low-cost and low-power to communicate untethered in short distances and collaborate as a group. These sensor nodes leverage the strength of collaborative effort to provide higher quality sensing in time and space as compared to traditional stationary sensors, which are deployed in the following two ways [18]:

- Sensors can be positioned far from the actual *phenomenon*, i.e., something known by sense perception. In this approach, large sensors that use some complex techniques to distinguish the targets from environmental noise are required.

- Several sensors that perform only sensing can be deployed. The positions of the sensors and communications topology are carefully engineered. They transmit time series of the sensed phenomenon to the central nodes where computations are performed and data are fused.

The sensor nodes are deployed either inside the phenomenon or very close to it. They may self-organize into clusters or collaborate together to complete a task that is issued by the users. In addition, the positions of these nodes do not need to be predefined. As a result, the sensor nodes are fit for many applications, e.g., location tracking and chemical detection in areas not easily accessible. Since sensing applications generate a large quantity of data, these data may be fused or aggregated together to lower the energy consumption. The sensor nodes use their processing abilities to locally carry out simple computations and transmit only the required and partially processed data. In essence, wireless sensor networks with these capabilities may provide the end users with intelligence and a better understanding of the environment. In the future, the wireless sensor networks may be an integral part of our lives, more so than the present-day personal computers.

Although many protocols and algorithms have been proposed for traditional wireless ad hoc networks, they are not well suited for the unique features and application requirements of wireless sensor networks. To illustrate this point, the differences between wireless sensor networks and ad-hoc networks [32] are outlined below:

- The number of sensor nodes in a sensor network can be several orders of magnitude higher than the nodes in an ad hoc network.

- Sensor nodes are densely deployed.

- Sensor nodes are prone to failures.

- The topology of a sensor network changes very frequently.

- Sensor nodes mainly use broadcast communication paradigm whereas most ad hoc networks are based on point-to-point communications.

- Sensor nodes are limited in power, computational capacities, and memory.

- Sensor nodes may not have global *identification* (ID) because of the large amount of overhead and large number of sensors.

- Sensor networks are deployed with a specific sensing application in mind whereas ad-hoc networks are mostly constructed for communication purposes.

Recently, as a result of the above differences and the potential application of wireless sensor networks, the sensor networks have attracted many interest in the research community. In this chapter, the challenges and essence of designing wireless sensor network protocols are described and discussed. We present some potential sensor network applications in Section 2.1. These applications may provide more insight into the usefulness of wireless sensor networks. Also, the challenges and essence of designing transport, network, datalink, and physical layer protocols and algorithms to enable these applications are described. In addition to these guidelines, we explore the different types of timing techniques and the issues that these techniques have to address in Section 2.6. Lastly, we conclude our chapter in Section 2.7.

2.1 APPLICATIONS OF SENSOR NETWORKS

Sensor networks may consist of many different types of sensors such as seismic, low sampling rate magnetic, thermal, visual, infrared, acoustic and radar, which are able to monitor a wide variety of ambient conditions that are listed in Table 2.1. An example of some MICA [27] motes is illustrated in Figure 2.1. The size of the MICA motes is small as compared to the size of a dime. These motes may be controlled by a computer through the sink, which is a MICA mote with a computer interface.

Sensor nodes can be used for continuous sensing, event detection, event identification, location sensing, and local control of actuators. The concept of microsensing and wireless connection of these nodes promise many new application areas, e.g., military, environment, health, home, commercial, space exploration, chemical processing and disaster relief, etc. Some of these application areas are described in Section 2.1.1. In addition, some application layer protocols are introduced in Section 2.1.2; they are used to realize these applications.

Table 2.1. Examples of ambient conditions.

Environmental Ambient Conditions
Temperature
Humidity
Vehicular movement
Lightning condition
Pressure
Soil makeup
Noise levels

Figure 2.1. Example of MICA motes.

2.1.1 WIRELESS SENSOR NETWORK APPLICATIONS

The number of potential applications for wireless sensor networks is huge. Actuators may also be included in the sensor networks, which makes the number of applications that can be developed much higher. In this section, some example applications are given to provide the reader with a better insight about the potentials of wireless sensor networks.

Military Applications: Wireless sensor networks can be an integral part of military *command, control, communications, computers, intelligence, surveillance, reconnaissance and tracking* (C4ISRT) systems. The rapid deployment, self organization and fault tolerance characteristics of sensor networks make them a very promising sensing technique for military C4ISRT. Since sensor networks are based on the dense deployment of disposable and low cost sensor nodes, destruction of some nodes by hostile actions does not affect a military operation as much as the destruction of a traditional sensor. Some of the military applications are monitoring friendly forces, equipment and ammunition; battlefield surveillance; reconnaissance of opposing forces and terrain; targeting; battle damage assessment; and nuclear, biological and chemical attack detection and reconnaissance.

Environmental Applications: Some environmental applications of sensor networks include tracking the movements of species, i.e., habitat monitoring; monitoring environmental conditions that affect crops and livestock; irrigation; macroinstruments for large-scale Earth monitoring and planetary exploration; and chemical/biological detection [1, 3–5, 16, 18, 20, 21, 43, 47].

Commercial Applications: The sensor networks are also applied in many commercial applications. Some of them are building virtual keyboards; managing inventory control; monitoring product quality; constructing smart office spaces; and environmental control in office buildings [1, 5, 11, 12, 21, 37, 38, 42, 47, 35].

2.1.2 APPLICATION LAYER PROTOCOLS FOR WIRELESS SENSOR NETWORKS

Although many application areas for wireless sensor networks are defined and proposed, potential application layer protocols for sensor networks remain largely unexplored. Three possible application layer protocols are introduced in this section; they are *Sensor Management Protocol, Task Assignment and Data Advertisement Protocol,* and *Sensor Query and Data Dissemination Protocol.* These protocols may require protocols at other stack layers that are explained in Sections 2.2, 2.3, 2.4, 2.5, and 2.6.

Sensor Management Protocol (SMP). Designing an application layer management protocol has several advantages. Sensor networks have many different application areas, and accessing them through networks such as Internet is aimed in some current projects [35]. An application layer management protocol makes the hardware and softwares of the lower layers transparent to the sensor network management applications.

System administrators interact with sensor networks by using *sensor management protocol* (SMP). Unlike many other networks, sensor networks consist of nodes that do not have global identification, and they are usually infrastructureless. Therefore, SMP needs to access the nodes by using attribute-based naming and location-based addressing, which are explained in detail in Section 2.3.

SMP is a management protocol that provides the software operations needed to perform the following administrative tasks:

- Introducing the rules related to data aggregation, attribute-based naming and clustering to the sensor nodes,

- Exchanging data related to the location finding algorithms,

- Time synchronization of the sensor nodes,

- Moving sensor nodes,

- Turning sensor nodes on and off,

- Querying the sensor network configuration and the status of nodes, and re-configuring the sensor network, and

- Authentication, key distribution and security in data communications.

The descriptions of some of these tasks are given in [8, 11, 33, 40, 41].

Task Assignment and Data Advertisement Protocol (TADAP). Another important operation in the sensor networks is interest dissemination. Users send their interest to a sensor node, a subset of the nodes or whole network. This interest may be about a certain attribute of the phenomenon or a triggering event. Another approach is the advertisement of available data in which the sensor nodes advertise the available data to the users, and the users query the data which they are interested in. An application layer protocol that provides the user software with efficient interfaces for interest dissemination is useful for lower layer operations, such as routing.

Sensor Query and Data Dissemination Protocol (SQDDP). The *Sensor Query and Data Dissemination Protocol* (SQDDP) provides user applications with interfaces to issue queries, respond to queries and collect incoming replies. Note that these queries are generally not issued to particular nodes. Instead, attribute-based or location-based naming is preferred. For instance, *"the locations of the nodes that sense temperature higher than* $70°F$ *"* is an attribute-based query. Similarly, *"temperatures read by the nodes in Region A"* is an example for location based naming.

Likewise, the *sensor query and tasking language* (SQTL) [41] is proposed as an application that provides even a larger set of services. SQTL supports three types of events, which are defined by keywords *receive, every*, and *expire. Receive* keyword defines events generated by a sensor node when the sensor node receives a message; *every* keyword defines events occurred periodically due to a timer time-out; and *expire* keyword defines the events occurred when a timer is expired. If a sensor node receives a message that is intended for it and contains a script, the sensor node then executes the script. Although SQTL is proposed, different types of SQDDP can be developed for various applications. The use of SQDDPs may be unique to each application.

2.2 TRANSPORT LAYER

The wireless sensor network is an event driven paradigm that relies on the collective effort of numerous sensor nodes. This collaborative nature brings several advantages over traditional sensing including greater accuracy, larger coverage area and extraction of localized features. The realization of these potential gains, however, directly depends on the efficient reliable communication between the wireless sensor network entities, i.e., the sensor nodes and the sink.

To accomplish this, in addition to robust modulation and media access, link error control and fault tolerant routing, a reliable transport mechanism is imperative. The functionalities and design of a suitable transport solution for the wireless sensor networks are the main issues addressed in this section.

The need for transport layer in the wireless sensor networks is pointed out in the literature [35, 38]. In general, the main objectives of the transport layer are (i) to bridge application and network layers by application multiplexing and demultiplexing; (ii) to provide data delivery service between the source and the sink with an error control mechanism tailored according to the specific reliability requirement of the application layer; and (iii) to regulate the amount of traffic injected to the network via flow and congestion control mechanisms. Although these objectives are still valid, the required transport layer functionalities to achieve these objectives in the wireless sensor networks are subject to significant modifications in order to accommodate unique characteristics of the wireless sensor network paradigm. The energy, processing, and hardware limitations of the wireless sensor nodes bring further constraints on the transport layer protocol design. For example, the conventional end-to-end retransmission-based error control and the window-based additive-increase multiplicative-decrease congestion control mechanisms adopted by the vastly used *Transport Control Protocol* (TCP) protocols may not be feasible for the wireless sensor network domain and hence, may lead to waste of scarce resources.

On the other hand, unlike the other conventional networking paradigms, the wireless sensor networks are deployed with a specific sensing application objective. For example, sensor nodes can be used within a certain deployment scenario to perform continuous sensing, event detection, event identification, location sensing, and local control of actuators for a wide range of applications such as military, environment, health, space exploration, and disaster relief.

The specific objective of a sensor network also influences the design requirements of the transport layer protocols. For example, the sensor networks deployed for different applications may require different reliability level as well as different congestion control approaches.

Consequently, the development of transport layer protocols is a challenging effort, because the limitations of the sensor nodes and the specific application requirements primarily determine the design principles of the transport layer protocols. With this respect, the main objectives and the desired features of the transport layer protocols that can address the unique requirements of the wireless sensor networks paradigm can be stated as follows:

- *Reliable Transport:* Based on the application requirements, the extracted event features should be reliably transferred to the sink. Similarly, the programming/retasking data for sensor operation, command and queries should be reliably delivered to the target sensor nodes to assure the proper functioning of the wireless sensor network.

- *Congestion Control:* Packet loss due to congestion can impair event detection at the sink even when enough information is sent out by the sources. Hence, congestion control is an important component of the transport layer to achieve reliable event detection. Furthermore, congestion control not only increases the network efficiency but also helps conserve scarce sensor network resources.

- *Self-configuration:* The transport layer protocols must be adaptive to dynamic topologies caused by node mobility/failure/temporary power-down, spatial variation of events and random node deployment.

- *Energy Awareness:* The transport layer functionalities should be energy-aware, i.e., the error and congestion control objectives must be achieved with minimum possible energy expenditure. For instance, if reliability levels at the sink are found to be in excess of that required for the event detection, the source nodes can conserve energy by reducing the amount of information sent out or temporarily powering down.

- *Biased Implementation:* The algorithms must be designed such that they mainly run on the sink with minimum functionalities required at sensor nodes. This helps conserve limited sensor resources and shifts the burden to the high-powered sink.

- *Constrained Routing/Addressing:* Unlike protocols such as TCP, the transport layer protocols for wireless sensor networks should not assume the existence of an end-to-end global addressing. It is more likely to have attribute-based naming and data-centric routing, which call for different transport layer approaches.

Due to the application-oriented and collaborative nature of the wireless sensor networks, the main data flow takes place in the *forward path*, where the source nodes transmit their data to the sink. The *reverse path*, on the other hand, carries the data originated from the sink such as programming/retasking binaries, queries and commands to the source nodes. Although the above objectives and the desired features are common for the transport layer protocols, different functionalities are required to handle the transport needs of the forward and reverse paths.

For example, the correlated data flows in the forward path are loss-tolerant to the extent that event features are reliably communicated to the sink. However, data flows in the reverse channel are mainly related to the operational communication such as dissemination of the new operating system binaries, which usually requires 100 % reliable delivery. Therefore, a reliability mechanism would not suffice to address the requirements of both forward and reverse paths. Hence, the transport layer issues pertaining to these distinct cases are studied separately in the following sections.

2.2.1 EVENT-TO-SINK TRANSPORT

In order to realize the potential gains of the collective effort of numerous sensor nodes, it is detrimental that extracted event features at the sensor nodes are reliably communicated to the sink. This necessitates a reliable transport layer mechanism that can assure the *event-to-sink reliability*.

The need for a transport layer for data delivery in the wireless sensor networks was questioned in [36] under the premise that data flows from source to sink are generally loss tolerant. While the need for end-to-end reliability may not exist due to the sheer amount of correlated data flows, an event in the sensor field needs to be tracked with a certain accuracy at the sink. Hence, unlike traditional communication networks, the sensor network paradigm necessitates an event-to-sink reliability notion at the transport layer [39]. This involves a reliable communication of the event features to the sink rather than conventional packet-based reliable delivery of the individual sensing reports/packets generated by each sensor node in the field. Such *event-to-sink reliable transport* notion based on collective identification of data flows from the event to the sink is illustrated in Figure 2.2.

In order to provide reliable event detection at the sink, possible congestion in the forward path should also be addressed by the transport layer. Once the event is sensed by a number of sensor nodes within the coverage of the phenomenon, i.e., event radius, significant amount of traffic is triggered by these sensor nodes, which may easily lead to congestion in the forward path. The need for transport layer congestion control to assure reliable event detection at the sink is revealed by the results in [19]. It has been shown in [19] that

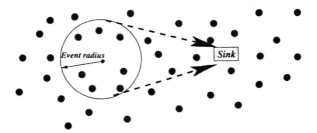

Figure 2.2. Typical sensor network topology with event and sink. The sink is only interested in collective information of sensor nodes within the event radius and not in their individual data.

exceeding network capacity can be detrimental to the observed goodput at the sink. Moreover, although the event-to-sink reliability may be attained even in the presence of packet loss due to network congestion thanks to the correlated data flows, a suitable congestion control mechanism can also help conserve energy while maintaining desired accuracy levels at the sink.

On the other hand, although the transport layer solutions in conventional wireless networks are relevant, they are simply inapplicable for the event-to-sink reliable transport in the wireless sensor networks. These solutions mainly focus on reliable data transport following end-to-end TCP semantics and are proposed to address the challenges posed by wireless link errors and mobility [2]. The primary reason for their inapplicability is their notion of end-to-end reliability which is based on acknowledgments and end-to-end retransmissions. Due to inherent correlation in the data flows generated by the sensor nodes, however, these mechanisms for strict end-to-end reliability are significantly energy-draining and superfluous. Furthermore, all these protocols bring considerable memory requirements to buffer transmitted packets until they are ACKed by the receiver. In contrast, sensor nodes have limited buffering space (<4KB in MICA motes [27]) and processing capabilities.

In contrast to the transport layer protocols for conventional end-to-end reliability, *Event-to-Sink Reliable Transport* (ESRT) protocol [39] is based on the event-to-sink reliability notion and provides reliable event detection without any intermediate caching requirements. ESRT is a novel transport solution developed to achieve reliable event detection in the wireless sensor networks with minimum energy expenditure. It includes a congestion control component that serves the dual purpose of achieving reliability and conserving energy. ESRT also does not require individual sensor identification, i.e., an event ID suffices. Importantly, the algorithms of ESRT mainly run on the sink, with minimal functionality required at resource constrained sensor nodes.

2.2.2 SINK-TO-SENSORS TRANSPORT

While the data flows in the forward path carry correlated sensed/detected event features, the flows in the reverse path mainly contain data transmitted by the sink for an operational or application-specific purposes. This may include the operating system binaries, programming/retasking configuration files, application-specific queries and commands. Dissemination of this type of data mostly requires 100 % reliable delivery. Therefore, the event-to-sink reliability approach introduced before would not suffice to address such tighter reliability requirement of the flows in the reverse paths.

Such strict reliability requirement for the sink-to-sensors transport of operational binaries and application-specific query and commands involves in certain level of retransmission as well as acknowledgment mechanisms. However, these mechanisms should be incorporated into the transport layer protocols cautiously in order not to totally compromise scarce sensor network resources. With this respect, local retransmissions and negative acknowledgment approaches would be preferable over the end-to-end retransmissions and acknowledgments to maintain minimum energy expenditure.

On the other hand, sink is involved more in the sink-to-sensor data transport on the reverse path. Hence, the sink with plentiful energy and communication resources can broadcast the data with its powerful antenna. This helps to reduce the amount of traffic forwarded in the multi-hop wireless sensor network infrastructure and hence, helps sensor nodes conserve energy. Therefore, data flows in the reverse path may experience less congestion in contrast to the forward path, which is totally based on multi-hop communication. This calls for less aggressive congestion control mechanisms for the reverse path as compared to the forward path in the wireless sensor networks.

The multi-hop and one-to-many nature of data flows in the reverse path of the wireless sensor networks prompt a review of reliable multicast solutions proposed in other wired/wireless networks. There exist many such schemes that address the reliable transport and congestion control for the case of single sender and multiple receivers [14]. Although the communication structure of the reverse path, i.e., from sink to sources, is an example of multicast, these schemes do not stand as directly applicable solutions; rather, they need significant modifications/improvements to address the unique requirements of the wireless sensor network paradigm.

In [36], the *Pump Slowly, Fetch Quickly* (PSFQ) mechanism is proposed for reliable retasking/reprogramming in the wireless sensor networks. PSFQ is based on slowly injecting packets into the network but performing aggressive hop-by-hop recovery in case of packet loss. The pump operation in PSFQ simply performs controlled flooding and requires each intermediate node to create and maintain a data cache to be used for local loss recovery and in-sequence data delivery. Although this is an important transport layer solution for the wireless sensor networks, PSFQ does not address packet loss due to congestion.

In summary, the transport layer mechanisms that can address the unique challenges posed by the wireless sensor network paradigm are essential to realize the potential gains of the collective effort of wireless sensor nodes. As discussed in Sections 2.2.1 and 2.2.2, there exist promising solutions for both event-to-sink and sink-to-sensors reliable transports. These solutions and the ones that are currently under development, however, need to be exhaustively evaluated under the real wireless sensor network deployment scenarios to reveal their shortcomings; hence, necessary modifications/improvements may be required to provide a complete transport layer solution for the wireless sensor networks.

2.3 NETWORK LAYER

Sensor nodes are scattered densely in a field either close to or inside the phenomenon. Since they are densely deployed, neighbor nodes may be very close to each other. As a result, multihop communication in the wireless sensor networks is expected to consume less power than the traditional single hop communication. Furthermore, the transmission power levels may be kept low, which is highly desired for covert operations. In addition, multihop communication may effectively overcome some of the signal propagation effects experienced in long distance wireless communication. As discussed in Section 2, the ad hoc routing techniques already proposed in the literature [32] do not usually fit the requirements of the wireless sensor networks. As a result, the networking layer of the sensor networks is usually designed according to the following principles:

- Power efficiency is always an important consideration.
- Sensor networks are mostly data-centric.
- An ideal sensor network has attribute-based addressing and location awareness.
- Data aggregation is useful only when it does not hinder the collaborative effort of the sensor nodes.
- The routing protocol is easily integrated with other networks, e.g., Internet.

The above principles serve as a guideline in designing a routing protocol for the wireless sensor networks. As discussed in Section 2.2, a transport protocol has to be energy aware. This criteria also applies to a routing protocol designed for the wireless sensor networks since the life-time of the networks depends on the longevity of each sensor node. In addition, a routing protocol may be data-centric. A data-centric routing protocol requires attribute-based naming [31, 41, 10, 8]. For attribute-based naming, the users are more interested in querying an attribute of the phenomenon, rather than querying an individual node. For instance, *"the areas where the temperature is over $70°F$"* is a more common query than *"the temperature read by a certain node"*. The attribute-based naming is used to carry out queries by using the attributes of the phenomenon. It also makes broadcasting, attribute-based multi-casting, geo-casting and any-casting important for sensor networks.

For example, if interest dissemination is based on data-centric, it is performed by assigning the sensing tasks to the sensor nodes. There are two approaches used for interest dissemination: (i) sinks broadcast the interest [18] and (ii) sensor nodes broadcast an advertisement for the available data [16] and wait for a request from the interested sinks.

As data-centric routing is important, it should also leverage the usefulness of data aggregation. Data-aggregation is a technique used to solve the implosion and overlap problems in data-centric routing [16]. In this technique, a sensor network is usually perceived as a reverse multicast tree as shown in Figure 2.3, where the sink asks the sensor nodes to report the ambient condition of the phenomena. Data coming from multiple sensor nodes are aggregated as if they are about the same attribute of the phenomenon when they reach the same routing node on the way back to the sink. For example, sensor node E aggregates the data from sensor nodes A and B while sensor node F aggregates the data from sensor nodes C and D as shown in Figure 2.3. Data aggregation can be perceived as a set of automated methods of combining the data that comes from many sensor nodes into a set of meaningful information [17]. With this respect, data aggregation is known as data fusion [16]. Also, care must be taken when aggregating data, because the specifics of the data, e.g., the locations of reporting sensor nodes, should not be left out. Such specifics may be needed by certain applications.

One other important function of the network layer is to provide internetworking with other networks such as other sensor networks, command and control systems and the Internet. In one scenario, the sink nodes can be used as a gateway to other networks while in another scenario they serve as a backbone to other networks. As shown in Figure 2.4, the sinks are the gateways to the sensor networks as well as the bases for the communication backbone between the user and the sensor nodes.

When developing a routing protocol with the design-principles in mind, one of the following approaches can be used to select an energy efficient route. Figure 2.5 is used to describe each of these approaches, and node T is the source node that senses the phenomena. The possible routes used to communicate with the sink in these approaches are given in Table 2.2.

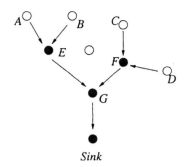

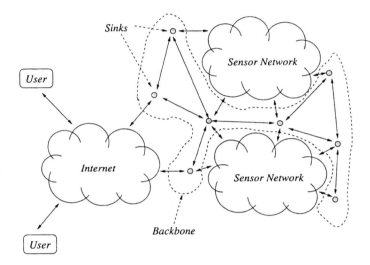

Figure 2.4. Internetworking between sensor nodes and user.

Table 2.2. Possible routes between the source and sink.[a]

Possible Routes	Description
Route 1	Sink-A-B-T, total PA=4, total $\alpha = 3$
Route 2	Sink-A-B-C-T, total PA=6, total $\alpha = 6$
Route 3	Sink-D-T, total PA=3, total $\alpha = 4$
Route 4	Sink-E-F-T, total PA=5, total $\alpha = 6$

[a] PA is the available power, and α_i is the energy required to transmit a data packet through the related link.

- *Maximum Available Power (PA) Route:* The route that has maximum total available power is preferred. The total PA is calculated by summing the PAs of each node along the route. Based on this approach, Route 2

Communication Protocols for Sensor Networks

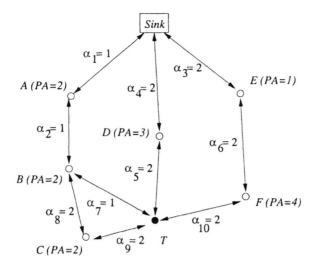

Figure 2.5. The power efficiency of the routes.

is selected in Figure 2.5. However, Route 2 includes the nodes in Route 1 and an extra node. Therefore, although it has a higher total PA, it is not a power efficient one. As a result, it is important not to consider the routes derived by extending the routes that can connect the sensor to the sink as an alternative route. Eliminating Route 2, Route 4 is selected as the power efficient route when the maximum PA scheme is used.

- *Minimum Energy (ME) Route:* The route that consumes minimum energy to transmit the data packets between the sink and the sensor node is the ME route. As shown in Figure 2.5, Route 1 is the ME route.

- *Minimum Hop (MH) Route:* The route that makes the minimum hop to reach the sink is preferred. Route 3 in Figure 2.5 is the most efficient route based on this scheme. Note that the ME scheme selects the same route as the MH when the same amount of energy, i.e., all α are the same, is used on every link. Therefore, when nodes broadcast with same power level without any power control, MH is then equivalent to ME.

- *Maximum Minimum PA Node Route:* The route along which the minimum PA is larger than the minimum PAs of the other routes is preferred. In Figure 2.5, Route 3 is the most efficient and Route 1 is the second efficient paths. This scheme precludes the risk of using up a sensor node with low PA much earlier than the others because they are on a route with nodes which has very high PAs.

A brief summary of the state-of-the-arts in the networking area is shown in Table 2.3. The *Small Minimum Energy Communication Network* (SMECN) [24] creates an energy efficient subgraph of the sensor networks. It tries to minimize the energy consumption while maintaining connectivity of the nodes in the network. Besides subgraph creation, the sensor nodes may form energy efficient clusters using the *Low Energy Adaptive Clustering Hierarchy* (LEACH) scheme [17]. In addition, QoS routing trees may be created with the *Sequential Assignment Routing* (SAR) protocol [45]; the sources send the collected data back to the sink through one of these routing trees. The collected data or queries may also be disseminated by flooding, gossiping [15], *Sensor Protocols for Information via Negotiation* (SPIN) [16], or directed diffusion protocol [18]. The directed diffusion protocol is a data-centric dissemination protocol, and the queries and collected data use the attribute-based naming schemes.

Although the protocols listed in Table 2.3 resolve some of the network layer issues, there are still room for more advanced data-centric routing protocols. In addition, different applications of the sensor networks may require different types of routing protocols. This is also a driving force for developing new transport protocols as described in Section 2.2 as well as data link schemes, which is discussed in the following section.

Table 2.3. An overview of network layer schemes.

Network Layer Scheme	Description
SMECN [24]	-Creates a sub graph of the sensor network that contains the minimum-energy path.
LEACH [17]	-Forms clusters to minimize energy dissipation.
SAR [45]	-Creates multiple trees where the root of each tree is one hop neighbor from the sink; select a tree for data to be routed back to the sink according to the energy resources and additive QoS Metric.
Flooding	-Broadcasts data to all neighbor nodes regardless if they receive it before or not.
Gossiping [15]	-Sends data to one randomly selected neighbor.
SPIN [16]	-Sends data to sensor nodes only if they are interested; has three types of messages, i.e., ADV, REQ, and DATA.
Directed Diffusion [18]	-Sets up gradients for data to flow from source to sink during interest dissemination.

2.4 DATA LINK LAYER

As discussed in Section 2.2, the wireless sensor networks are deployed with an objective of reliable event detection at the sink based on the collective effort of numerous sensor nodes spread in the sensor field. Although the transport layer is essential to achieve higher level error and congestion control, it is still imperative to have the data link layer functionalities in the wireless sensor networks.

In general, the data link layer is primarily responsible for the multiplexing of data streams, data frame detection, medium access and error control. It ensures reliable point-to-point and point-to-multipoint connections in a communication network. Nevertheless, the collaborative and application-oriented nature of the wireless sensor networks and the physical constraints of the sensor nodes such as energy and processing limitations determine the way these responsibilities are fulfilled. In the following two subsections, the data link layer issues are explored within the discussion of the medium access and error control strategies in the wireless sensor networks.

2.4.1 MEDIUM ACCESS CONTROL

The *Medium Access Control* (MAC) layer protocols in a wireless multi-hop self-organizing sensor network must achieve two objectives. The first one is to establish communication links for data transfer; this is necessary for creating a basic network infrastructure that is needed for multi-hop wireless communication in a densely scattered sensor field. This also provides the sensor network with self-organizing ability. The second objective is to regulate the access to the shared media such that communication resources are fairly and efficiently shared between the wireless sensor nodes.

The unique resource constraints and application requirements of sensor networks, however, denounce the MAC protocols for the conventional wireless networks inapplicable to the wireless sensor network paradigm. For example, the primary goal of a MAC protocol in an infrastructure-based cellular system is the provision of high QoS and bandwidth efficiency mainly with dedicated resource assignment strategy. Power conservation assumes only secondary importance as base stations have unlimited power supply and the mobile user can replenish exhausted batteries in the handset. Such an access scheme is impractical for sensor networks as there is no central controlling agent like the base station. Moreover, power efficiency directly influences network lifetime in a sensor network and hence, is of prime importance.

While Bluetooth and the *Mobile Ad-Hoc Network* (MANET) show similarities to the wireless sensor networks in terms of communication infrastructure, both of them consist of the nodes that have portable battery-powered devices, which can be replaced by the user. Hence, unlike the wireless sensor networks, power consumption is only of secondary importance in both of these systems. For example, the transmission power of a Bluetooth device is typically around 20 dBm, and the transmission range is of the order of 10s of meters. However, the transmission power of a sensor node is around 0 dBm, and hence, the radio range is much less than the one of a Bluetooth or MANET device. Therefore, none of the existing Bluetooth or MANET MAC protocols can be directly used in the wireless sensor networks due to the network lifetime concerns in a sensor network.

It is evident that the MAC protocol for sensor networks must have built-in power conservation, mobility management, and failure recovery strategies. Thus far, both *fixed allocation* and *random access* versions of medium access have been proposed [45, 49]. *Demand-based* MAC schemes may be unsuitable for sensor networks due to their large messaging overhead and link setup delay. Furthermore, contention-based channel access is deemed unsuitable due to their requirement to monitor the channel at all times, which is an energy-draining task. A qualitative overview of some MAC protocols proposed for wireless sensor networks are summarized in Table 2.4. The applicability of the fundamental MAC schemes in the wireless sensor networks is discussed along with some proposed MAC solutions using that access method as follows:

Table 2.4. Qualitative overview of MAC protocols for sensor networks.

MAC Protocol	Channel Access	Features and Advantages
SMACS [45]	Fixed allocation of duplex time slots at fixed frequency	- Exploits large available bandwidth compared to sensor data rate - Random wake up during setup and turning radio off while idle
Hybrid TDMA/FDMA [42]	Centralized frequency and time division	- Optimum number of channels for minimum system energy - Hardware based approach for system energy minimization
CSMA based [49]	Contention based random access	- Application phase shift and pre-transmit delay - Constant listening time for energy efficiency

- *TDMA-Based Medium Access:* In a *time-division multiple-access* (TDMA) scheme, a channel is granted to a source for a certain time duration. TDMA-based access schemes are inherently more energy-conserving compared to contention-based schemes since the duty cycle of the radio is reduced, and there is no contention-introduced overhead and collisions. It has been reasoned in [35] that MAC scheme for energy-constrained sensor networks should include a variant of TDMA since radios must be turned off during idling for precious power savings. The *Self-organizing Medium Access Control for Sensor networks* (SMACS)

[45] is such a time-slot based access protocol where each sensor node maintains a TDMA-like frame, called super frame, in which the node schedules different time slots to communicate with its known neighbors. SMACS achieves power conservation by using a random wake-up schedule during the connection phase and by turning the radio off during idle time slots. However, while TDMA-based access scheme minimizes the transmit-on time, it is not always preferred due to the associated time synchronization costs.

- *Hybrid TDMA/FDMA Based Medium Access:* While a pure TDMA-based access scheme dedicates the entire channel to a single sensor node, a pure *Frequency-Division Multiple Access* (FDMA) scheme allocates minimum signal bandwidth per node. Such contrast brings the tradeoff between the access capacity and the energy consumption. An analytical formula is derived in [42] to find the optimum number of channels, which gives the minimum system *power consumption*. This determines the hybrid TDMA-FDMA scheme to be used. The optimum number of channels is found to depend on the ratio of the power consumption of the transmitter to that of the receiver. If the transmitter consumes more power, a TDMA scheme is favored, while the scheme leans toward FDMA when the receiver consumes greater power. Such centrally controlled hybrid TDMA/FDMA based MAC scheme is already developed [42] for a time-sensitive machine monitoring application of the energy-constrained sensor network.

- *CSMA-Based Medium Access:* The traditional *Carrier-Sense Multiple Access* (CSMA) based schemes, which are based on carrier sensing and backoff mechanism, are deemed inappropriate since they all make the fundamental assumption of stochastically distributed traffic and tend to support independent point-to-point flows. On the contrary, the MAC protocol for sensor networks must be able to support variable, but highly correlated and dominantly periodic traffic. Any CSMA-based medium access scheme has two important components, the *listening mechanism* and the *backoff scheme*. A CSMA-based MAC scheme for sensor networks is presented in [49]. As reported and based on simulations in [49], the constant listen periods are energy efficient and the introduction of random delay provides robustness against repeated collisions.

2.4.2 ERROR CONTROL

In addition to the medium access control, error control of the transmitted data in the wireless sensor networks is another extremely important function of the data link layer. Error control is critical especially in some sensor network applications such as mobile tracking and machine monitoring. In general, the error control mechanisms in communication networks can be categorized into two main approaches, i.e., *Forward Error Correction* (FEC) and *Automatic Repeat reQuest* (ARQ).

ARQ-based error control mainly depends on the retransmission for the recovery of the lost data packets/frames. It is clear that such ARQ-based error control mechanism incurs significant additional retransmission cost and overhead. Although ARQ-based error control schemes are utilized at the data link layer for the other wireless networks, the usefulness of ARQ in sensor network applications is limited due to the scarcity of the energy and processing resources of the wireless sensor nodes. On the other hand, FEC schemes have inherent decoding complexity which require relatively considerable processing resources in the wireless sensor nodes. In this respect, simple error control codes with low-complexity encoding and decoding might present the best solutions for error control in the wireless sensor networks. In the following sections, the motivation and basic design requirements for FEC in the wireless sensor networks are explored.

Forward Error Correction. Due to the unpredictable and harsh nature of channels encountered in various wireless sensor network application scenarios, link reliability is detrimental to the performance of the entire sensor network. Some of the applications like mobile tracking and machine monitoring require high data precision. It is important to have good knowledge of the channel characteristics and implementation techniques for the design of efficient FEC schemes.

Channel *bit error rate* (BER) is a good indicator of link reliability. The BER is directly proportional to the symbol rate and inversely proportional to both the received signal-to-noise ratio and the transmitter power level. For a given error coding scheme, the received energy per symbol decreases if the data symbol rate and the transmission power remain unchanged. This corresponds to a higher BER at the decoder input than the one without coding. The decoder equipped with the coding scheme can then utilize the received redundant bits to correct the transmission errors to a certain degree. In fact, a good choice of the error correcting code can result in several orders of magnitude reduction in BER and an overall gain. The coding gain is generally expressed in terms of the additional transmit power needed to obtain the same BER without coding. For instance, a simple (15,11) Hamming code reduces BER by almost 10^3 and achieves a coding gain of 1.5 dB for binary phase shift keying modulated data and additive white gaussian noise model [48].

Therefore, the link reliability can be achieved either by increasing the output transmit power or the use of suitable FEC scheme. Due to the energy constraints of the wireless sensor nodes, increasing the transmit power is not a feasible option. Therefore, use of FEC is still the most efficient solution given the constraints of the sensor nodes. Although the FEC can achieve significant reduction in the BER for any given value of the transmit power, the additional processing power that is consumed during encoding and decoding must be considered when designing an FEC scheme. If the additional processing power is greater than the coding gain, then the whole process is not energy efficient and hence, the system is better off without coding. On the other hand, the FEC is a valuable asset to the sensor networks if the additional processing power is less than the transmission power savings. Thus, the tradeoff between this additional processing power and the associated coding gain need to be optimized in order to have powerful, energy-efficient and low-complexity FEC schemes for the error control in the wireless sensor networks.

In summary, it is evident that the performance of the entire wireless sensor network directly depends on the performance of the medium access and error control protocols used for the data link layer. Much additional research effort will be required to ultimately obtain the complete data link layer solutions that can efficiently address the unique challenges posed by the wireless sensor network paradigm.

2.5 PHYSICAL LAYER

The data link layer discussed in Section 2.4 multiplex the data streams and pass them to the the lowest layer in the communication architecture, i.e., the physical layer, for transmission. The physical layer is responsible for the conversion of bit streams into signals that are best suited for communication across the channel. More specifically, the physical layer is responsible for frequency selection, carrier frequency generation, signal detection, modulation and data encryption.

In a multi-hop sensor network, communicating nodes are linked by a wireless medium. These links can be formed by radio, infrared or optical media. To enable global operation of these networks, the chosen transmission medium must be available worldwide. One option for radio links is the use of *Industrial, Scientific and Medical* (ISM) bands, which offer license-free communication in most countries. The *International Table of Frequency Allocations*, contained in Article S5 of the Radio Regulations (volume 1), specifies some frequency bands that may be made available for ISM applications. These frequency bands and the corresponding center frequencies are shown in Table 2.5.
Some of these frequency bands are already being used for communication in cordless phone systems and wireless local area networks. For sensor networks, a small sized, low cost, ultralow power transceiver is required. According to the authors of [34], certain hardware constraints and the tradeoff between antenna efficiency and power consumption limit the choice of a carrier frequency for such transceivers to the ultra high frequency range. They also propose

Table 2.5. Frequency bands available for ISM applications.

Frequency Band	Center Frequency
6765-6795 kHz	6780 kHz
13553-13567 kHz	13560 kHz
26957-27283 kHz	27120 kHz
40.66-40.70 MHz	40.68 MHz
433.05-434.79 MHz	433.92 MHz
902-928 MHz	915 MHz
2400-2500 MHz	2450 MHz
5725-5875 MHz	5800 MHz
24-24.25 GHz	24.125 GHz
61-61.5 GHz	61.25 GHz
122-123 GHz	122.5 GHz
244-246 GHz	245 GHz

the use of the 433 MHz ISM band in Europe and the 917 MHz ISM band in North America. Transceiver design issues in these two bands are addressed in [13, 26]. The main advantages of using the ISM bands are the free radio, huge spectrum allocation and global availability. They are not bound to a particular standard, thereby giving more freedom for the implementation of power saving strategies in sensor networks. On the other hand, there are various rules and constraints, like power limitations and harmful interference from existing applications. These frequency bands are also referred to as unregulated frequencies in literature.

Much of the current hardware for sensor nodes is based upon *radio frequency* (RF) circuit design. The μAMPS wireless sensor node [42] uses a Bluetooth-compatible 2.4 GHz transceiver with an integrated frequency synthesizer. In addition, the low-power sensor device [49] uses a single channel RF transceiver operating at 916 MHz. The *Wireless Integrated Network Sensors* architecture [35] also uses radio links for communication.

Another possible mode of inter-node communication in sensor networks is by infrared. Infrared communication is license-free and robust to interference from electrical devices. Infrared based transceivers are cheaper and easier to build. Many of today's laptop's, PDAs, and mobile phones offer an *Infrared Data Association* interface. The main drawback is the requirement of a line-of-sight between the sender and receiver. This makes infrared a reluctant choice for transmission medium in the sensor network scenario.

An interesting development is the *Smart Dust* mote [21], which is an autonomous sensing, computing, and communication system that uses optical medium for transmission. Two transmission schemes, passive transmission using a *corner-cube retroreflector* and active communication using a laser diode and steerable mirrors, are examined in [47]. In the former, the mote does not require an on-board light source. A configuration of three mirrors is used to communicate a digital high or low. The latter uses an on-board laser diode and an active-steered laser communication system to send a tightly collimated light beam toward the intended receiver.

The unusual application requirements of the wireless sensor networks make the choice of transmission media more challenging. For instance, marine applications may require the use of the aqueous transmission medium. Here, one would like to use long-wavelength radiation that can penetrate the water surface. Inhospitable terrain or battlefield applications might encounter error prone channels and greater interference. Moreover, the antenna of the sensor nodes might not have the height and radiation power of those in traditional wireless devices. Hence, the choice of transmission medium must be supported by robust coding and modulation schemes that efficiently model these vastly different channel characteristics.

The choice of a good modulation scheme is critical for reliable communication in a sensor network. Binary and M-ary modulation schemes are compared in [42]. While an M-ary scheme can reduce the transmit on-time by sending multiple bits per symbol, it results in complex circuitry and increased radio power consumption. The authors formulate these trade-off parameters and conclude that under startup power dominant conditions, the binary modulation scheme is more energy efficient. Hence, M-ary modulation gains are significant only for low startup power systems. A low-power direct-sequence spread-spectrum modem architecture for sensor networks is presented in [6]. This low power architecture can be mapped to an application-specific integrated circuit technology to further improve efficiency.

The *Ultra Wideband* (UWB) or impulse radio has been used for baseband pulse radar and ranging systems and has recently drawn considerable interest for communication applications, especially in indoor wireless networks [30]. The UWB employs baseband transmission and thus, requires no intermediate or radio carrier frequencies. Generally, pulse position modulation is used. The main advantage of UWB is its resilience to multipath [7, 22, 25]. Low transmission power and simple transceiver circuitry make UWB an attractive candidate for the wireless sensor networks.

It is well known that long distance wireless communication can be expensive, both in terms of energy and cost. While designing the physical layer for sensor networks, energy minimization assumes significant importance, over and above the decay, scattering, shadowing, reflection, diffraction, multipath and fading effects. In general, the minimum output power required to transmit a signal over a distance d is proportional to d^n, where $2 <= n < 4$. The exponent n is closer to four for low-lying antennae and near-ground channels [44], as is typical in sensor network communication. This can be attributed to the partial signal cancellation by a ground-reflected ray. While trying to resolve these problems, it is important that the designer is aware of inbuilt diversities and exploits this to the fullest. For instance, multihop communication in a sensor network can effectively overcome shadowing and path loss effects, if the node density is high enough. Similarly, while propagation losses and channel capacity limit data reliability, this very fact can be used for spatial frequency re-use. Energy efficient physical layer solutions are currently being pursued by researchers. Although some of these topics have been addressed in literature, it still remains a vastly unexplored domain of the wireless sensor network.

2.6 TIME SYNCHRONIZATION

Instead of time synchronization between just the sender and receiver during an application like in the Internet, the sensor nodes in the sensor field have to maintain a similar time within a certain tolerance throughout the lifetime of the network. Combining with the criteria that sensor nodes have to be energy efficient, low-cost, and small in a multi-hop environment as described in Section 2, this requirement makes a challenging problem to solve. In addition, the sensor nodes may be left unattended for a long period of time, e.g. in deep space or on an ocean floor. For short distance multi-hop broadcast, the data processing time and the variation of the data processing time may contribute the most in time fluctuations and differences in the path delays. Also, the time difference between two sensor nodes is significant over time due to the wandering effect of the local clocks.

Small and low-end sensor nodes may exhibit device behaviors that may be much worst than large systems such as *personal computers (PCs)*. Some of the factors influencing time synchronization in large systems also apply to sensor networks [23]; they are *temperature, phase noise, frequency noise, asymmetric delays,* and *clock glitches*.

- *Temperature:* Since sensor nodes are deployed in various places, the temperature variation throughout the day may cause the clock to speed up or slow down. For a typical PC, the clock drifts few parts per million during the day [29]. For low end sensor nodes, the drifting may be even worst.

- *Phase Noise:* Some of the causes of phase noise are due to access fluctuation at the hardware interface, response variation of the operating system to interrupts, and jitter in the network delay. The jitter in the network delay may be due to medium access and queueing delays.

- *Frequency Noise:* The frequency noise is due to the unstability of the clock crystal. A low-end crystal may experience large frequency fluctuation, because the frequency spectrum of the crystal has large sidebands on adjacent frequencies.

- *Asymmetric Delay:* Since sensor nodes communicate with each other through the wireless medium, the delay of the path from one node to another may be different than the return path. As a result, an asymmetric delay may cause an offset to the clock that can not be detected by a variance type method [23]. If the asymmetric delay is static, the time offset between any two nodes is also static. The asymmetric delay is bounded by one-half the round trip time between the two nodes [23].

- *Clock Glitches:* Clock glitches are sudden jumps in time. This may be caused by hardware or software anomalies such as frequency and time steps.

Table 2.6. Three types of timing techniques.

Type	Description
(1) Relies on fixed time servers to synchronize the network	-The nodes are synchronized to time servers that are readily available. These time servers are expected to be robust and highly precise.
(2) Translates time throughout the network	-The time is translated hop-by-hop from the source to the sink. In essence, it is a time translation service.
(3) Self-organizes to synchronize the network	-The protocol does not depend on specialized time servers. It automatically organizes and determines the master nodes as the temporary time-servers.

There are three types of timing techniques as shown in Table 2.6, and each of these types has to address the challenges mentioned above. In addition, the timing techniques have to be energy aware since the batteries of the sensor nodes are limited. Also, they have to address the mapping between the sensor network time and the Internet time, e.g., universal coordinated time. In the following, examples of these types of timing techniques are described, namely the *Network Time Protocol* (NTP) [28], the *Reference-Broadcast Synchronization* (RBS) [9], and the *Time-Diffusion Synchronization Protocol* (TDP) [46].

In Internet, the NTP is used to discipline the frequency of each node's oscillator. It may be useful to use NTP to disciple the oscillators of the sensor nodes, but the connection to the time servers may not be possible because of frequent sensor node failures. In addition, disciplining all the sensor nodes in

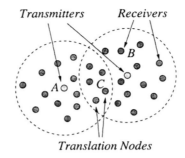

Figure 2.6. Illustration of the RBS.

the sensor field may be a problem due to interference from the environment and large variation of delay between different parts of the sensor field. The interference can temporarily disjoint the sensor field into multiple smaller fields causing undisciplined clocks among these smaller fields. The NTP protocol may be considered as type (1) of the timing techniques. In addition, it has to be refined to address the timing challenges in the wireless sensor networks.

As for type (2) of the timing techniques, the RBS provides an instantaneous time synchronization among a set of receivers that are within the reference broadcast of the transmitter. The transmitter broadcasts m reference packets. Each of the receivers that are within the broadcast range records the time-of-arrival of the reference packets. Afterwards, the receivers communicate with each other to determine the offsets. To provide multi-hop synchronization, it is proposed to use nodes that are receiving two or more reference broadcasts from different transmitters as translation nodes. These translation nodes are used to translate the time between different broadcast domains. As shown in Figure 2.6, nodes A, B, and C are the transmitter, receiver, and translation nodes, respectively.

Another emerging timing technique is the TDP. The TDP is used to maintain the time throughout the network within a certain tolerance. The tolerance level can be adjusted based on the purpose of the sensor networks. The TDP automatically self-configures by electing master nodes to synchronize the sensor network. In addition, the election process is sensitive to energy requirement as well as the quality of the clocks. The sensor network may be deployed in unattended areas, and the TDP still synchronizes the unattended network to a common time. It is considered as a type (3) of the timing techniques.

In summary, these timing techniques may be used for different types of applications as discussed in Section 2.1; each of these types has its benefits. A time-sensitive application has to choose not only the type of timing techniques but also the type of transport, network, datalink, and physical schemes as described in Sections 2.2, 2.3, 2.4, and 2.5, respectively. This is because different protocols provide different features and services to the time-sensitive application.

2.7 CONCLUSION

The design-principles of developing application, transport, network, datalink, and physical schemes as well as timing techniques are described. They are to guide and encourage new developments in the wireless sensor network areas. As the technologies for the wireless sensor networks advanced, the pervasive daily usage of the wireless sensor networks is foreseeable in the near future.

ACKNOWLEDGMENTS

The authors wish to thank Dr. Ian F. Akyildiz for his encouragement and support.

REFERENCES

[1] Agre, J., and Clare, L., "An Integrated Architecture for Cooperative Sensing Networks," *IEEE Computer Magazine*, pp.106-108, May 2000.

[2] Balakrishnan, H., Padmanabhan, V. N., Seshan, S., and Katz, R. H., "A Comparison of Mechanisms for Improving TCP Performance over Wireless Links", *IEEE/ACM Trans. Networking*, vol. 5, no. 6, pp. 756-769, December 1997.

[3] Bhardwaj, M., Garnett, T., and Chandrakasan, A. P., "Upper Bounds on the Lifetime of Sensor Networks," *IEEE International Conference on Communications '01*, Helsinki, Finland, June 2001.

[4] Bonnet, P., Gehrke J., and Seshadri, P., "Querying the Physical World," *IEEE Personal Communications*, pp. 10-15, October 2000.

[5] Bulusu, N., Estrin, D., Girod, L., and Heidemann, J., "Scalable Coordination for Wireless Sensor Networks: Self-Configuring Localization Systems," *International Symposium on Communication Theory and Applications (ISCTA 2001)*, Ambleside, UK, July 2001.

[6] Chien, C., Elgorriaga, I., and McConaghy, C., "Low-Power Direct-Sequence Spread-Spectrum Modem Architecture For Distributed Wireless Sensor Networks," in *ISLPED '01*, Huntington Beach, California, USA, August 2001.

[7] Cramer, R.J., Win, M. Z., and Scholtz, R. A., "Impulse radio multipath characteristics and diversity reception," *IEEE International Conference on Communications '98*, vol. 3, pp. 1650-1654, 1998.

[8] Elson, J., and Estrin, D., "Random, Ephemeral Transaction Identifiers in Dynamic Sensor Networks," *Proceedings 21st International Conference on Distributed Computing Systems*, pp. 459-468, Phoenix, Arizona, USA, April 2001.

[9] Elson, J., Girod, L., and Estrin, D., "Fine-Grained Network Time Synchronization using Reference Broadcasts," in *Proceedings of the Fifth Symposium on Operating Systems Design and Implementation (OSDI 2002)*, Boston, MA, USA, December 2002.

[10] Estrin, D., Girod, L., Pottie, G., and Srivastava, M., "Instrumenting the World With Wireless Sensor Networks," *International Conference on Acoustics, Speech, and Signal Processing (ICASSP 2001)*, Salt Lake City, Utah, USA, May 2001.

[11] Estrin, D., Govindan, R., Heidemann, J., and Kumar, S., "Next Century Challenges: Scalable Coordination in Sensor Networks," *ACM Mobicom'99*, pp.263-270, Seattle, Washington, USA, August 1999.

[12] Estrin, D., Govindan R., and Heidemann J., "Embedding the Internet," *Commun. ACM*, vol. 43, pp. 38-41, May 2000.

[13] Favre, P. and et al., "A 2V, 600µA, 1 GHz BiCMOS Super Regenerative Receiver for ISM Applications," *IEEE J. Solid-State Circuits*, vol. 33, pp.2186-2196, December 1998.

[14] Floyd, S., Jacobson, V., Liu, C., Macanne, S., and Zhang, L., "A Reliable Multicast Framework for Lightweight Sessions and Application Level Framing," *IEEE/ACM Trans. Networking*, vol. 5, no. 6, pp.784-803, December 1997.

[15] Hedetniemi, S., Hedetniemi, S., and Liestman, A., "A Survey of Gossiping and Broadcasting in Communication Networks," *Networks*, vol. 18, no. 4, pp. 319-349, 1988.

[16] Heinzelman, W. R., Kulik, J., and Balakrishnan, H., "Adaptive Protocols for Information Dissemination in Wireless Sensor Networks," *ACM Mobicom'99*, pp. 174-185, Seattle, Washington, USA, August 1999.

[17] Heinzelman, W. R., Chandrakasan, A., and Balakrishnan, H., "Energy-Efficient Communication Protocol for Wireless Microsensor Networks," *IEEE Proceedings of the Hawaii International Conference on System Sciences*, pp. 1-10, Maui, Hawaii, USA, January 2000.

[18] Intanagonwiwat, C., Govindan, R., and Estrin, D., "Directed Diffusion: A Scalable and Robust Communication Paradigm for Sensor Networks," *ACM Mobicom'00*, pp. 56-67, Boston, MA, USA, August 2000.

[19] Tilak, S., Abu-Ghazaleh, N. B., and Heinzelman, W., "Infrastructure Tradeoffs for Sensor Networks," *In Proc. WSNA 2002*, Atlanta, GA, USA, September 2002.

[20] Jaikaeo, C., Srisathapornphat, C., and Shen, C., "Diagnosis of Sensor Networks," *IEEE International Conference on Communications '01*, Helsinki, Finland, June 2001.

[21] Kahn, J. M., Katz, R. H., and Pister, K. S. J., "Next Century Challenges: Mobile Networking for Smart Dust," *ACM Mobicom'99*, pp.271-278, Seattle, Washington, USA, August 1999.

[22] Lee, H., Han, B., Shin, Y., and Im, S., "Multipath characteristics of impulse radio channels," *Vehicular Technology Conference Proceedings 2000*, vol. 3, pp. 2487-2491, Tokyo, Japan, May 2000.

[23] Levine, J., "Time Synchronization Over the Internet Using an Adaptive Frequency-Locked Loop," *IEEE Transaction on Ultrasonics, Ferroelectrics, and Frequency Control*, vol. 46, no. 4, pp. 888-896, July 1999.

[24] Li, L., and Halpern, J. Y., "Minimum-Energy Mobile Wireless Networks Revisited," *IEEE International Conference on Communications ICC'01*, Helsinki, Finland, June 2001.

[25] J. Le Martret, C. and Giannakis, G. B., "All-Digital Impulse radio for MUI/ISI-resilient multiuser communications over frequency-selective multipath channels," *MILCOM 2000. 21st Century Military Communications Conference Proceedings*, vol. 2, pp. 655-659, Los Angeles, CA, USA, October 2000.

[26] Melly, T., Porret, A., Enz, C. C., and Vittoz, E. A., "A 1.2 V, 430 MHz, 4dBm Power Amplifier and a 250 µW Front-End, using a Standard Digital CMOS Process" *IEEE International Symposium on Low Power Electronics and Design Conf.*, pp.233-237, San Diego, CA, USA, August 1999.

[27] MICA Motes and Sensors, http://www.xbow.com.

[28] Mills, D. L. (1994). "Internet Time Synchronization: The Network Time Protocol," *In Z. Yang and T. A. Marsland, editors, Global States and Time in Distributed Systems.* IEEE Computer Society Press.

[29] Mills, D. L., "Adaptive Hybrid Clock Discipline Algorithm for the Network Time Protocol," *IEEE/ACM Trans. on Networking,* vol. 6, no. 5, pp. 505-514, October 1998.

[30] Mireles, F. R. and Scholtz, R. A., "Performance of equicorrelated ultra-wideband pulse-position-modulated signals in the indoor wireless impulse radio channel," *IEEE Conference on Communications, Computers and Signal Processing '97,* vol. 2, pp 640-644, Victoria, BC, Canada, August 1997.

[31] Mirkovic, J., Venkataramani, G. P., Lu, S., and Zhang, L., "A Self-Organizing Approach to Data Forwarding in Large-Scale Sensor Networks," *IEEE International Conference on Communications ICC'01,* Helsinki, Finland, June 2001.

[32] Perkins, C. (2000). *Ad Hoc Networks.* Addison-Wesley.

[33] Perrig, A., Szewczyk, R., Wen, V., Culler, D., and Tygar, J. D., "SPINS: Security Protocols for Sensor Networks," *Proc. of ACM MobiCom'01,* pp. 189-199, Rome, Italy, July 2001.

[34] Porret, A., Melly, T., Enz, C. C., and Vittoz, E. A., "A Low-Power Low-Voltage Transceiver Architecture Suitable for Wireless Distributed Sensors Network," *IEEE International Symposium on Circuits and Systems '00,* vol. 1, pp.56-59, Geneva, Switzerland, May 2000.

[35] Pottie, G.J. and Kaiser, W.J., "Wireless Integrated Network Sensors," *Communications of the ACM,* vol. 43, no. 5, pp. 551-8, May 2000.

[36] Wan, C. Y., Campbell, A. T., and Krishnamurthy, L., "PSFQ: A Reliable Transport Protocol for Wireless Sensor Networks," *In Proc. WSNA 2002,* Atlanta, GA, USA, September 2002.

[37] Rabaey, J., Ammer, J., L. da Silva Jr., J., and Patel, D., "PicoRadio: Ad-hoc Wireless Networking of Ubiquitous Low-Energy Sensor/Monitor Nodes," *Proceedings of the IEEE Computer Society Annual Workshop on VLSI (WVLSI'00),* pp. 9-12, Orlando, Florida, USA, April 2000.

[38] Rabaey, J. M., Ammer, M. J., L. da Silva Jr., J., Patel, D., and Roundy, S., "PicoRadio Supports Ad Hoc Ultra-Low Power Wireless Networking," *IEEE Computer Magazine,* vol. 33, pp. 42-48, July 2000.

[39] Sankarasubramaniam, Y., Akan, O. B., and Akyildiz, I. F., "ESRT: Event-to-Sink Reliable Transport for Wireless Sensor Networks," in *Proc. ACM MOBIHOC 2003,* pp. 177-188, Annapolis, Maryland, USA, June 2003.

[40] Savvides, A., Han, C., and Srivastava, M., "Dynamic Fine-Grained Localization in Ad-Hoc Networks of Sensors," *Proc. of ACM MobiCom'01,* pp. 166-179, Rome, Italy, July 2001.

[41] Shen, C., Srisathapornphat, C., and Jaikaeo, C., "Sensor Information Networking Architecture and Applications," *IEEE Personal Communications,* pp. 52-59, August 2001.

[42] Shih, E., Cho, S., Ickes, N., Min, R., Sinha, A., Wang, A., and Chandrakasan, A., "Physical layer Driven Protocol and Algorithm Design for Energy-Efficient Wireless Sensor Networks," *ACM Mobicom'01,* pp. 272-286, Rome, Italy, July 2001.

[43] Slijepcevic, S. and Potkonjak, M., "Power Efficient Organization of Wireless Sensor Networks," *IEEE International Conference on Communications '01,* Helsinki, Finland, June 2001.

[44] Sohrabi, K., Manriquez, B., and Pottie, G. J., "Near-ground Wideband Channel Measurements in 800-1000 MHz," *IEEE Proc.of 49th Vehicular Technology Conference,* Houston, TX, USA, May 1999.

[45] Sohrabi, K., Gao, J., Ailawadhi, V., and Pottie, G. J., "Protocols for Self-Organization of a Wireless Sensor Network," *IEEE Personal Communications,* pp. 16-27, October 2000.

[46] Su, W. and Akyildiz, I. F., "Time-Diffusion Synchronization Protocol for Sensor Networks," *Georgia Tech Technical Report,* 2003.

[47] Warneke, B., Liebowitz, B., and Pister, K. S. J., "Smart Dust: Communicating with a Cubic-Millimeter Computer," *IEEE Computer Magazine,* pp. 2-9, January 2001.

[48] Wicker, S. (1995). *Error Control Coding for Digital Communication and Storage.* Prentice-Hall.

[49] Woo, A. and Culler, D., "A Transmission Control Scheme for Media Access in Sensor Networks," *ACM Mobicom'01,* pp.221-235, Rome, Italy, July 2001.

Chapter 3

ENERGY EFFICIENT DESIGN OF WIRELESS SENSOR NODES

Vijay Raghunathan[1], Curt Schurgers[2], Sung Park[3] and Mani B. Srivastava[1]
[1]*Department of Electrical Engineering, University of California Los Angeles*
[2]*Department of Electrical and Computer Engineering, University of California San Diego*
[3]*Raytheon Inc., Los Angeles*

Abstract: The battery driven nature of wireless sensor networks, combined with the need for operational lifetimes of months to years, mandates that energy efficiency be treated as a metric of utmost priority while designing these distributed sensing systems. This chapter presents an overview of energy-centric sensor node design techniques that enable designers to significantly extend system and network lifetime. Such extensions to battery life can only be obtained by eliminating energy inefficiencies from all aspects of the sensor node, ranging from the hardware platform to the operating system, network protocols, and application software.

Key words: Wireless Sensor Networks, Low Power Design, Energy Efficient Design, Dynamic Power Management, Battery Aware Design

3.1 INTRODUCTION

Self-configuring wireless sensor networks can be invaluable in many civil and military applications for collecting, processing, and disseminating wide ranges of complex environmental data. They have therefore, attracted considerable research attention in the last few years. The WINS [1] and SmartDust [2] projects for instance, aim to integrate sensing, computing, and wireless communication capabilities into a small form factor to enable low-cost production of these tiny nodes in large numbers. Several other groups are investigating efficient hardware/software system architectures, signal processing algorithms, and network protocols for wireless sensor networks [3], [4], [5].

Sensor nodes are battery-driven, and hence operate on an extremely frugal energy budget. Further, they must have a lifetime on the order of months to years, since battery replacement is not an option for networks with thousands of physically embedded nodes. In some cases, these networks may be required to operate solely on energy scavenged from the environment through seismic, photovoltaic, or thermal conversion. This transforms energy consumption into the most important factor that determines sensor node lifetime.

Conventional low-power design techniques [6] and hardware architectures only provide point solutions which are insufficient for these highly energy constrained systems. Energy optimization, in the case of sensor networks, is far more complex, since it involves not only reducing the energy consumption of a single sensor node, but also maximizing the lifetime of an entire network. The network lifetime can be maximized only by incorporating energy-awareness into every stage of wireless sensor network design and operation, thus empowering the system with the ability to make dynamic tradeoffs between energy consumption, system performance, and operational fidelity. This new networking paradigm, with its extreme focus on energy efficiency, poses several system and network design challenges that need to be overcome to fully realize the potential of the wireless sensor systems.

A quite representative application in wireless sensor networks is event tracking, which has widespread use in applications such as security surveillance and wildlife habitat monitoring. Tracking involves a significant amount of collaboration between individual sensors to perform complex signal processing algorithms such as Kalman Filtering, Bayesian Data Fusion, and Coherent Beamforming. This collaborative signal processing nature of sensor networks offers significant opportunities for energy management. For example, just the decision of whether to do the collaborative signal processing at the user end-point or somewhere inside the network has significant implication on energy and lifetime. We will use tracking as the driver to illustrate many of the techniques presented in this chapter.

Chapter overview

This chapter describes architectural and algorithmic techniques that designers can use to design energy aware wireless sensor networks. We start off with an analysis of the power consumption characteristics of typical sensor node architectures, and identify the various factors that affect system lifetime. We then present a suite of techniques that perform aggressive energy optimization while targeting all the individual sensor node components. Designing a sensor node that has a high battery life requires the use of a well-structured design methodology, which enables energy aware design and operation of all aspects of the sensor node, from the underlying hardware platform, to the application software and network protocols. Adopting such a holistic approach also ensures that energy awareness is incorporated not only into individual sensor nodes, but also into groups of communicating nodes, and the entire sensor network (although the latter two categories are not discussed in this chapter). By following an energy-aware design methodology based on techniques such as in this chapter, designers can enhance sensor node lifetime by orders of magnitude.

3.2 WHERE DOES THE POWER GO?

The first step in designing energy aware sensor systems involves analyzing the power dissipation characteristics of a wireless sensor node. Systematic power analysis of a sensor node is extremely important to identify power bottlenecks in the system, which can then be the target of aggressive optimization. We analyze two popular sensor nodes from a power consumption perspective, and discuss how decisions taken during node design can significantly impact the system energy consumption.

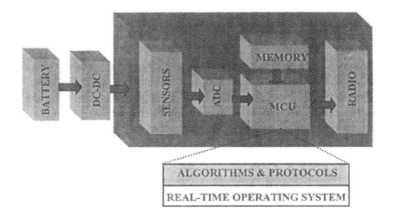

Figure 3.1 System architecture of a canonical wireless sensor node

The system architecture of a canonical wireless sensor node is shown in Figure 3.1. The node is comprised of four subsystems: (i) a computing subsystem consisting of a microprocessor or microcontroller, (ii) a communication subsystem consisting of a short range radio for wireless communication, (iii) a sensing subsystem that links the node to the physical world and consists of a group of sensors and actuators, and (iv) a power supply subsystem, which houses the battery and the DC-DC converter, and powers the rest of the node. The sensor node shown in Figure 3.1 is representative of commonly used node architectures such as [1], [2].

3.2.1 Microcontroller unit (MCU)

Providing intelligence to the sensor node, the MCU is responsible for control of the sensors, and execution of communication protocols and signal processing algorithms on the gathered sensor data. Commonly used MCUs are Intel's StrongARM microprocessor and Atmel's AVR microcontroller. The power-performance characteristics of MCUs have been studied extensively, and several techniques have been proposed to estimate the power consumption of these embedded processors [7], [8]. While the choice of MCU is dictated by the required performance levels, it can also significantly impact the node's power dissipation characteristics. For example, the StrongARM microprocessor from Intel, used in high end sensor nodes, consumes around 400mW of power while executing instructions, whereas the ATmega103L AVR microcontroller from Atmel consumes only around 16.5mW, but provides much lower performance. Thus, the choice of MCU should be dictated by the application scenario, to achieve a close match between the performance level offered by the MCU, and that demanded by the application. Further, MCUs usually support various operating modes, including *Active, Idle,* and *Sleep* modes, for power management purposes. Each mode is characterized by a different amount of power consumption. For example, the StrongARM consumes 50mW of power in the *Idle* mode, and just 0.16mW in the *Sleep* mode. However, transitioning between operating modes involves a power and latency overhead. Thus, the power consumption levels of the various modes, the transition costs, and the amount of time spent by the MCU in each mode, all have a significant bearing on the total energy consumption (battery lifetime) of the sensor node.

3.2.2 Radio

The sensor node's radio enables wireless communication with neighboring nodes and the outside world. There are several factors that affect the power consumption characteristics of a radio, including the type of modulation scheme used, data rate, transmit power (determined by the transmission distance), and the operational duty cycle. In general, radios can operate in four distinct modes of operation, namely *Transmit, Receive, Idle,* and *Sleep* modes. An important observation in the case of most radios is that, operating in *Idle* mode results in significantly high power consumption, almost equal to the power consumed in the *Receive* mode [11]. Thus, it is important to completely shutdown the radio rather than transitioning to *Idle* mode, when it is not transmitting or receiving data. Another influencing factor is that, as the radio's operating mode changes, the transient activity in the radio electronics causes a significant amount of power dissipation. For example, when the radio switches from sleep mode to transmit mode to send a packet, a significant amount of power is consumed for starting up the transmitter itself [9].

3.2.3 Sensors

Sensor transducers translate physical phenomena to electrical signals, and can be classified as either analog or digital devices depending on the type of output they produce. There exists a diversity of sensors that measure environmental parameters such as temperature, light intensity, sound, magnetic fields, image *etc.* There are several sources of power consumption in a sensor, including (i) signal sampling and conversion of physical signals to electrical ones, (ii) signal conditioning, and (iii) analog to digital conversion. Given the diversity of sensors there is no typical power consumption number. In general, however, passive sensors such as temperature, seismic *etc.*, consume negligible power relative to other components of the sensor node. However, active sensors such as sonar rangers, array sensors such as imagers, and narrow field-of-view sensors that require repositioning such as cameras with pan-zoom-tilt can be large consumers of power. A significant source of power consumption in the sensing subsystem is the analog-to-digital converter (ADC) that converts an analog sensor reading into a digital value. The power consumption of ADCs is related to the data conversion speed and the resolution of the ADC. It has been shown that there exists a fundamental limit on the speed-resolution product per unit power consumption of ADCs, and that state-of-the-art ADCs have almost reached this fundamental limit. For example, seismic sensors often use 24-bit converters that operate at a conversion rate of on the order of thousands of samples per second. This results in a high speed-resolution product and, therefore, high power consumption. Therefore, node designers should perform a detailed study of applications' resolution and speed requirements and design the ADC to closely match this, without over provisioning and hence, introducing significant power overhead.

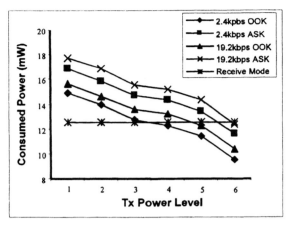

Figure 3.2 Power consumption of an RFM radio in various modes of operation

3.2.4 Power analysis of sensor nodes

Table 3.1 shows the power consumption characteristics of Rockwell's WINS node [10], which represents a high-end sensor node, and is equipped with a powerful StrongARM SA-1100 processor from Intel, a radio module from Conexant Systems, and several sensors including acoustic and seismic ones. Table 3.2 gives the characteristics of the MEDUSA-II, an experimental sensor node developed at the Networked and Embedded Systems Lab, UCLA. The MEDUSA node, designed to be ultra low power, is a low-end sensor node similar to the COTS Motes developed as part of the SmartDust project [2]. It is equipped with an AVR microcontroller from ATMEL, a low-end RFM radio module, and a few sensors. As can be seen from the tables, the power dissipation characteristics of the two nodes differ significantly. There are several inferences that can be drawn from these tables:

- Using low-power components and trading off unnecessary performance for power savings during node design, can have a significant impact, up to a few orders of magnitude.
- The node power consumption is strongly dependent on the operating modes of the components. For example, as Table 3.1 shows, the WINS node consumes only around one sixth the power when the MCU is in *Sleep* mode, than when it is in *Active* mode.
- Due to extremely small transmission distances, the power consumed while receiving data can often be greater than the

power consumed while transmitting packets. This is illustrated in Figure 3.2, which plots the power consumed by an RFM radio in *Receive* mode as well as in various *Transmit* modes, as a function of the output power level. Thus, conventional network protocols which usually assume the receive power to be negligible, are no longer efficient for sensor networks, and customized protocols that explicitly account for receive power have to be developed.

- The power consumed by the node with the radio in *Idle* mode is approximately the same with the radio in *Receive* mode. Thus, operating the radio in *Idle* mode does not provide any advantage in terms of power. Several network protocols have often ignored this fact, leading to fallacious savings in power consumption, as pointed out in [11]. Therefore, the radio should be completely shut off whenever possible, to obtain energy savings.

Table 3.1. Power analysis of Rockwell's WINS nodes

MCU Mode	Sensor Mode	Radio Mode	Power (mW)
Active	On	Tx (Power: 36.3 mW)	1080.5
		Tx (Power: 19.1 mW)	986.0
		Tx (Power: 13.8 mW)	942.6
		Tx (Power: 3.47 mW)	815.5
		Tx (Power: 2.51 mW)	807.5
		Tx (Power: 0.96 mW)	787.5
		Tx (Power: 0.30 mW)	773.9
		Tx (Power: 0.12 mW)	771.1
Active	On	Rx	751.6
Active	On	Idle	727.5
Active	On	Sleep	416.3
Active	On	Removed	383.3
Sleep	On	Removed	64.0
Active	Removed	Removed	360.0

3.2.5 Battery issues

The battery supplies power to the complete sensor node, and hence plays a vital role in determining sensor node lifetime. Batteries are complex devices whose operation depends on many factors including battery dimensions, type of electrode material used, and diffusion rate of the active materials in the electrolyte. In addition, there can be several non-idealities that can creep in during battery operation, which adversely affect system lifetime. We describe the various battery non-idealities, and discuss system level design approaches that can be used to prolong battery lifetime.

Table 3.2. Power analysis of UCLA's Medusa nodes

MCU Mode	Sensor Mode	Radio Mode	Mod. Scheme	Data Rate	Power (mW)
Active	On	Tx (0.7368 mW)	OOK	2.4 kbps	24.58
		Tx (0.0979 mW)	OOK	2.4 kbps	19.24
		Tx (0.7368 mW)	OOK	19.2 kbps	25.37
		Tx (0.0979 mW)	OOK	19.2 kbps	20.05
		Tx (0.7368 mW)	ASK	2.4 kbps	26.55
		Tx (0.0979 mW)	ASK	2.4 kbps	21.26
		Tx (0.7368 mW)	ASK	19.2 kbps	27.46
		Tx (0.0979 mW)	ASK	19.2 kbps	22.06
Active	On	Rx	Any	Any	22.20
Active	On	Idle	Any	Any	22.06
Active	On	Off	Any	Any	9.72
Idle	On	Off	Any	Any	5.92
Sleep	Off	Off	Any	Any	0.02

- **Rated capacity effect**

The most important factor that affects battery lifetime is the discharge rate or the amount of current drawn from the battery. Every battery has a rated current capacity, specified by the manufacturer. Drawing higher current than the rated value leads to a significant reduction in battery life. This is because, if a high current is drawn from the battery, the rate at which active ingredients diffuse through the electrolyte falls behind the rate at which they are consumed at the electrodes. If the high discharge rate is maintained for a long time, the electrodes run out of active materials, resulting in battery death even though active ingredients are still present in the electrolyte. Hence, to avoid battery life degradation, the amount of current drawn from the battery should be kept under tight check. Unfortunately, depending on the battery type (Lithium Ion, NiMH, NiCd, Alkaline, *etc.*), the minimum required current consumption of sensor nodes often exceeds the rated current capacity, leading to sub-optimal battery lifetime.

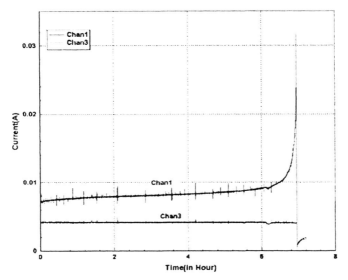

Figure 3.3 Current drawn from the battery (Chan 1) and current supplied to the sensor node (Chan 3)

- **Relaxation effect**

The effect of high discharge rates can be mitigated to a certain extent through battery relaxation. If the discharge current from the battery is cut off or reduced, the diffusion and transport rate of active materials catches up with the depletion caused by the discharge. This phenomenon is called the relaxation effect, and enables the battery to recover a portion of its lost capacity. Battery lifetime can be significantly increased if the system is operated such that the current drawn from the battery is frequently reduced to very low values, or is completely shut off [12].

3.2.6 DC-DC converter

The DC-DC converter is responsible for providing a constant supply voltage to the rest of the sensor node while utilizing the complete capacity of the battery. The efficiency factor associated with the converter plays a big role in determining battery lifetime [13]. A low efficiency factor leads to significant energy loss in the converter, reducing the amount of energy available to other sensor node components. Also, the voltage level across the battery terminals constantly decreases as it gets discharged. The converter therefore draws increasing amounts of current from the battery to maintain a constant supply voltage to the sensor node. As a result, the current drawn from the battery becomes progressively higher than the current that actually gets supplied to the rest of the sensor node. This leads to depletion in battery life due to the rated capacity effect, as explained earlier. Figure 3.3 shows the difference in current drawn from the battery and the current delivered to the sensor node for a Lithium-Ion coin cell battery.

3.3 NODE LEVEL ENERGY OPTIMIZATION

Having studied the power dissipation characteristics of wireless sensor nodes, we now focus our attention to the issue of minimizing the power consumed by these nodes. In order to incorporate energy awareness into the network, it is necessary to first develop hardware/software design methodologies and system architectures that enable energy-aware design and operation of individual sensor nodes in the network.

3.3.1 Power aware computing

Advances in low-power circuit and system design [6] have resulted in the development of several ultra low power microprocessors, and microcontrollers. In addition to using low-power hardware components during sensor node design, operating the various system resources in a power-aware manner through the use of dynamic power management (DPM) [14] can reduce energy consumption further, increasing battery lifetime. A commonly used power management scheme is based on idle component shutdown, in which the sensor node, or parts of it, is shutdown or sent into one of several low-power states if no interesting events occur. Such event-driven power management is extremely crucial in maximizing node lifetime. The core issue in shutdown based DPM is deciding the state transition policy [14], since different states are characterized by different amounts of power consumption, and state transitions have a non-negligible power and time overhead.

While shutdown techniques save energy by turning off idle components, additional energy savings are possible in active state through the use of dynamic voltage scaling (DVS) [15]. Most microprocessor-based systems have a time varying computational load, and hence peak system performance is not always required. DVS exploits this fact by dynamically adapting the processor's supply voltage and operating frequency to just meet the instantaneous processing requirement, thus trading off unutilized performance for energy savings. DVS based power management, when applicable, has been shown to have significantly higher energy efficiency compared to shutdown based power management due to the convex nature of the energy- speed curve [15]. Several modern processors such as Intel's StrongARM and Transmeta's Crusoe support scaling of voltage and frequency, thus providing control knobs for energy-performance management.

For example, consider the target-tracking application discussed earlier. The duration of node shutdown can be used as a control knob to trade off tracking fidelity against energy. A low operational duty cycle for a node reduces energy consumption at the cost of a few missed detections. Further, the target update rate varies, depending on the Quality of Service requirements of the user. A low update rate implies more available latency to process each sensor data sample, which can be exploited to reduce energy through the use of DVS.

3.3.2 Energy aware software

Despite the higher energy efficiency of application specific hardware platforms, the advantage of flexibility offered by microprocessor and DSP based systems has resulted in the increasing use of programmable solutions during system design. Sensor network lifetime can be significantly enhanced if the system software, including the operating system (OS), application layer, and network protocols are all designed to be energy aware.

The OS is ideally poised to implement shutdown-based and DVS-based power management policies, since it has global knowledge of the performance and fidelity requirements of all the applications, and can directly control the underlying hardware resources, fine tuning the available performance-energy control knobs. At the core of the OS is a task scheduler, which is responsible for scheduling a given set of tasks to run on the system while ensuring that timing constraints are satisfied. System lifetime can be increased considerably by incorporating energy awareness into the task scheduling process [16], [17].

The energy aware real-time scheduling algorithm proposed in [16] exploits two observations about the operating scenario of wireless systems, to provide an adaptive power *vs.* fidelity tradeoff. The first observation is that these systems are inherently designed to operate resiliently in the presence of varying fidelity in the form of data losses, and errors over wireless links. This ability to adapt to changing fidelity is used to trade off against energy. Second, these systems exhibit significant correlated variations in computation and communication processing load due to underlying time-varying physical phenomena. This observation is exploited to proactively manage energy resources by predicting processing requirements. The voltage is set according to predicted computation requirements of individual task instances, and adaptive feedback control is used to keep the system fidelity (timing violations) within specifications.

The energy-fidelity tradeoff can be exploited further by designing the application layer to be energy scalable. This can be achieved by transforming application software such that the most significant computations are performed first. Thus, terminating the algorithm prematurely due to energy constraints, does not impact the result severely. For example, the target tracking application described earlier involves the extensive use of signal filtering algorithms such as Kalman filtering. Transforming the filtering algorithms to be energy scalable trades off computational precision (and hence, tracking precision) for energy consumption. Several transforms to enhance the energy scalability of DSP algorithms are presented in [18].

3.3.3 Power management of radios

While power management of embedded processors has been studied extensively, incorporating power awareness into radio subsystems has remained relatively unexplored. Power management of radios is extremely important since wireless communication is a major power consumer during system operation. One way of characterizing the importance of this problem is in terms of the ratio of the energy spent in sending one bit to the energy spent in executing one instruction. While it is not quite fair to compare this ratio across nodes without normalizing for transmission range, bit error probability, and the complexity of instruction (8-bit vs. 32-bit), this ratio is nevertheless useful. Example values are from 1500 to 2700 for Rockwell's WIN nodes, 220 to 2900 for the MEDUSA II nodes, and is around 1400 for the WINS NG 2.0 nodes from the Sensoria Corporation that are used by many researchers.

The power consumed by a radio has two main components to it, an RF component that depends on the transmission distance and modulation parameters, and an electronics component that accounts for the power consumed by the circuitry that performs frequency synthesis, filtering, up-converting, *etc*. Radio power management is a non-trivial problem, particularly since the well-understood techniques of processor power management may not be directly applicable. For example, techniques such as dynamic voltage and frequency scaling reduce processor energy consumption at the cost of an increase in the latency of computation. However, in the case of radios, the electronics power can be comparable to the RF component (which varies with the transmission distance). Therefore, slowing down the radio may actually lead to an increase in energy consumption. Other architecture specific overheads like the startup cost of

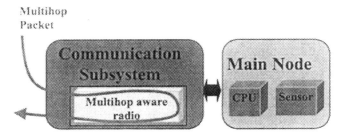

Figure 3.4 Energy aware packet forwarding architecture

the radio can be quite significant [9], making power management of radios a complex problem. The various tradeoffs involved in incorporating energy awareness into wireless communication will be discussed in detail in Section 3.4.

3.3.4 Energy aware packet forwarding

In addition to sensing and communicating its own data to other nodes, a sensor node also acts as a router, forwarding packets meant for other nodes. In fact, for typical sensor network scenarios, a large portion (around 65%) of all packets received by a sensor node need to be forwarded to other destinations [19]. Typical sensor node architectures implement most of the protocol processing functionality on the main computing engine. Hence, every received packet, irrespective of its final destination, travels all the way to the computing subsystem and gets processed, resulting in a high energy overhead. The use of intelligent radio hardware, as shown in Figure 3.4, enables packets that need to be forwarded to be identified and re-directed from the communication subsystem itself, allowing the computing subsystem to remain in *Sleep* mode, saving energy [19].

3.4 ENERGY AWARE WIRELESS COMMUNICATION

While power management of individual sensor nodes reduces energy consumption, it is important for the communication between nodes to be conducted in an energy efficient manner as well. Since the wireless transmission of data accounts for a major portion of the total energy consumption, power management decisions that take into account the effect of inter-node communication yield significantly higher energy savings. Further, incorporating power management into the communication process enables the diffusion of energy awareness from an individual sensor node to

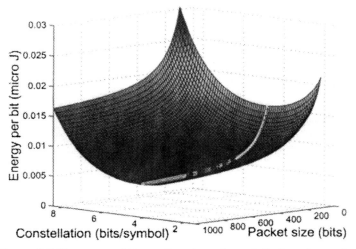

Figure 3.5 Energy per bit as a function of constellation size and packet size

a group of communicating nodes, thereby enhancing the lifetime of entire regions of the network. To achieve power-aware communication it is necessary to identify and exploit the various performance-energy trade-off knobs that exist in the communication subsystem.

3.4.1 Modulation schemes

Besides the hardware architecture itself, the specific radio technology used in the wireless link between sensor nodes plays an important role in energy considerations. The choice of modulation scheme greatly influences the overall energy versus fidelity and latency tradeoff that is inherent to a wireless communication link. Equation (1) expresses the energy cost for transmitting one bit of information, as a function of the packet payload size L, the header size H, the fixed overhead E_{start} associated with the radio startup transient, and the symbol rate R_S for an M-ary modulation scheme [9], [20]. P_{elec} represents the power consumption of the electronic circuitry for frequency synthesis, filtering, modulating, up converting, etc. The power delivered by the power amplifier, P_{RF}, needs to go up as M increases, in order to maintain the same error rate.

$$Ebit = \frac{E_{start}}{L} + \frac{P_{elec} + P_{RF}(M)}{R_S * \log_2 M} * (1 + \frac{H}{L}) \qquad (1)$$

Figure 3.5 plots the communication energy per bit as a function of the packet size and the modulation level M. This curve was obtained using the parameters given in Table 3.3, which are representative for sensor networks, and choosing Quadrature Amplitude Modulation (QAM) [9], [20]. The markers in Figure 3.5 indicate the optimal modulation setting for each packet size, which is independent of L. In fact, this optimal modulation level is relatively high, close to 16-QAM for the values specified in Table 3.3. Higher modulation levels might be unrealistic in low-end wireless systems, such as sensor nodes. In these scenarios, a practical guideline for saving energy is to transmit as fast as possible, at the optimal setting [9]. However, if for reasons of peak-throughput, higher modulation levels than the optimal one need to be provided, adaptively changing the modulation level can lower the overall energy consumption. When the instantaneous traffic load is lower than the peak value, transmissions can be slowed down, possibly all the way to the optimal operating point. This technique of dynamically adapting the modulation level to match the instantaneous traffic load, as part of the radio power management, is called modulation scaling [20]. It is worth noting that dynamic modulation scaling is the exact counterpart of dynamic voltage scaling, which has been shown to be extremely effective for processor power management, as described earlier.

Table 3.3. Typical radio parameters for wireless sensor nodes

E_{start}	1 µJ
P_{elec}	12 mW
P_{RF}	1 mW for 4-QAM
R_S	1 Mbaud
H	16 bits

The above conclusions are expected to hold for other implementations of sensor network transceivers as well. Furthermore, since the startup cost is significant in most radio architectures [9], it is beneficial to operate with as large a packet size as possible, since it amortizes this fixed overhead over more bits. However, aggregating more data into a single packet has the downside of increasing the overall latency of information exchange. The discussion up until now has focused on the links between two sensor nodes, which are characterized by their short distance. However, when external users interact with the network, they often times do so via specialized gateway nodes [22], [23]. These gateway nodes offer long-haul communication services, and are therefore in a different regime where P_{RF} dominates P_{elec}. In this case, the optimal M shifts to the lowest possible

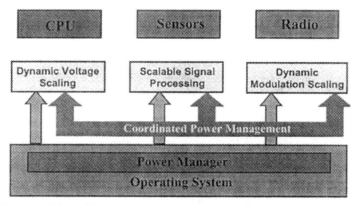

Figure 3.6 Coordinated power management of the computation, communication, and sensing subsystems

value, such that it becomes beneficial to transmit as slow as possible, subject to the traffic load. In this regime, modulation scaling is clearly effective [20].

3.4.2 Coordinated power management to exploit computation communication tradeoff

Sensor networks involve several node-level and network-wide computation-communication tradeoffs, which can be exploited for energy management. At the individual node level, power management techniques such as DVS and modulation scaling reduce the energy consumption at the cost of increased latency. Since both the computation and communication subsystems take from the total acceptable latency budget, exploiting the inherent synergy between them to perform coordinated power management will result in far lower energy consumption. For example, the relative split up of the available latency for the purposes of dynamic voltage scaling and dynamic modulation scaling significantly impacts the energy savings obtained. Figure 3.6 shows a system power management module that is integrated into the OS, and performs coordinated power management of the computing, communication and sensing subsystems.

The computation-communication tradeoff manifests itself in a powerful way due to the distributed nature of these sensor networks. The network's inherent capability for parallel processing offers further energy optimization potential. Distributing an algorithm's computation among multiple sensor nodes enables the computation to be performed in parallel. The increased allowable latency per computation enables the use of voltage scaling, or other energy-latency tradeoff techniques. Distributed computing algorithms however demand more inter-node collaboration, thereby increasing the amount of communication that needs to take place.

These computation-communication tradeoffs extend beyond individual nodes to the network level too. As we will discuss in the next section, the high redundancy present in the data gathering process, enables the use of data combining techniques to reduce the amount of data to be communicated, at the expense of extra computation at individual nodes to perform data aggregation.

3.4.3 Low level network protocol optimizations

While exploring energy-performance-quality tradeoffs, reliability constraints also have to be considered, which are related to the interplay of communication packet losses and sensor data compression. Reliability decisions are usually taken at the link layer, which is responsible for some form of error detection and correction. Adaptive error correction schemes were proposed in [24] to reduce energy consumption, while maintaining the bit error rate (BER) specifications of the user. For a given BER requirement, error control schemes reduce the transmit power required to send a packet, at the cost of additional processing power at the transmitter and receiver. This is especially useful for long-distance transmissions to gateway nodes, which involve large transmit power. Link layer techniques also play an indirect role in reducing energy consumption. The use of a good error control scheme minimizes the number of times a packet is retransmitted, thus reducing the power consumed at the transmitter as well as the receiver.

3.5 CONCLUSIONS

Sensor networks have emerged as a revolutionary technology for querying the physical world and hold promise in a wide variety of applications. However, the extremely energy constrained nature of these networks necessitate that their design and operation be done in an energy-aware manner, enabling the system to dynamically make tradeoffs between performance, fidelity, and energy consumption. We presented several energy optimization and management techniques at the node level, leveraging which can lead to significant enhancement in sensor network lifetime.

ACKNOWLEDGEMENTS

This chapter is based in part on research funded through DARPA's PAC/C and SensIT programs under AFRL contracts F30602-00-C-0154 and F30602-99-1-0529 respectively, and through NSF Grants ANI-0085773 and MIPS-9733331. Any opinions, findings and conclusions or recommendations expressed in this chapter are those of the author(s), and do not necessarily reflect the views of DARPA, AFRL, or NSF. The authors would like to acknowledge their colleagues at the Networked and Embedded Systems Laboratory, UCLA for several interesting and stimulating technical discussions.

REFERENCES

[1] Wireless Integrated Network Sensors, University of California, Los Angeles. (http://www.janet.ucla.edu/WINS)
[2] J. M. Kahn, R. H. Katz, and K. S. J. Pister, "Next century challenges: mobile networking for smart dust", in Proc. Mobicom, pp. 483-492, 1999.
[3] D. Estrin and R. Govindan, "Next century challenges: scalable coordination in sensor networks", in *Proc. Mobicom*, pp. 263-270, 1999.
[4] A. P. Chandrakasan, et al., "Design considerations for distributed microsensor systems", in *Proc. CICC*, 1999, pp. 279-286.
[5] J. Rabaey, et al., "PicoRadio supports ad hoc ultra low power wireless networking", in *IEEE Computer*, July 2000, pp. 42-48.
[6] A. P. Chandrakasan and R. W. Brodersen, Low Power CMOS Digital Design, Kluwer Academic Publishers, Norwell, MA, 1996.
[7] V. Tiwari, S. Malik, A. Wolfe, and M. T. C. Lee, "Instruction level power analysis and optimization of software", in *Jrnl. VLSI Signal Processing*, Kluwer Academic Publishers, pp. 1-18, 1996.
[8] A. Sinha and A. P. Chandrakasan, "Jouletrack: A web based tool for software energy profiling", in *Proc. Design Automation Conf.*, 2001.
[9] A. Wang, S-H. Cho, C. G. Sodini, and A. P. Chandrakasan, "Energy-efficient modulation and MAC for asymmetric microsensor systems", in *Proc. ISLPED*, 2001.
[10] WINS project, Rockwell Science Center, (http://wins.rsc.rockwell.com).
[11] Y. Xu, J. Heidemann and D. Estrin, "Geography-informed energy conservation for ad hoc routing", in *Proc. Mobicom*, 2001.
[12] C. F. Chiasserini and R. R. Rao, "Pulsed battery discharge in communication devices", in *Proc. Mobicom*, 1999.
[13] S. Park, A. Savvides, and M. Srivastava, "Battery capacity measurement and analysis using lithium coin cell battery", in *Proc. ISLPED*, 2001.
[14] L. Benini and G. DeMicheli, Dynamic Power Management: Design Techniques & CAD Tools, Kluwer Academic Publishers, Norwell, MA, 1997.
[15] T. A. Pering, T. D. Burd, and R. W. Brodersen, "The simulation and evaluation of dynamic voltage scaling algorithms", in *Proc. ISLPED*, pp. 76-81, 1998.

[16] V. Raghunathan, P. Spanos, and M. Srivastava, "Adaptive power-fidelity in energy aware wireless embedded systems", to be presented at *IEEE Real Time Systems Symposium*, 2001.
[17] F. Yao, A. Demers, and S. Shenker, "A scheduling model for reduced CPU energy", in *Proc. Annual Symp. on Foundations of Computer Science*, pp.374-382, 1995.
[18] A. Sinha, A. Wang, and A. P. Chandrakasan, "Algorithmic transforms for efficient energy scalable computation", in *Proc. ISLPED*, 2000.
[19] V. Tsiatsis, S. Zimbeck, and M. Srivastava, "Architectural strategies for energy efficient packet forwarding in wireless sensor networks", in *Proc. ISLPED*, 2001.
[20] C. Schurgers, O. Aberthorne, and M. Srivastava, "Modulation scaling for energy aware communication systems", in *Proc. ISLPED*, 2001.
[21] C. Schurgers, G. Kulkarni, and M. Srivastava, "Distributed assignment of encoded MAC addresses in wireless sensor networks", in *Proc.MobiHoc*, 2001.
[22] K. Sohrabi, J. Gao, V. Ailawadhi, and G. Pottie, "Protocols for self-organization of a wireless sensor network", in *IEEE Personal Comm. Magazine*, vol.7, no.5, pp. 16-27, Oct. 2000.
[23] W. Heinzelman, A. Chandrakasan, and H. Balakrishnan, "Energy-efficient communication protocol for wireless sensor networks", in *Proc. Hawaii Intl. Conf. on System Sciences*, Hawaii, 2000.
[24] P. Lettieri, C. Fragouli, and M. Srivastava, "Low power error control for wireless links", in *Proc. Mobicom*, pp. 139-150, 1997.
[25] J.-H. Chang and L. Tassiulas, "Energy conserving routing in wireless ad-hoc networks", in *Proc. INFOCOM*, 2000.
[26] C. Schurgers and M. Srivastava, "Energy efficient routing in sensor networks", in *Proc. Milcom*, 2001.
[27] B. Chen, K. Jamieson, H. Balakrishnan, and R Morris, "SPAN: An energy-efficient coordination algorithm for topology maintenance in ad hoc wireless networks, in *Proc. Mobicom*, 2001.
[28] C. Schurgers, V. Tsiatsis, and M. Srivastava, "STEM: Topology management for energy efficient sensor networks", Proceedings of the 2002 IEEE Aerospace Conference, March 2002.

II

NETWORK PROTOCOLS

Chapter 4

MEDIUM ACCESS CONTROL IN WIRELESS SENSOR NETWORKS

Wei Ye
Information Sciences Institute
University of Southern California
weiye@isi.edu

John Heidemann
Information Sciences Institute
University of Southern California
johnh@isi.edu

Abstract
This chapter reviews medium access control (MAC), an enabling technology in wireless sensor networks. MAC protocols control how sensors access a shared radio channel to communicate with neighbors. Battery-powered wireless sensor networks with many nearby nodes challenge traditional MAC design. This chapter discusses design trade-offs with an emphasis on energy efficiency. It classifies existing MAC protocols and compares their advantages and disadvantages in the context of sensor networks. Finally, it presents S-MAC as an example of a MAC protocol designed specifically for a sensor network, illustrating one combination of design trade-offs.

Keywords: Medium access control, wireless sensor networks, energy efficiency

4.1 INTRODUCTION

A wireless sensor network is a special network with large numbers of nodes equipped with embedded processors, sensors and radios. These nodes collaborate to accomplish a common task such as environment monitoring or asset tracking. In many applications, sensor nodes will be deployed in an ad hoc fashion without careful planning. They must organize themselves to form a multi-hop, wireless communication network.

A common challenge in wireless networks is collision, resulting from two nodes sending data at the same time over the same transmission medium or channel. Medium access control (MAC) protocols have been developed to assist each node to decide when and how to access the channel. This problem is also known as channel allocation or multiple access problem. The MAC layer is normally considered as a sublayer of the data link layer in the network protocol stack.

MAC protocols have been extensively studied in traditional areas of wireless voice and data communications. Time division multiple access (TDMA), frequency division multiple access (FDMA) and code division multiple access (CDMA) are MAC protocols that are widely used in modern cellular communication systems [1]. Their basic idea is to avoid interference by scheduling nodes onto different sub-channels that are divided either by time, frequency or orthogonal codes. Since these sub-channels do not interfere with each other, MAC protocols in this group are largely collision-free. We refer to them as scheduled protocols.

Another class of MAC protocols is based on contention. Rather than pre-allocate transmissions, nodes compete for a shared channel, resulting in probabilistic coordination. Collision happens during the contention procedure in such systems. Classical examples of contention-based MAC protocols include ALOHA [2] and carrier sense multiple access (CSMA) [3]. In ALOHA, a node simply transmits a packet when it is generated (pure ALOHA) or at the next available slot (slotted ALOHA). Packets that collide are discarded and will be retransmitted later. In CSMA, a node listens to the channel before transmitting. If it detects a busy channel, it delays access and retries later. The CSMA protocol has been widely studied and extended; today it is the basis of several widely-used standards including IEEE 802.11 [4].

Sensor networks differ from traditional wireless voice or data networks in several ways. First of all, most nodes in sensor networks are likely to be battery powered, and it is often very difficult to change batteries for all the nodes. Second, nodes are often deployed in an ad hoc fashion rather than with careful pre-planning; they must then organize themselves into a communication network. Third, many applications employ large numbers of nodes, and node density will vary in different places and times, with both sparse networks and nodes with many neighbors. Finally, most traffic in the network is triggered by sensing events, and it can be extremely bursty. All these characteristics suggest that traditional MAC protocols are not suitable for wireless sensor networks without modifications.

This chapter reviews MAC protocols for wireless sensor networks. After discussing the attributes of MAC protocols and design trade-offs for sensor networks (Section 4.2), we present TDMA protocols (Section 4.3) and contention-based protocols (Section 4.4). We then examine S-MAC as a case study of a sensor-net specific MAC protocol (Section 4.5).

4.2 TRADE-OFFS IN MAC DESIGN FOR WIRELESS SENSOR NETWORKS

This section discusses important attributes of MAC protocols and how design trade-offs can be made to meet the challenges of the sensor network and its applications. Because sensor networks are often battery constrained, we emphasize energy efficiency in MAC design.

4.2.1 MAC ATTRIBUTES AND TRADE-OFFS

MAC protocols are influenced by a number of constraints. A protocol designer needs to make trade-offs among different attributes. This section examines MAC attributes and trade-offs in detail, and how their importance varies in the context of wireless sensor networks.

Collision avoidance is the basic task of all MAC protocols. It determines when and how a node can access the medium and send its data. Collisions are not always completely avoided in regular operation; contention-based MAC protocols accept some level of collisions. But all MAC protocols avoid frequent collisions.

Energy efficiency is one of the most important attributes for sensor-net MAC protocols. As stated above, with large numbers of battery powered nodes, it is very difficult to change or recharge batteries for these nodes. In fact, some design goals of sensor networks are to build nodes that are cheap enough to be discarded rather than recharged, or that are efficient enough to operate only on ambient power sources. In all cases, prolonging the lifetime of each node is a critical issue. On many hardware platforms, the radio is a major energy consumer. The MAC layer directly controls radio activities, and its energy savings significantly affect the overall node lifetime. We explore energy conservation in more detail below.

Scalability and adaptivity are closely related attributes of a MAC protocol that accommodate changes in network size, node density and topology. Some nodes may die over time; some new nodes may join later; some nodes may move to different locations. A good MAC protocol should accommodate such changes gracefully. Scalability and adaptivity to changes in size, density, and topology are important attributes, because sensor networks are deployed in an ad hoc manner and often operate in uncertain environments.

Channel utilization reflects how well the entire bandwidth of the channel is utilized in communications. It is also referred to as bandwidth utilization or channel capacity. It is an important issue in cell phone systems or wireless local area networks (LANs), since the bandwidth is the most valuable resource in such systems and service providers want to accommodate as many users as possible. In contrast, the number of active nodes in sensor networks is primarily determined by the application. Channel utilization is normally a secondary goal in sensor networks.

Latency refers to the delay from when a sender has a packet to send until the packet is successfully received by the receiver. In sensor networks, the importance of latency depends on the application. In applications such as surveillance or monitoring, nodes will be vigilant for long time, but largely inactive until something is detected. These applications can often tolerate some additional messaging latency, because the network speed is typically orders of magnitude faster than the speed of a physical object. The speed of the sensed object places a bound on how rapidly the network must react. During a period that there is no sensing event, there is normally very little data flowing in the network. Sub-second latency for an initial message after an idle period may be less important than potential energy savings and longer operational lifetime. By contrast, after a detection, low-latency operation becomes more important.

Throughput (often measured in bits or bytes per second) refers to the amount of data successfully transfered from a sender to a receiver in a given time. Many factors affect the throughput, including efficiency of collision avoidance, channel utilization, latency and control overhead. As with latency, the importance of throughput depends on the application. Sensor applications that demand long lifetime often accept longer latency and lower throughput. A related attribute is *goodput*, which refers to the throughput measured only by data received by the receiver without any errors.

Fairness reflects the ability of different users, nodes, or applications to share the channel equally. It is an important attribute in traditional voice or data networks, since each user desires an equal opportunity to send or receive data for their own applications. However, in sensor networks, all nodes cooperate for a single common task. At any particular time, one node may have dramatically more data to send than some other nodes. Thus, rather than treating each node equally, success is measured by the performance of the application as a whole, and per-node or per-user fairness becomes less important.

In summary, the above attributes reflects the characteristics of a MAC protocol. For wireless sensor networks, the most important factors are effective collision avoidance, energy efficiency, scalability and adaptivity to densities and numbers of nodes. Other attributes are normally secondary.

4.2.2 ENERGY EFFICIENCY IN MAC PROTOCOLS

Energy efficiency is one of the most important issues in wireless sensor networks. To design an energy-efficient MAC protocol, we must consider the

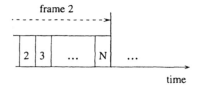

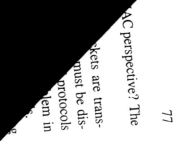

following question: what causes energy waste from the MAC perspective? The following sources are major causes of energy waste.

Collision is a first source of energy waste. When two packets are transmitted at the same time and collide, they become corrupted and must be discarded. Follow-on retransmissions consume energy too. All MAC protocols try to avoid collisions one way or another. Collision is a major problem in contention protocols, but is generally not a problem in scheduled protocols.

A second source is *idle listening*. It happens when the radio is listening to the channel to receive possible data. The cost is especially high in many sensor network applications where there is no data to send during the period when nothing is sensed. Many MAC protocols (such as CSMA and CDMA protocols) always listen to the channel when active, assuming that the complete device would be powered off by the user if there is no data to send.

The exact cost of idle listening depends on radio hardware and mode of operation. For long-distance radios (0.5km or more), transmission power dominates receiving and listening costs. By contrast, several generations of short-range radios show listening costs of the same order of magnitude as receiving and transmission costs, often 50–100% of the energy required for receiving. For example, Stemm and Katz measure that the power consumption ratios of idle:receiving:transmission are 1:1.05:1.4 [5] on the 915MHz Wavelan card, while the Digitan wireless LAN module (IEEE 802.11/2Mbps) specification shows the ratios are 1:2:2.5 [6]. On the Mica2 mote [7], the ratios for radio power draw are 1:1:1.41 at 433MHz with RF signal power of 1mW in transmission mode. Most sensor networks are designed to operate over long time, and the nodes will be in idle state for long time. In such cases, idle listening is a dominant factor of radio energy consumption.

A third source is *overhearing*, which occurs when a node receives packets that are destined to other nodes. Overhearing unnecessary traffic can be a dominant factor of energy waste when traffic load is heavy and node density is high.

The last major source that we consider is *control packet overhead*. Sending, receiving, and listening for control packets consumes energy. Since control packets do not directly convey data, they also reduce the effective goodput.

A MAC protocol achieves energy savings by controlling the radio to avoid or reduce energy waste from the above sources. Turning off the radio when it is not needed is an important strategy for energy conservation. A complete energy management scheme must consider all sources of energy consumption, not just the radio. In laptop computers, for example, display back-lighting can dominate costs [8].

On a tiny sensor node such as the Berkeley mote [9], the radio and the CPU are two major energy consumers. For example, on the Mica2 mote, the 433MHz radio consumes 22.2mW [10] when idle or receiving data, about the

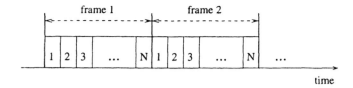

Figure 4.1. TDMA divides the channel into N time slots.

same power draw as the CPU when active [11], and is much higher than other components. From a system point-of-view, MAC energy control must be integrated with control of the CPU and other components.

4.3 SCHEDULED PROTOCOLS

According to the underlying mechanism for collision avoidance, MAC protocols can be broadly divided into two groups: scheduled and contention-based. Among protocols in the first group, TDMA has attracted attentions of sensor network researchers.

TDMA divides the channel into N time slots, as shown in Figure 4.1. In each slot, only one node is allowed to transmit. The N slots comprises a frame, which repeats cyclically. TDMA is frequently used in cellular wireless communication systems, such as GSM [1]. Within each cell, a base station allocates time slots and provides timing and synchronization information to all mobile nodes. Typically, mobile nodes communicate only with the base station; there is no direct, peer-to-peer communications between mobile nodes. The major advantage of TDMA is its energy efficiency, because it directly supports low-duty-cycle operations on nodes.

However, TDMA has some disadvantages that limits its use in wireless sensor networks. TDMA normally requires nodes to form clusters, analogous to the cells in the cellular communication systems. One of the nodes within the cluster is selected as the cluster head, and acts as the base station. This hierarchical organization has several implications. Nodes are normally restricted to communicate with the cluster head within a cluster; peer-to-peer communication is not directly supported. (If nodes communicate directly, then they must listen during all slots, reducing energy efficiency.) Inter-cluster communications and interference need to be handled by other approaches, such as FDMA or CDMA. More importantly, TDMA protocols have limited scalability and adaptivity to the changes on number of nodes. When new nodes join or old nodes leave a cluster, the base station must adjust frame length or slot allocation. Frequent changes may be expensive or slow to take effect. Also, frame length and static slot allocation can limit the available throughput for any given node, and the the maximum number of active nodes in any cluster may be limited. Finally, TDMA protocols depend on distributed, fine-grained time synchronization to align slot boundaries.

Many variations on this basic TDMA protocol are possible. Rather than scheduling slots for node transmissions, slots may be assigned for reception with some mechanism for contention within each slot. The base station may dynamically allocate slot assignments on a frame-by-frame basis. In ad hoc settings, regular nodes may assume the role of base station, and this role may rotate to balance energy consumption.

4.3.1 EXAMPLES OF SCHEDULED PROTOCOLS

This subsection shows some examples of scheduled protocols for sensor networks. (We do not consider cellular communication systems here. Interested readers can refer to [1])

Sohrabi and Pottie proposed a self-organization protocol for wireless sensor networks [12]. The protocol assumes that multiple channels are available (via FDMA or CDMA), and any interfering links select and use different sub-channels. During the time that is not scheduled for transmission or reception, a node turns off its radio to conserve energy. Each node maintains its own time slot schedules with all its neighbors, which is called a superframe. Time slot assignment is only decided by the two nodes on a link, based on their available time. It is possible that nodes on interfering links will choose the same time slots. Although the superframe looks like a TDMA frame, it does not prevent collisions between interfering nodes, and this task is actually accomplished by FDMA or CDMA. This protocol supports low-energy operation, but a disadvantage is the relatively low utilization of available bandwidth. A sub-channel is dedicated to two nodes on a link, but is only used for a small fraction of time, and it cannot be re-used by other neighboring nodes.

LEACH (Low-Energy Adaptive Clustering Hierarchy), proposed by Heinzelman *et al.* [13] is an example of utilizing TDMA in wireless sensor networks. LEACH organizes nodes into cluster hierarchies, and applies TDMA within each cluster. The position of cluster head is rotated among nodes within a cluster depending on their remaining energy levels. Nodes in the cluster only talk to their cluster head, which then talks to the base station over a long-range radio. LEACH is an example that directly extends the cellular TDMA model to sensor networks. The advantages and disadvantages of LEACH are summarized above.

Bluetooth [14, 15] is designed for personal area networks (PAN) with target nodes as battery-powered PDAs, cell phones and laptop computers. Its design for low-energy operation and inexpensive cost make it attractive for use in wireless sensor networks. As with LEACH, Bluetooth also organizes nodes into clusters, called *piconets*. Frequency-hopping CDMA is adopted to handle inter-cluster communications and interference. Within a cluster, a TDMA-based protocol is used to handle communications between the cluster head (master) and other nodes (slaves). The channel is divided into time slots for alternate master transmission and slave transmission. The master uses *polling* to decide which slave has the right to transmit. Only the communication between the master and one or more slaves is possible. The maximum number of active nodes within a cluster is limited to eight, an example of limited scalability. Larger networks can be constructed as *scatternets*, where one node bridges two piconets. The bridge node can temporarily leave one piconet and join another, or operate two radios.

4.3.2 ENERGY CONSERVATION IN SCHEDULED PROTOCOLS

Scheduled protocols such as TDMA are very attractive for applications in sensor networks because of their energy efficiency. Since slots are pre-allocated to individual nodes, they are collision-free. There is no energy wasted on collisions due to channel contention. Second, TDMA naturally supports low-duty-cycle operation. A node only needs to turn on its radio during the slot that it is assigned to transmit or receive. Finally, overhearing can be easily avoided by turning off the radio during the slots of other nodes.

In general, scheduled protocols can provide good energy efficiency, but they are not flexible to changes in node density or movement, and lack of peer-to-peer communication.

4.4 CONTENTION-BASED PROTOCOLS

Unlike scheduled protocols, contention protocols do not divide the channel into sub-channels or pre-allocate the channel for each node to use. Instead, a common channel is shared by all nodes and it is allocated on-demand. A contention mechanism is employed to decide which node has the right to access the channel at any moment.

Contention protocols have several advantages compared to scheduled protocols. First, because contention protocols allocate resources on-demand, they can scale more easily across changes in node density or traffic load. Second, contention protocols can be more flexible as topologies change. There is no requirement to form communication clusters, and peer-to-peer communication is directly supported. Finally, contention protocols do not require fine-grained time synchronizations as in TDMA protocols.

The major disadvantage of a contention protocol is its inefficient usage of energy. It normally has all the sources of energy waste we discussed in Section 4.2: nodes listen at all times and collisions and contention for the media

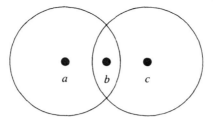

Figure 4.2. Hidden terminal problem: nodes a and c are hidden to each other.

can waste energy. Overcoming this disadvantage is required if contention-based protocols are to be applied to long-lived sensor networks.

4.4.1 EXAMPLES OF CONTENTION PROTOCOLS

As mentioned in Section 4.1, CSMA [3] is an important contention protocol. Its central idea is listening before transmitting. The purpose of listening is to detect if the medium is busy, also known as carrier sense. There are several variants of CSMA, including non-persistent, 1-persistent, and p-persistent CSMA. In non-persistent CSMA, if a node detects an idle medium, it transmits immediately. If the medium is busy, it waits a random amount of time and start carrier sense again. In 1-persistent CSMA, a node transmit if the medium is idle. Otherwise it continues to listen until the medium becomes idle, and then transmits immediately. In p-persistent CSMA, a node transmits with probability p if the medium is idle, and with probability $(1 - p)$ to back-off and restart carrier sense. Woo and Culler examined the performance of CSMA with various configurations when it is used in wireless sensor networks [16].

In a multi-hop wireless network, however, CSMA alone is not sufficient due to the hidden terminal problem [17]. Figure 4.2 illustrates the hidden terminal problem on a two-hop network with three nodes. Suppose nodes a, b and c can only hear from their immediate neighbors. When node a is sending to b, node c is not aware of this transmission, and its carrier sense still indicates that the medium is idle. If c starts transmitting now, b will receive collided packets from both a and c.

CSMA/CA, where CA stands for collision avoidance, was developed to address the hidden terminal problem, and is adopted by the wireless LAN standard, IEEE 802.11 [4]. The basic mechanism in CSMA/CA is to establish a brief handshake between a sender and a receiver before the sender transmits data. The handshake starts from the sender by sending a short Request-to-Send (RTS) packet to the intended receiver. The receiver then replies with a Clear-to-Send (CTS) packet. The sender starts sending data after it receives the CTS packet. The purpose of RTS-CTS handshake is to make an announcement to ind the receiver. In the example of Figure 4.2, RTS from a, it can hear the CTS from b. If destined to other nodes, it should back-off SMA/CA does not completely eliminate the the collisions are mainly on RTS packets. the cost of collisions is greatly reduced. osed MACA [18], which added a duration indicating the amount of data to be transw long they should back-off. Bharghavan eir protocol MACAW [19]. MACAW pro- ...itions to MACA, including use of an acknowledgment (ACK) packet after each data packet, allowing rapid link-layer recovery from transmission errors. The transmission between a sender and a receiver follows the sequence of RTS-CTS-DATA-ACK.

IEEE 802.11 adopted all these features of CSMA/CA, MACA and MACAW in its distributed coordination function (DCF), and made various enhancement, such as virtual carrier sense, binary exponential back-off, and fragmentation support [4]. DCF is designed for ad hoc networks, while the point coordination function (PCF, or infrastructure mode) adds support where designated access points (or base-stations) manage wireless communication.

Woo and Culler proposed a MAC protocol for wireless sensor networks [16], which combined CSMA with an adaptive rate control mechanism. This protocol is based on a specific network setup where there is a base station that tries to collect data equally from all sensors in the field. The major problem faced by the network is that nodes that are closer to the base station carry more traffic, since they have to forward more data from nodes down to the network. The MAC protocol aims to fairly allocate bandwidth to all nodes in the network. Each node dynamically adjusts its rate of injecting its original packets to the network: linearly increases the rate if it successfully injects a packet; otherwise multiplicatively decreases the rate. This protocol does not use RTS and CTS packets to address the hidden terminal problem. Instead, a node relies on overhearing the transmissions of the next-hop node and longer back-off time in CSMA to reduce the effect of the hidden terminal problem.

4.4.2 ENERGY CONSERVATION IN CONTENTION PROTOCOLS

Various techniques have been proposed to improve energy consumption of contention-based protocols for sensor networks. The basic approach is to put the radio into sleep state when it is not needed. For example, the Chipcon radio used on a Mica2 mote only consumes $15\mu W$ in sleep mode [10], three orders of magnitude less than that in idle/receive mode.

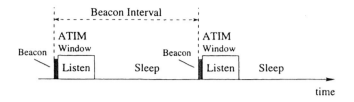

Figure 4.3. The power-save (PS) mode in IEEE 802.11 DCF.

However, uncoordinated sleeping can make it difficult for adjacent nodes to communicate with each other. TDMA protocols provide structure by scheduling when nodes can communicate. Contention-based MAC protocols have explored similar but less restrictive sleep/wake schedules to improve energy consumption. Some examples are described in this subsection.

Piconet is a low-power ad hoc wireless network developed by Bennett et al. [20]. (It is not the same piconet in Bluetooth.) The basic MAC protocol used in Piconet is the 1-persistent CSMA protocol. To reduce energy consumption, each node sleeps autonomously. Since nodes do not know when their neighbors are listening, they beacon their ID each time they wake up. Neighbors with data for a particular destination must listen until they hear the destination's beacon. They then coordinate using CSMA.

In IEEE 802.11, both PCF and DCF have power-save (PS) modes that allow nodes to periodically sleep to conserve energy. Figure 4.3 shows the diagram of the PS mode in DCF. A basic assumption here is that all nodes can hear each other—the network consists of only a single hop. One node periodically broadcasts a beacon to synchronize all nodes clocks. All nodes participate in beacon generation, and if one node sends it out first, others will suppress their transmissions. Following each beacon, there is an ATIM (ad hoc traffic indication message) window, in which all nodes are awake. If a sender wants to transmit to a receiver in power save-mode, it first sends out an ATIM packet to the receiver. After the receiver replies to the ATIM packet, the sender starts sending data.

The above PS mode in 802.11 DCF is designed for a single-hop network. Generalizing it to a multi-hop network is not easy, since problems may arise in clock synchronization, neighbor discovery and network partitioning, as pointed out by Tseng et al. [21]. They designed three sleep patterns to enable robust operation of 802.11 power-saving mode in a multi-hop network. Their schemes do not synchronize the listen time of each node. Instead, the three sleep patterns guarantee that the listen intervals of two nodes periodically overlap. Thus it resolves the problems of 802.11 in multi-hop networks. The cost is the increased control overhead and longer delay. For example, to send a broadcast

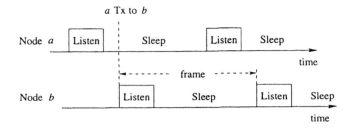

Figure 4.4. Node a transmit to b when b starts listening.

packet, the sender has to explicitly wake up each individual neighbor before it sends out the actual packet. Without synchronization, each node has to send beacons more frequently than the original 802.11 PS mode to prevent long-term clock drift.

Both Piconet and the 802.11 PS mode try to save energy by reducing the time of idle listening. They do not address the overhearing problem. PAMAS, proposed by Singh and Raghavendra [22], avoids overhearing by putting nodes into sleep state when their neighbors are in transmission. PAMAS uses two channels, one for data and one for control. All control packets are transmitted in the control channel. After a node wakes up from sleep, it also probes in the control channel to find any possible ongoing transmissions and their durations. If any neighbor answers the probe, the node will sleep again for the specified duration. Probing in the control channel avoids interfering a neighbor's transmission in the data channel, and the neighbor is able to answer the probe in the control channel without interrupting its data transmission. However, the requirement of separate control and data channels makes PAMAS more difficult to deploy, since multiple channels require multiple radios or additional complex channel allocation. Also, PAMAS does not reduce idle listening.

4.5   CASE STUDY: S-MAC

S-MAC is a MAC protocol specifically designed for wireless sensor networks, proposed by Ye *et al.* [23, 24]. Building on contention-based protocols like 802.11, S-MAC strives to retain the flexibility of contention-based protocols while improving energy efficiency in multi-hop networks. S-MAC includes approaches to reduce energy consumption from all the major sources of energy waste described in Section 4.2: idle listening, collision, overhearing and control overhead.

4.5.1 S-MAC DESIGN APPROACHES

At a high-level, S-MAC uses a coarse-grained sleep/wakeup cycle to allow nodes to spend most of their time asleep, as shown in Figure 4.4. We call a complete listen/sleep cycle a frame, after the TDMA frame. Each frame begins with a listen period for nodes that have data to send to coordinate. A sleep period follows, during which nodes sleep if they have no data to send or receive, and nodes remain awake and exchange data if they are involved in communication. We briefly describe how S-MAC establishes schedules in a multi-hop network, how nodes contend for the channel during listen periods, and how several optimizations improve throughput.

Scheduling: The first technique in S-MAC is to establish low-duty-cycle operation on nodes in a multi-hop network. For long-lived sensor networks, we expect duty cycles of 1–10%. The basic scheme is similar to the 802.11 PS mode, but without assuming all nodes can hear each other, or a designated base-station.

In S-MAC, all nodes are free to choose their own listen/sleep schedules. They share their schedules with their neighbors so that communication between all nodes is possible. Nodes then schedule transmissions during the listen time of their destination nodes. For example, nodes a and b in Figure 4.4 follow different schedules. If a wants to send to b, it just wait until b is listening. S-MAC enables multi-hop operation by accommodating multiple schedules in the network.

To prevent timing errors due to long-term clock drift, each node periodically broadcasts its schedule in a SYNC packet, which provides simple clock synchronization. The period for a node to send a SYNC packet is called a synchronization period. Combined with relatively long listen time and short guard time in waking up, S-MAC does not require tight clock synchronization of a TDMA protocol.

On the other hand, to reduce control overhead, S-MAC encourages neighboring nodes to adopt identical schedules. When a node first configures itself, it listens for a synchronization period and adopts the first schedule it hears. In addition, nodes periodically perform neighbor discovery, listening for an entire frame, allowing them to discover nodes on different schedules that may have moved within range.

Data transmission: The collision avoidance mechanism in S-MAC is similar to that in the IEEE 802.11 DCF [4]. Contention only happens at a receiver's listen interval. S-MAC uses both virtual and physical carrier sense. Unicast packets combine CSMA with an RTS-CTS-DATA-ACK exchange between the sender and the receiver, while broadcast packet use only CSMA procedure.

S-MAC puts a duration field in each packet, which indicates the time needed in the current transmission. If a neighboring node receives any packet from the sender or the receiver, it knows how long it needs to keep silent. In this case, S-MAC puts the node into sleep state for this amount of time, avoiding energy waste on overhearing. Ideally the node goes to sleep after receiving a short RTS or CTS packet destined to other nodes, and it avoids overhearing subsequent data and ACK packets. Compared with PAMAS, S-MAC only uses in-channel signaling for overhearing avoidance.

An important feature of wireless sensor networks is the in-network data processing, since data aggregation or other techniques can greatly reduces energy consumption by largely reducing the amount of data to be transmitted [25–27]. In-network processing requires store-and-forward processing of application-level *messages*, not just individual MAC-layer packets or fragments. While traditional MAC protocols emphasize fairness and interleave communication from concurrent senders, S-MAC utilizes *message-passing*, an optimization that allows multiple fragments from a message to be sent in a burst. It reduces message-level latency by disabling fragment-level interleaving of multiple messages.

In message passing, only one RTS and one CTS are used to reserve the medium for the time needed to transmit all fragments. Each fragment is separately acknowledged (and retransmitted if necessary). Besides RTS and CTS, each fragment or ACK also includes the duration of the remaining transmission, allowing nodes that wake up in the middle of the transmission to return to sleep. This is unlike 802.11's fragmentation mode where each fragment only indicates the presence of an additional fragment, not all of them.

With the low-duty-cycle operation, nodes must delay sending a packet until the next listen period of a destination, which increases latency. In addition, by limiting the opportunity to content for the channel, throughput can be reduced to one message per frame. These costs can accumulate at each hop of a multi-hop network. As an optimization to reduce this delay, S-MAC uses *adaptive listening*. Rather than waiting until the next scheduled listen interval after an RTS-CTS-DATA-ACK sequence, neighbors wake up immediately after the exchange completes. This allows immediate contention for the channel, either by another node with data to send, or for the next hop in a multi-hop path. With adaptive listen, the overall multi-hop latency can be reduced by at least half [24].

4.5.2   S-MAC PERFORMANCE

S-MAC has been implemented on Berkeley motes [28, 7]. Motes use an 8-bit embedded CPU and short-range, low-power radios: either an RFM TR1000 [29] or TR3000 [30], or a Chipcon CC1000 [10]. The following measurements use Mica motes with RFM TR3000 and 20kb/s bandwidth. An attractive fea-

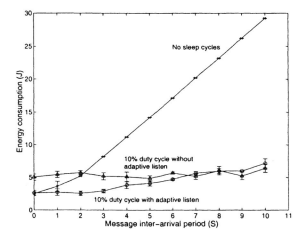

Figure 4.5. Aggregate energy consumption on radios in the entire 10-hop network using three S-MAC modes. (From [24], ©2004 IEEE)

ture of the mote for MAC research is that it provides very low-level access to the radio.

S-MAC implementation allows a user to configure it into different modes. This subsection shows some measurement results of S-MAC over Mica motes with the following configurations:

- 10% duty cycle without adaptive listen
- 10% duty cycle with adaptive listen
- No sleep cycles (100% duty cycle), but with overhearing avoidance

The topology in the measurement is a linear network of 11 nodes with the first node as the source and the last node as the sink. For complete details of these experiments, see [24].

Energy consumption. Energy consumption is measured in the ten-hop network with S-MAC configured in the above modes. In each test, the source node sends a fixed amount of data, 20 messages of 100-bytes each. Figure 4.5 shows how energy consumption on all nodes in the network changes as traffic load varies from heavy (on the left) to light (on the right).

Figure 4.5 shows that, at light load, operating at a low duty cycle can save significant amounts of energy compared to not sleeping, a factor of about 6 in this experiment. It also shows the importance of adaptive listening when traffic becomes heavy. Without adaptive listening, a 10% duty cycle consumes *more* energy than always listening because fewer opportunities to send require

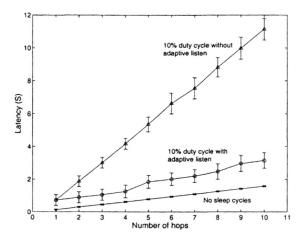

Figure 4.6. Mean message latency on each hop under the lightest traffic load. (From [24], ©2004 IEEE)

a longer time to send the same amount of data. By contrast, adaptive sending allows S-MAC to be as efficient as a MAC that never sleeps, even when there is always data to send.

Latency. A disadvantage of S-MAC is that the latency of sending a message can be increased. In this example, latency is measured by the time a message takes to travel over a varying number of hops when there is only one message in the network at a time.

Figure 4.6 shows the measured latency as a function of distance. In all three S-MAC modes, the latency increases linearly with the number of hops. However, S-MAC at 10% duty cycle without adaptive listen has much higher latency than the other two. The reason is that each message has to wait for one sleep cycle on each hop. In comparison, the latency of S-MAC with adaptive listen is very close to that of the MAC without any periodic sleep, because adaptive listen often allows S-MAC to immediately send a message to the next hop. On the other hand, for either low-duty-cycle mode, the variance in latency is much larger than that in the fully active mode, and it increases with the number of hops. The reason is that messages can miss sleep cycles in the path, and different parts of the network may be on different schedules.

Energy vs. Latency and Throughput. Now we look at the trade-offs that S-MAC has made on energy, latency and throughput. S-MAC reduces energy consumption, but it increases latency, and thus has a reduced throughput. To evaluate the overall performance, we compare the combined effect of energy

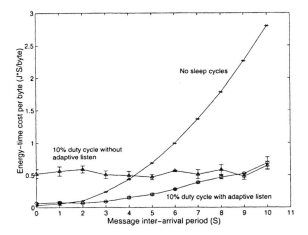

Figure 4.7. Energy-time cost per byte on passing data from source to sink under different traffic load. (From [24], ©2004 IEEE)

consumption and reduced throughput by calculating the per-byte cost of energy and time to pass data from the source to the sink.

Figure 4.7 shows the results under different traffic loads. We can see that when traffic load is very heavy (on the left), adaptive listen and the no-sleep mode both show statistically equivalent performances that are significantly better than sleeping without adaptive listen. Without adaptive listen, the sleep delay at each hop lowers overall energy-time efficiency. At lighter traffic load, the energy-time cost without sleeping quickly exceeds the cost of sleep modes.

In summary, periodic sleep provides excellent performance at light traffic load. Adaptive listen is able to adjust to traffic, and achieves performance as good as the no-sleep mode at heavy load. Therefore, S-MAC with adaptive listen is a good choice for sensor networks.

ACKNOWLEDGMENTS

This work is in part supported by NSF under grant ANI-0220026 as the MACSS project, by DARPA under grant DABT63-99-1-0011 as the SCADDS project, and a grant from the Intel Corporation.

4.6 SUMMARY

This chapter reviews MAC protocols for wireless sensor networks. Large scale, battery powered wireless sensor networks put serious challenges to the MAC design. We have discussed important MAC attributes and possible design trade-offs, with an emphasis on energy efficiency. It described both scheduled and contention-based MAC protocols and evaluated their advantages and disadvantages when applied to sensor networks. Finally, we presented S-MAC as an example of sensor-net MAC protocols, illustrating the design trade-offs for energy conservation.

REFERENCES

[1] T. S. Rappaport. *Wireless Communications, Principles and Practice.* Prentice Hall, 1996.

[2] Norman Abramson. Development of the ALOHANET. *IEEE Transactions on Information Theory*, 31(2):119–123, March 1985.

[3] Leonard Kleinrock and Fouad Tobagi. Packet switching in radio channels: Part I - carrier sense multiple access modes and their throughput delay characteristics. *IEEE Transactions on Communications*, 23(12):1400–1416, December 1975.

[4] LAN MAN Standards Committee of the IEEE Computer Society. *Wireless LAN medium access control (MAC) and physical layer (PHY) specification.* IEEE, New York, NY, USA, IEEE Std 802.11-1999 edition, 1999.

[5] Mark Stemm and Randy H Katz. Measuring and reducing energy consumption of network interfaces in hand-held devices. *IEICE Transactions on Communications*, E80-B(8):1125–1131, August 1997.

[6] Oliver Kasten. Energy consumption. http://www.inf.ethz.ch/~kasten/research/bathtub/energy_consumption.html. Eldgenossische Technische Hochschule Zurich.

[7] Crossbow Technology Inc. http://www.xbow.com/.

[8] Brian Marsh and Bruce Zenel. Power measurements of typical notebook computers. Technical Report MITL-TR-110-94, Matsushita Information Technology Laboratory, May 1994.

[9] Jason Hill, Robert Szewczyk, Alec Woo, Seth Hollar, David Culler, and Kristofer Pister. System architecture directions for networked sensors. In *Proceedings of the 9th International Conference on Architectural Support for Programming Languages and Operating Systems*, pages 93–104, Cambridge, MA, USA, November 2000. ACM.

[10] Chipcon Inc., http://www.chipcon.com/. *Chipcon CC1000 Data Sheet.*

[11] Atmel Corporation, http://www.atmel.com/. *AVR Microcontroller ATmega128L Reference Manual.*

[12] Katayoun Sohrabi and Gregory J. Pottie. Performance of a novel self-organization protocol for wireless ad hoc sensor networks. In *Proceedings of the IEEE 50th Vehicular Technology Conference*, pages 1222–1226, 1999.

[13] Wendi Rabiner Heinzelman, Anantha Chandrakasan, and Hari Balakrishnan. Energy-efficient communication protocols for wireless microsensor networks. In *Proceedings of the Hawaii International Conference on Systems Sciences*, January 2000.

[14] Bluetooth SIG Inc. Specification of the Bluetooth system: Core. http://www.bluetooth.org/, 2001.

[15] Jaap C. Haartsen. The Bluetooth radio system. *IEEE Personal Communications Magazine*, pages 28–36, February 2000.

[16] Alec Woo and David Culler. A transmission control scheme for media access in sensor networks. In *Proceedings of the ACM/IEEE International Conference on Mobile Computing and Networking*, pages 221–235, Rome, Italy, July 2001. ACM.

[17] Fouad Tobagi and Leonard Kleinrock. Packet switching in radio channels: Part II - the hidden terminal problem in carrier sense multiple access and the busy-tone solution. *IEEE Transactions on Communications*, 23(12):1417–1433, December 1975.

[18] Phil Karn. MACA: A new channel access method for packet radio. In *Proceedings of the 9th ARRL Computer Networking Conference*, pages 134–140, London, Ontario, Canada, September 1990.

[19] V. Bharghavan, A. Demers, S. Shenker, and L. Zhang. MACAW: A media access protocol for wireless lans. In *Proceedings of the ACM SIGCOMM Conference*, pages 212–225, London, UK, September 1994.

[20] Frazer Bennett, David Clarke, Joseph B. Evans, Andy Hopper, Alan Jones, and David Leask. Piconet: Embedded mobile networking. *IEEE Personal Communications Magazine*, 4(5):8–15, October 1997.

[21] Yu-Chee Tseng, Chih-Shun Hsu, and Ten-Yueng Hsieh. Power-saving protocols for IEEE 802.11-based multi-hop ad hoc networks. In *Proceedings of the IEEE Infocom*, pages 200–209, New York, NY, June 2002.

[22] S. Singh and C.S. Raghavendra. PAMAS: Power aware multi-access protocol with signalling for ad hoc networks. *ACM Computer Communication Review*, 28(3):5–26, July 1998.

[23] Wei Ye, John Heidemann, and Deborah Estrin. An energy-efficient mac protocol for wireless sensor networks. In *Proceedings of the IEEE Infocom*, pages 1567–1576, New York, NY, June 2002.

[24] Wei Ye, John Heidemann, and Deborah Estrin. Medium access control with coordinated, adaptive sleeping for wireless sensor networks. Technical Report ISI-TR-567, USC Information Sciences Institute, January 2003. To appear in the IEEE/ACM Transactions on Networking, April, 2004.

[25] Gregory J. Pottie and William J. Kaiser. Embedding the internet: wireless integrated network sensors. *Communications of the ACM*, 43(5):51–58, May 2000.

[26] Chalermek Intanagonwiwat, Ramesh Govindan, and Deborah Estrin. Directed diffusion: A scalable and robust communication paradigm for sensor networks. In *Proceedings of the ACM/IEEE International Conference on Mobile Computing and Networking*, pages 56–67, Boston, MA, USA, August 2000. ACM.

[27] John Heidemann, Fabio Silva, Chalermek Intanagonwiwat, Ramesh Govindan, Deborah Estrin, and Deepak Ganesan. Building efficient wireless sensor networks with low-level naming. In *Proceedings of the Symposium on Operating Systems Principles*, pages 146–159, Lake Louise, Banff, Canada, October 2001.

[28] http://webs.cs.berkeley.edu/tos.

[29] RF Monolithics Inc., http://www.rfm.com/. *ASH Transceiver TR1000 Data Sheet*.

[30] RF Monolithics Inc., http://www.rfm.com/. *ASH Transceiver TR3000 Data Sheet*.

Chapter 5

A SURVEY OF MAC PROTOCOLS FOR SENSOR NETWORKS

Piyush Naik and Krishna M. Sivalingam
Dept. of CSEE, University of Maryland, Baltimore County, Baltimore, MD 21250
piyush.naik@umbc.edu,krishna@umbc.edu

Abstract In this chapter, we continue the discussion on medium access control protocols designed for wireless sensor networks. This builds upon the background material and protocols presented in Chapter 4. We first present protocols that are based on random access techniques such as Carrier Sense Multiple Access. These include the Sift protocol, the T-MAC protocol and other protocols. The second set of protocols are based on static access and scheduling mechanisms. These include the UNPF framework, T-RAMA protocols and other related protocols.

Keywords: Medium access control, sensor networks, wireless ad-hoc networks.

5.1 INTRODUCTION

Wireless sensor networks have attracted a considerable attention from the researchers in the recent past as described in previous chapters in this book. Though the initial impetus came from military applications, the advancements in the field of pervasive computing have led to possibilities of a wide range of civilian, environmental, bio-medical, industrial and other applications. In order to practically realize such networks, Medium access control (MAC) is one of the basic protocol functionality that has to be appropriately defined.

The previous chapter presented some of the fundamental issues underlying the design of MAC protocols for sensor networks. In this chapter, we continue this discussion and present a comprehensive survey of other MAC protocols studied for sensor networks. We first present protocols that are based on random access techniques such as Carrier Sense Multiple Access. These include the Sift protocol [1], the T-MAC protocol [2] and other protocols presented in [3, 4]. The second set of protocols are based on static access and scheduling mechanisms. These include the UNPF framework [5], T-RAMA protocols [6] and the work presented in [7].

5.2 RANDOM ACCESS BASED PROTOCOLS

This section presents MAC protocols based on a random access mechanism.

5.2.1 CSMA-BASED EXPERIMENTS

One of the first experimental results for sensor networks based on the Berkeley motes was presented in [3]. The protocol is based on CSMA and its variants based on tuning many system parameters such as: (i) whether random delay is used before transmission, (ii) whether the listening time is constant or random and (iii) whether fixed or exponential window backoff mechanisms are used. An experimental testbed consisting of 10 sensor nodes and a base station was used for the analysis.

A detailed analysis of each of these CSMA schemes is performed through simulations and actual experiments. The 802.11 CSMA with ACK scheme was used as the baseline for comparison. A simple single-hop star topology and a more complex multi-hop tree topology were used in the analysis. The performance metrics considered include the average energy consumed per packet and the fraction of packets delivered to the base station. It was observed that a combination of random delay, constant listening and backoff with radio powered off provided the best results, for the metrics of interest. Interestingly the performance was found to be almost insensitive to the backoff mechanism.

The paper also presents the Adaptive Rate Control (ARC) mechanism. This mechanism tries to balance the originating traffic with the route-through traffic. It is similar to the congestion control scheme of TCP and works as follows. The transmission rate of the either traffic is given by Sp, where S is the original transmission rate and p is the probability of transmission. The factor p is governed by linear increase and multiplicative decrease: p is incremented by adding a constant α ($0 < \alpha < 1$) on a successful transmission and decremented by multiplying with β ($0 < beta < 1$) in the case of a failure. In short $alpha$ is a reward while $beta$ is a penalty. Naturally, a large α makes the scheme aggressive in acquiring the channel while a small β makes it very conservative.

Given that the network has invested more resources in the route-through traffic, it is given more consideration. The penalty (in case of a failure) for the route-through traffic is set to 50% less than that for the originating traffic. Also in order to provide a fair proportion of bandwidth to each node routing through it, the α_{route} is given by $\alpha_{route} = \frac{\alpha_{originate}}{n+1}$, where n is the number of nodes routing through that node.

A Survey of MAC Protocols for Sensor Networks

Experiments and Results. Studies through simulation and actual experiment support the analytical claims and expectations. Delivered bandwidth per node is nearly constant for all the nodes with ARC mechanism as compared to the IEEE 802.11 mechanisms or simple RTS/CTS mechanism where the variance is quite high. This observation clearly proves that the protocol is fair to all the network nodes. Average energy cost per packet is lower than IEEE 802.11 for smaller values of α. This is to be expected because larger values of α tends to be aggressive and injects more originating traffic.

To summarize, Adaptive Rate Control provides a good balance between energy efficiency and fairness of the network.

5.2.2 SIFT: AN EVENT-DRIVEN MAC PROTOCOL

The *Sift* protocol [1] exploits the event driven nature of the sensor networks for MAC protocol design. This work points out that the contention among the sensor nodes is often spatially correlated. This means that at a given time, only a set of adjacent ($R|R \leq N$) sensors have data to transmit and this is most likely to be after detection of some specific event. Thus, contention resolution may be limited to these R sensors rather than the entire set of N sensors. The protocol adopts a typical random access protocol such as CSMA or CSMA/CA and uses a fixed size contention window with a non-uniform probability distribution for choosing the contention slot for a node.

Protocol Details. At system initialization, every node uses a large estimate of a node population and hence correspondingly small transmission probability, p_r. Each node also continuously monitors the contention slots and reduces the node count estimate after every contention slot that goes free. That is, a free slot is taken as an indication of fewer number of sensors than assumed. Likewise, the node increases its transmission probability multiplicatively after each free slot. Thus the contention is reduced to geometrically decreasing number of nodes for the same number of contention slots. This is the core idea of Sift wherein the protocol *sifts* the slot-winner node from the others.

The geometrical distribution is chosen to be:

$$p_r = \frac{(1-\alpha)\alpha^{CW}}{1-\alpha^{CW}}\alpha^{-r}, r = 1, ..., CW$$

where α is the distribution parameter and CW is the fixed size of the contention window.

Note that p_r increases exponentially with increasing r. This means that later slots have higher probability of transmission than the earlier ones. When a node successfully transmits in a slot or when there is a collision, the other nodes select new random contention slots and repeat this backoff procedure. An example slot probability allocation is shown in Table 5.1.

Table 5.1. Example slot probabilities with the Sift protocol.

Slot	1	2	3	4	5	6	7	8
p_r	0.02	0.03	0.04	0.06	0.1	0.15	0.24	0.37
N_r	24	15	10	6	4	3	2	1

This protocol is designed on the premise that often only a subset of all the nodes need to send their reports. Thus the goal is to minimize the latency for the first R sensors sending their reports and suppress the rest. The scheme indeed works well for this purpose. The detailed probability analysis presented in [1] determines the optimal value of alpha to be 0.82 for a system with 512 sensors. The merit of the protocol lies in the fact that the performance degrades gracefully for more than 512 sensors. The probability of success in both the cases is almost the same until the performance of Sift starts degrading marginally for $N > 512$.

Experiments and Results. A set of experiments is conducted to observe the protocol performance with respect to different metrics viz. latency, throughput and fairness. To capture the burstiness of the sensor data traffic realistically, the work considers a motion-sensor video camera focused on a street and logging every motion event. The log contains the time and x, y coordinates of the motion event. This trace is mapped to an imaginary sensor field with randomly placed sensors. Sensors near the given x, y position at a given time from the log send the reports. The experiments are run to compare Sift with IEEE 802.11 [8]. Latency experiments show a seven-fold latency reduction compared to 802.11. Furthermore, Sift is found to be the least susceptible to the changes in latency with changes in number of reporting sensors and the variation in report times. Throughput analyses show that Sift shows promise under both event-driven and non-event-driven workloads.

Under constant bit rate workload, Sift lags behind 802.11 for a small number of flows (≤ 2). This is due to the higher delay per slot as compared to 802.11 which wins slot early. However as the number of flows increases, Sift performs better and surpasses 802.11 for 8 or more number of flows.

However, the work does acknowledge that Sift does not focus on the energy consumption issue since it constantly listens during the backoff period like 802.11. However, it is possible to integrate the Sift mechanisms with other wireless MAC protocols that focus on minimizing the energy consumption [9].

5.2.3 THE T-MAC PROTOCOL

The T-MAC protocol presented in [2] attempts to improve upon the performance of the S-MAC protocol [10]. It proposes using a dynamic duty cycle as against the fixed one in S-MAC to further reduce the idle listening periods. It also introduces some additional features described below.

Protocol details. Since idle listening is a major source of overhead, T-MAC, similar to S-MAC, maintains a sleep-sense cycle. However instead of having a fixed duty cycle like in S-MAC (say 10% sense and 90% sleep), it has a variable duty cycle. The idea is similar to that of a screen-saver. Just as the screen-saver starts after a certain period of inactivity, the node switches itself to a sleep mode when no activation event has occurred for a predetermined time period. The activation event can be a reception of some data, expiration of some timer, sensing of the communication, knowledge of an impending data reception through neighbors' RTS/CTS and so on. Synchronization of the schedules is achieved in an exactly similar manner as S-MAC through the scheme dubbed as virtual clustering.

T-MAC uses a fixed contention interval to send an RTS. A special case arises in the RTS transmission due to the dynamic duty cycle. When a node sends an RTS, it may not get a CTS back if that RTS was lost or if the receiving node could not transmit because one or more of its neighbors were communicating. In this case, the sender node might go to sleep if it does not hear a CTS for the predetermined time, resulting in a reduced throughput. To correct this problem, T-MAC specifies that the RTS be sent twice before the sender gives up.

The paper also describes another type of problem, called the *early sleeping problem*. Consider a scenario where a node X may constantly lose the contention to transmit an RTS to its neighbor (say N). This can happen if another neighbor of X (say Z), which is not a neighbor of N, is communicating with its own neighbor (say A). As a result the node X has to remain silent either because of an RTS transmitted to it by Z or because of an overheard CTS of the neighbor of Z. The situation is illustrated in the Figure 5.1. Ultimately, the active period of node N ends and it goes to sleep. Now node X can only transmit to N in its next active period. This plight of node X affects the throughput badly. This is termed as the *Early Sleeping problem* because a node (N in this case) goes to sleep even when another node (X here) has data to send to it.

T-MAC offers two solutions to this problem. In the first solution, the blocking node (X) sends a Future-RTS packet to its intended receiver (N), with the information about the blocking duration. The receiving node now knows that it is the future receiver at a particular time and must be awake by then. However this solution is found to increase the throughput and the energy requirements considerably. In the other solution, the node gives itself a higher priority if its

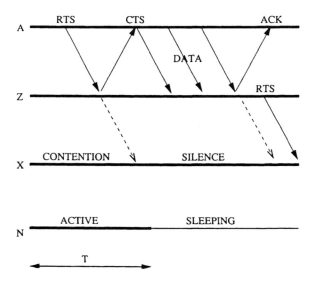

Figure 5.1. The Early Sleeping Problem.

transmission buffers are full. In other words, it will not send a CTS reply to an RTS but would rather send its own RTS for its own contention for the medium. This solution is termed as *full-buffer-priority*. Although this reduces the probability of the early sleeping problem, it is not advisable under high traffic load conditions because it will increase the number of collisions dramatically due to increased aggression. To avoid this, T-MAC puts a constraint that a node may only use the priority if it has lost the contention at least twice.

Experiments and results. Simulations are carried out with energy consumption as the primary metric. In a comparison of T-MAC with CSMA and S-MAC with different duty cycles for a homogeneous local unicast traffic, T-MAC is shown to perform far better than CSMA and at least as well as S-MAC. In a Nodes-To-Sink type of communication, T-MAC again outperforms S-MAC and CSMA especially at higher traffic loads. However it results in lower throughput as compared to S-MAC because of the early sleeping problem even with the FRTS and full-buffer-priority solutions. In a more realistic scenario with event based local unicast where nodes send unicast messages to their neighbors upon the occurrence of certain events, T-MAC is shown to perform the best. Once again, the early sleeping problem limits the overall throughput of T-MAC.

In a separate comparison of the solutions to the early sleeping problem, FRTS provides higher throughput at a higher energy cost while the full-buffer-priority scheme provides a slightly lesser throughput than FRTS but with no additional energy costs. Lastly in a combined simulation of the event based unicast and Nodes-To-Sink reporting, T-MAC has the least energy consumption.

T-MAC has been experimentally implemented on the EYES nodes [11]. Through this implementation, extensive experiments are carried out which provide important power usage characteristics of the sensor nodes.

5.2.4 MEDIUM ACCESS CONTROL WITH CHANNEL STATE INFORMATION

In [4], the authors develop a protocol based on the hypothesis that a sophisticated physical layer model can help improve the MAC protocol. A MAC protocol that uses the channel state information (CSI) in presented in this work. The authors propose a variation of the slotted ALOHA protocol in which the nodes transmit with a probability that is a function of the observed channel state in a particular slot. At the end of the time slot, base station transmits the indices of the nodes from which it received the packets successfully. The channel state is assumed to be identically and independently distributed from slot to slot and from node to node.

A metric termed Asymptotic Stable Throughput (AST) is introduced, where AST is defined as the maximum stable throughput achieved as the number of users goes to infinity while keeping the total packet rate constant [4]. It is evident that given a scheduler which uses the channel state information, AST can be significantly improved. The channel state is chosen to be proportional to the transmit power and propagation channel gain.

This concludes the discussion on the CSMA-based MAC protocols for sensor networks.

5.3 STATIC ACCESS BASED PROTOCOLS

This section present the different static access schemes such as those based on Time Division Multiplexed Access (TDMA).

5.3.1 UNPF PROTOCOLS

In [5], a unified protocol framework denoted *UNPF*, that comprises a network organization protocol, a MAC protocol and a routing protocol, are presented for a large-scale wireless sensor network architecture. In this network architecture, the network nodes are organized into layers where the layers are formed based on a node's hop-count to the base station [12]. For example, a node two hops away from the base station belongs to layer 2 and so on. Fig. 5.2 illustrates how the 10 sensor nodes are organized into three layers. When the network is first created, the nodes are organized into layers by using periodic *beacon* packets. The details of the network organization phase are explained in

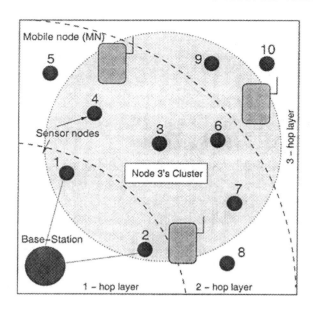

Figure 5.2. An example of MINA with 10 nodes organized into 3 layers. The shaded region represents the broadcast region of node 3.

[5]. The layered organization leads to a simple multi-hop routing mechanism well aligned with the limited memory and computation capabilities of the sensor node. A source node in layer i selects one of its inward neighbors in layer $(i - 1)$) as its forwarding node, and sends the data packet to that node. This node then forwards the packet to *its* forwarding node in layer $(i - 2)$, and so on until the packet reaches the BS. Thus, a node k in layer $i - 1$ acts as a forwarding node for a set of nodes in layer i and transmissions to node k need to be coordinated with a MAC protocol as explained below.

Protocol details. A Time Division protocol is proposed as the MAC protocol. The MAC protocol assumes the availability of a number of channels either in the form of a code or a frequency. The receiver of a forwarding node receiver (in layer i) is assigned a unique channel (with spatial reuse possible) and the MAC protocol is designed to share this channel among the transmitters of the forwarding node's client nodes in layer $i - 1$. A simple scheduling scheme is used for this purpose. This protocol is termed as Distributed TDMA Receiver Oriented Channeling (DTROC). Thus, each forwarding node can define its own schedule for its clients. Also, a node may be able to reach several nodes in its inward layer and chooses one among those as its forwarding node based on criteria such as available buffer space and remaining energy.

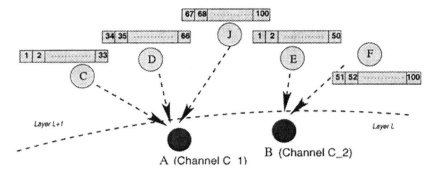

Figure 5.3. Slot allocation for forwarding nodes A and B, in a specific frame.

Fig. 5.3 shows a sample slot allocation and channel assignment for a sensor network with $\beta = 100$ slots in each data frame. In the figure, node A uses channel C_1 and $\mathcal{S}_A = \{C, D, J\}$; similarly, node B uses channel C_2 and $\mathcal{S}_B = \{J, E, F\}$. Note that in the figure, node **J** can choose between node **A** and node **B** as a forwarding node. In this frame, it has selected node A.

Experiments and results. With the layered architecture, the number of layers and the transmission range play an important role. In fact the latter influences the former. The simulations are therefore geared towards studying the effect of variation in the transmission range on different parameters. These parameters include packet latency, average energy consumed per packet, energy-delay product, time to first node death and time to network partition. All the simulations are carried out for varying node density viz. 200, 400, 600, and 800 nodes for a given field size.

With the increase in the transmission range, the number of layers decreases and so does the hop-count. This results in a lesser average delay. However for the transmission range greater than 40-60m, delay actually increases because there are lesser intermediate layers available which causes more queuing delays.

Similarly the energy per packet decreases for the transmission range of up to 60m. This again can be attributed to the lesser hop-count. However for range greater than 60m, the transmit power required is higher resulting in higher energy per packet. Energy-delay product too predictably follows the same trend. However the optimal value of the range varies with the number of nodes in this case.

Network lifetime is quantified by two metrics namely time to first node death and time to network partition. Both the metrics increase with the increase in transmission range. However for the range greater than 120m, the network lifetime drops significantly due to the added cost of transmission.

Thus from these simulations, the optimal range of the transmission is identified to be between 40m and 60m. Thus the MAC protocol for MINA provides a clean way to perform medium access control and complements the other protocols of the suite perfectly.

5.3.2 TRAFFIC ADAPTIVE MEDIUM ACCESS PROTOCOL (TRAMA)

The goal of the TRAMA protocol [6] is to provide a completely collision free medium access and thus achieve significant energy savings. It is primarily a scheduled based MAC protocol with a random access component for establishing the schedules. TRAMA relies on switching the nodes to a low power mode to realize the energy savings. The Protocol has different phases or components namely: Neighbor Protocol (NP), Schedule Exchange Protocol (SEP) and Adaptive Election Algorithm (AEA). NP uses the random access period to gather the one-hop and two-hop neighbor information. SEP helps establishing the schedules for a given interval among the one-hop and two-hop neighbors. Finally, AEA decides the winner of a given time slot and also facilitates the reuse of unused slots.

Protocol details. TRAMA derives from the idea proposed in the Neighbor-Aware Contention Resolution (NCR) [13] to select the winner of the given time slot in a two-hop neighborhood. For every one-hop and two-hop neighbor, a node calculates a MD5 hash of the concatenation of the node-id and the time slot t. This gives the priority of a node for a given time slot. The node with the highest priority is chosen to be the slot winner. After the Neighbor Protocol has gathered the neighbor information using the signaling packets in a random access mode, the node computes a certain SCHEDULE_INTERVAL. This is the duration in which a node may transmit data and is based on the rate at which packets are generated from the application layer. The node further pre-computes the priorities to identify its own winning slots for the duration of SCHEDULE_INTERVAL. These schedules are announced in a schedule packet. Instead of including the receiver addresses in the schedule packet, a bitmap is included for its every winning slot. Each bit in the bitmap corresponding to its every one-hop neighbor; 1 if it is the intended receiver and 0 otherwise. This also simplifies the broadcast and multicast mechanisms.

Broadcast involves a bitmap with all 1's while for the multicast, specific bits corresponding to the intended receivers are set. Looking at the schedule packet, a node may go into sleep mode if it is not the intended receiver of any of its neighbors. This is helpful from the energy efficiency point of view. Also a node may not have enough data to transmit during all of its winner slots. In order not to waste these vacant slots and allow their re-use, the node announces these slots with a zero bitmap. However the last of the winning slots is reserved for the node to announce its schedule for the next SCHEDULE_INTERVAL. Due to this provision for the re-use of slots, schedules for the nodes may no longer remain synchronized. This is because a node may use some other node's unused slot to transmit. To alleviate the problem, schedules are timed out after a certain time period. Furthermore, schedule summaries are sent piggybacked along with the data packets to help maintain the synchronization.

The Adaptive Election Algorithm (AEA) determines the state of the node at a given time and facilitates the slot re-use. For every node, the protocol keeps track of the nodes in its neighborhood that need extra slots to transmit. The set of such nodes is called as *Need Contender Set*. Every slot that is owned but unused by a node X is given to the node with the highest priority in the Need Contender Set of node X. However, inconsistencies may arise as shown in Figure 5.4. For node B, node D is the winner since it has the highest priority in its two-hop neighborhood. But for node A, node D is not visible and hence it assumes itself as the winner. Thus both nodes A and D may transmit. Suppose node B is not the intended receiver for node D and goes to sleep. However the node A may transmit to B in which case the transmission will be lost. In order to deal with such a problem, a node also keeps track of an Alternate winner along with the Absolute winner. The node has to account for the Alternate winner as well if the Absolute winner does not have any data to send. In this case, D is the Absolute winner and A is the Alternate winner for B.

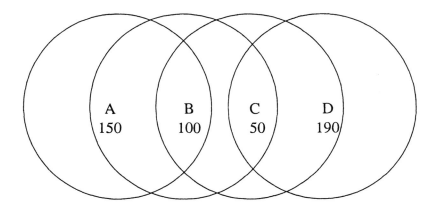

Figure 5.4. The slot inconsistency problem.

Experiments and Results. The performance of TRAMA has been studied using detailed discrete-event simulation based experiments. The performance of TRAMA is compared with both contention based protocols (IEEE 802.11 and S-MAC) as well as a scheduled based protocol (NAMA [13]). One set of experiments is conducted using with an exponential inter-arrival time for the data. In this case, a neighbor is randomly selected to perform either the unicast or the broadcast. TRAMA is shown to achieve higher throughput than the contention based protocols. This is to be expected given the collision-free nature of TRAMA and the fact that contention based protocols perform poorly due to the collisions. As to NAMA, the other scheduled access based protocol, the throughput achieved is comparable to that of TRAMA. Broadcasts are also found to be more feasible in case of scheduled based protocols, in particular TRAMA. This again can be attributed to the collision freedom guaranteed by TRAMA.

For the study of the energy efficiency, detailed investigations are performed by comparing the performance of TRAMA with that of S-MAC [10] and IEEE 802.11. The metrics considered are *sleep time percentage*, defined as the ratio of number of sleep slots to the total slots and the *average sleep interval*, used to measure the number of radio mode switches. In case of average sleep time percentage, S-MAC with 10% duty cycle (10% sense, 90% sleep) fares better than TRAMA. However TRAMA has better average sleep interval than S-MAC. This means that the switching between radio modes is more frequent in the case of S-MAC. The price to pay for the scheduled access based protocols is the higher latency. TRAMA incurs higher average queuing delays than the IEEE 802.11 and S-MAC. However it performs better in this respect than its other counterpart, NAMA.

The simulations are also performed under different sensor scenarios by varying the position of the sink node in the field (edge, corner or center). The results are almost similar to the ones observed before with the notable exception in case of the percentage sleep time. TRAMA exhibits higher energy savings than S-MAC in all the scenarios.

5.3.3 ENERGY AWARE TDMA BASED MAC

Another approach based on TDMA is considered in [7]. It assumes the presence of "gateway" nodes which act as the cluster heads for clusters of sensors. The gateway assigns the time slots to the sensor nodes in its cluster. Naturally this TDMA scheme eliminates majority of the potential collisions. Marginal possibility of collisions still exists in the case that a node does not hear the slot assignment. However this is highly limited.

Protocol details. The protocol consists of four phases namely, data transfer, refresh, event-triggered rerouting and refresh-based rerouting. Data transfer phase, understandably is the longest of all. Refresh phase is used by nodes to update the gateway about their current state (energy level, position etc.) This information is used by the gateway to perform rerouting if necessary. This is done during the event-triggered rerouting phase. Another form of rerouting occurs during refresh-based rerouting, which is scheduled periodically after the refresh phase. In both the rerouting phases, the gateway runs the routing algorithms and sends the new routes to the sensors.

The paper presents two algorithms for the slot assignment based on Breadth First Search (BFS) and Depth First Search (DFS). These graph-parsing strategies specify the order in which slot numbers are assigned to the nodes starting from the outermost active sensors. In BFS, the numbering starts from the outermost nodes giving them contiguous slots. On the other hand, the DFS strategy assigns contiguous time slots for the nodes on the routes from outermost sensors to the gateway. Figure 5.5 illustrates the two ideas.

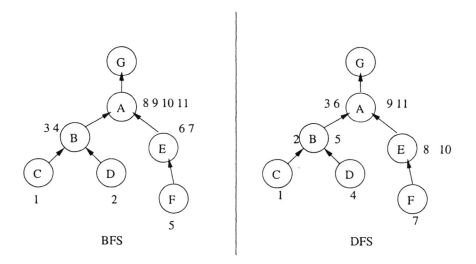

Figure 5.5. Slot Assignments for BFS and DFS Strategies. Note that Node E serves both as a sensor as well as a relay while nodes A and B only serve as relays.

With the BFS strategy, the relay nodes need only turn on once to route their children's packets. If the cost of turning the sensor nodes ON and OFF is high, this scheme offers a good economical option. On the other hand, the relay nodes need to have sufficient buffer capacity to store the data packets until it is time to forward them. This makes it susceptible to the buffer overflows and associated packet drops. The DFS strategy, on the other hand, does not demand any buffer capacity for the relays. However the relays have to switch on and off multiple times, which makes it a less attractive choice from the energy savings point of view.

Experiments and Results. The simulations are performed for a 1000 × 1000 square meter field with 100 randomly deployed nodes. The effects of buffer size on various parameters (e.g. end-to-end delay, throughput, energy consumed per packet, node lifetime, packet drop count etc.) are considered for both BFS and DFS mechanisms. Because there are no sensor state changes, BFS consumes less energy per packet and hence offers a higher node lifetime. However, DFS offers lesser end-to-end delay and lesser packet drop count and thus higher throughput. This is because there are no overheads associated with the buffers.

5.4 SUMMARY

This section presented a survey of the some of the recent medium access control protocols specifically designed for wireless sensor networks. The protocols were categorized based on the random access or static access nature. Further research is necessary in this topic to address highly scalable MAC protocols for networks involving a very large number (say 10,000) of nodes. Also, a comprehensive qualitative evaluation of the various protocols for different scenarios and traffic patterns is necessary. In addition, more experimental MAC-level protocol results especially for scheduling-based protocols will be useful.

ACKNOWLEDGMENTS

We would like to thank DAWN lab members Aniruddha Rangnekar and Uttara Korad for their valuable suggestions and help in preparing this document.

REFERENCES

[1] K. Jamieson, H. Balakrishnan, and Y. C. Tay, "Sift: A MAC protocol for event-driven wireless sensor networks," Tech. Rep. MIT-LCS-TR-894, Massachusetts Institute of Technology, May 2003.

[2] T. van Dam and K. Langendoen, "An adaptive energy-efficient MAC protocol for wireless sensor networks," in *ACM Intl. Conference on Embedded Networked Sensor Systems (SenSys)*, (Los Angeles, CA), Nov. 2003.

[3] A. Woo and D. Culler, "A transmission control scheme for media access in sensor networks," in *Proc. of ACM Mobicom*, (Rome, Italy), Aug 2001.

[4] S. Adireddy and L. Tong, "Medium access control with channel state information for large sensor networks," in *Proc. of IEEE Intl. Workshop on Multimedia Signal Processing*, (St. Thomas, US Virgin Islands), Dec. 2002.

[5] J. Ding, K. M. Sivalingam, R. Kashyapa, and L. J. Chuan, "A multi-layered architecture and protocols for large-scale wireless sensor networks," in *Proc. IEEE Semiannual Vehicular Technology Conference – Fall*, (Orlando, FL), Oct. 2003.

[6] V. Rajendran, K. Obraczka, and J. Garcia-Luna-Aceves, "Energy-efficient collision-free medium access control for wireless sensor networks," in *ACM Intl. Conference on Embedded Networked Sensor Systems (SenSys)*, (Los Angeles, CA), Nov. 2003.

[7] K. Arisha, M. Youssef, and M. Younis, "Energy-aware TDMA based MAC for sensor networks," in *IEEE Workshop on Integrated Management of Power Aware Communications Computing and Networking (IMPACCT 2002)*, 2002.

[8] IEEE, "Wireless LAN medium access control (MAC) and physical layer (PHY) Spec." IEEE 802.11 standard, 1998.

[9] C. E. Jones, K. M. Sivalingam, P. Agrawal, and J.-C. Chen, "A survey of energy efficient network protocols for wireless networks," *ACM/Baltzer Wireless Networks*, vol. 7, no. 4, pp. 343–358, 2001.

[10] W. Ye, J. Heidemann, and D. Estrin, "An energy-efficient MAC protocol for wireless sensor networks," in *Proc. IEEE INFOCOM*, (New York, NY), June 2002.

[11] P. Havinga, "Energy efficient sensor networks: European research project (ist-2001-34734)." http://eyes.eu.org, 2003.

[12] R. Kashyapa, "Medium access control and routing protocols for data gathering using wireless sensor networks: Design and analysis," Master's thesis, Washington State University, Pullman, Aug. 2001.

[13] L. Bao and J. Garcia-Luna-Aceves, "A new approach to channel access scheduling for ad hoc networks," in *Proc. ACM MobiCom*, (Rome, Italy), July 2001.

Chapter 6

DISSEMINATION PROTOCOLS FOR LARGE SENSOR NETWORKS

Fan Ye, Haiyun Luo, Songwu Lu and Lixia Zhang
Department of Computer Science
UCLA
Los Angeles, CA 90095-1596
{ yefan,hluo,slu,lixia } @cs.ucla.edu

Abstract   Dissemination protocols in a large sensor network typically take a data-centric paradigm in which the communication primitives are organized around the sensing data instead of the network nodes. A user expresses his interest in a given data type by specifying a query message, which then propagates through the network. When detecting a nearby stimulus that matches the query type, sensors generate data reports which traverse the network to reach the interested user. The large scale of a sensor network, the resource constraints of each node in terms of energy, computation and memory, and potentially frequent node failures all present formidable challenges to reliable data dissemination. The protocols must be energy efficient, robust against failures and channel dynamics, and be able to scale to large networks. In this chapter, we provide a brief overview and critique of the state-of-art dissemination protocols and discuss future research directions.

In this chapter, we focus on the design of data dissemination protocols for a large-scale sensor network. In such networks data flows from potentially multiple sources to potentially multiple, static or mobile sinks. We define a data source as a sensor node that detects a *stimulus*, which is a target or *an event of interest*, and generates data to report the event. A *sink* is defined as a user that collects data reports from the sensor network. In the example of battlefield surveillance, a group of soldiers, each of them serves as a sink, may be interested in collecting tank location information from a sensor network deployed in the battlefield. The sensor nodes surrounding a tank, which acts as a stimulus, detect it and generate data reports. These reports will be forwarded by intermediate sensors to each of the sinks.

The unique characteristics of wireless sensor networks make the problem of data dissemination challenging and different from conventional wired Internet or wireless ad-hoc networks. Three main design challenges are:

- **Network scale** A sensor network may have thousands or even hundreds of thousands of nodes to cover a vast geographical area. Each sensor node is a potential data source, and multiple sinks may exist. In contrast, a conventional ad-hoc network has no more than a few hundred nodes in general.

- **Constrained resources at each sensor node** To reduce cost and increase spatial sensing coverage, sensor nodes are typically simple and inexpensive devices manufactured according to the economy of scale. Compared to the typical notebook/desktop computers that serve a conventional ad-hoc network or wired Internet, sensors tend to have limited battery energy, relatively low-end CPUs, and constrained memory size. Once the battery is depleted, a node may be considered useless. Recharging nodes is difficult due to the ad hoc deployment and the large node population.

- **Unpredictable node failures and unreliable wireless channel** The limited energy supply and small size make it infeasible to equip nodes with powerful radios and large antennas for high communication channel quality. Potential environmental obstacles and interferences further exacerbate the situation, making the wireless communication channel used by the sensor nodes prone to errors. In a harsh environment, sensor nodes themselves may fail or be destroyed unexpectedly. These operating conditions can be much worse than those of conventional ad hoc wireless networks.

In the rest of this chapter, we first identify main design goals for dissemination protocols in large sensor networks. We next describe the state-of-the-art protocol designs by organizing them into two categories: approaches without hierarchy and the hierarchical approach. We compare their main differences, and provide critiques on their possible limitations. We then elaborate on related issues of exploiting data naming and location awareness and providing real-time delivery. We further identify possible future directions and finally conclude the chapter. It should be noted that there may exist other small-scale, human-planned sensor networks, which are not the focus in this chapter. We are mainly concerned with large sensor networks made of simple and inexpensive nodes.

6.1 DESIGN GOALS AND SOLUTION ROADMAP

Generally speaking, there are three desired properties for data dissemination protocols in a wireless sensor network:

- **Scalable and distributed solution** The dissemination protocol should scale with a large number of nodes, multiple sources and sinks.

- **Energy efficiency** Since energy is the most critical resource, the protocol should minimize computation and especially communication operations.

- **Robustness** The protocol should be able to work with highly dynamic, error-prone wireless communication channels, as well as fallible nodes that may fail due to damage or energy depletion.

The design of data dissemination protocols often take into account other desired properties as well, possibly with lesser degree of importance. For example, mission-critical applications may require that the report delivery be timely and reliable.

Note that some of the above goals may be in partial conflict with each other. For example, a distributed design, which can scale well, may not achieve the optimal energy efficiency. The robustness of the protocol is often realized through inter-node communications which consumes additional energy. Given an application scenario and network setting, the protocol design seeks the best tradeoff between conflicting metrics.

6.1.1 SOLUTION ROADMAP

Dissemination protocols in wireless sensor networks typically take a *data-centric* design paradigm, where the communication primitives of sensor networks are organized around the sensing data instead of network nodes as in either wired or conventional wireless networks. The operations of the dissemination protocol typically take a two-phase, "query-reply" process. The process is usually initiated by a sink, which sends query messages towards a geographical region of interest or the entire network, rather than a specific node. The query carries the sink's interests such as stimulus types and possibly the frequency it intends the data to be delivered back. The query message informs sensor nodes in the target region to send data reports back to the sink when the interested stimuli are detected. Data reports are tagged with the location, type and detection time of the stimulus, not necessarily with the detecting node's identifier. During the process, the protocol may install dissemination states in intermediate sensor nodes in order to guide future data delivery.

Such application-driven design distinct sensor networks from conventional wired and and wireless network routing. The unique characteristics of a sensor network also offer several unique opportunities for its protocol design:

- **Application semantics** Because the data dissemination protocol is data centric, it can exploit application semantics to improve its performance. For example, to reduce communication cost it is possible to aggregate data reports based on their common semantics and to do in-network data processing. A data centric design further allows to collapse multiple layers of the protocol stack and reduce inter-layer communication overhead.

- **Location awareness** Sensing application typically requires that each node know its (possibly approximate) location through GPS or other location services. The protocol design can exploit this extra information to achieve better scalability and efficiency.

- **Stationary nodes** Sensor nodes typically remain stationary after deployment. Therefore a dissemination protocol does not need to incorporate sophisticated mobility management component and may greatly simplify its operation compared to that of ad-hoc network routing protocols.

- **Dense deployment** To enhance reliability and system life time the initial deployment of a sensor network may have high node density. The scale and dense deployment of a sensor network also offer new opportunities for protocol design. Though each individual sensor has stringent resource constraints, collectively multiple sensors help to achieve better system robustness by harnessing the node population. The protocol can prolong the system operation lifetime via maintaining a small set of working nodes and letting others doze off.

Data dissemination protocols in sensor networks can be broadly classified into two categories: those protocols that operate without building a hierarchy among the sensor nodes in the network, and those protocols that build and maintain a communication hierarchy to help scale data dissemination. In the first category of protocols, two popular approaches are reverse-path-based forwarding and cost-field-driven dissemination. In the second category, the protocol builds a communication hierarchy by applying clustering or exploiting location information. We will discuss all these approaches in the following sections.

Table 6.1 summarizes the different data dissemination protocols in sensor networks. It shows the basic approach adopted by each protocol and the functionalities achieved.

Table 6.1. **Summary of Data Dissemination Protocols for Sensor Networks**
(Supported properties or functionalities: 1. multiple sinks 2. sink mobility 3. load balancing 4. robustness 6. real time 7. in-network processing 8. query dissemination to targeted regions)

Protocol	Properties	Basic Approach	Details
DRP [1]	7	Reverse path forwarding	per-sink tree; use semantic information
Directed Diffusion [8]	1,7	Reverse path forwarding	shared mesh; use semantic information
GRAd [13]	4	cost-field	receiver-decided
GRAB [16]	4	cost-field	receiver-decided; use credit to control forwarding redundancy
ReInForm [2]	4	cost-field	receiver-decided; sender calculates forwarding probabilities for receivers
EAR [14]	3	cost-field	sender-decided
TTDD [12]	1,2	hierarchy	virtual geographic grid per source
LEACH [6]	3	hierarchy	two-hop data delivery, a cluster head sends to the sink directly
GPSR [9]	1	Geographical forwarding	uses planar graph to bypass "holes"
GEAR [17]	3,8	Geographical forwarding	combines location and energy to choose next hop
SPIN [7]	7	Flooding	use semantic data descriptor to reduce redundant transmissions
RAP [11]	6	Geographical forwarding, virtual "velocity"	schedule packets using priority queues at each node
SPEED [4]	6	Geographical forwarding, virtual "velocity"	explicitly control the delay at each hop

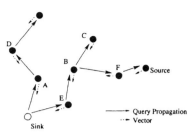
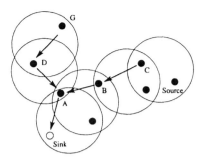

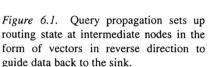

Figure 6.1. Query propagation sets up routing state at intermediate nodes in the form of vectors in reverse direction to guide data back to the sink.

Figure 6.2. Routing tree building by query propagation. Big circles denote neighborhoods. Only one forwarding node is needed in each neighborhood.

6.2 REVERSE PATH FORWARDING

Reserve path forwarding is one general approach to data dissemination in sensor networks and representative protocols include Declarative Routing Protocol (DRP), and Directed Diffusion (DD) protocol. In this approach, data reports flow in the reverse direction of query propagation to reach a sink. The sink sends out a query message that expresses its interest, possibly using flooding, throughout the network. Whenever a node receives a query from a neighbor, it sets up a forwarding state in the form of a *vector*, pointing from itself to the neighbor, i.e., indicating the reverse data path. When a data report is generated by a source, it goes along the direction of the vector from one node to the next, until it reaches the sink. The protocols in this category differ in their techniques to achieve scalability, energy efficiency and robustness. In the following, we will discuss three such protocols: DRP, DD and its variant – resilient multipath directed diffusion.

6.2.1 DECLARATIVE ROUTING PROTOCOL

Declarative Routing Protocol (DRP) [1] achieves energy efficiency through in-network aggregation. It uses reverse path forwarding to establish a routing tree for each sink to receive data reports for all other nodes. It also exploits the application-specific property of sensor networks to build a cross-layer declarative service for better efficiency.

DRP organizes nodes in the communication range of each other into *neighborhoods*. One node may belong to multiple neighborhoods at the same time. A sink sends out queries using a distributed algorithm, similar to an expand-ring search, which ensures loop-free forwarding. As the query is disseminated from one neighborhood to the next, a weighted sum of several factors is used to select to which node of the next neighborhood the query should be sent. The factors include *reachability* (the number of nodes that can be reached), *diversity* (the uniqueness of the set of nodes that can be reached), *directionality* (the proximity to the path to the target region), *remaining energy* and *link quality* in terms of SNR (Signal-to-Noise Ratio).

After the query traverses all neighborhoods, the sink obtains a routing tree rooted at itself, with vectors pointing in the reverse direction of query propagation. Figure 6.2 shows an example of such a tree. The routing vectors are maintained at selected nodes A, B, C, D and G. A data source broadcasts its data report in its local neighborhood. Any neighbor that finds a match between the data and the user's interest should forward data back along the reverse path to the sink.

DRP maintains the consistency of routing state at different nodes using a sequence number. Each time a routing record is updated, the associated sequence number is incremented. Neighbors detects inconsistency by comparing the sequence numbers of their cached records and requesting the latest version through NACK. To improve efficiency, DRP proposes information sharing across multiple layers in the protocol stack. A new entity called *declaration service* spans the medium access, network and transport layers, and interacts with the sensor applications directly. Such a cross-layer design allows sharing of state and reduces cross-layer function calls, thus trading off system modularity for performance.

6.2.2 DIRECTED DIFFUSION AND ITS VARIANTS

Similar to DRP, Directed Diffusion [8] also uses reserve-path to guide data and follows the data-centric paradigm. However, Diffusion focuses more on scaling to multiple sinks and robustness under network dynamics.

Diffusion builds a shared mesh to deliver data from multiple sources to multiple sinks, rather than per tree each sink. To scale to multiple sinks, nodes do not keep per sink state. When receiving a query from a neighbor, a node does not differentiate whether the neighbor is a sink or one that merely forwards the query. It always sets up a vector (called "gradient") pointing from itself to the neighbor to direct future data flow, without associating the vector to any particular sink. This is different from the per sink state in DRP.

After a sink floods its query across the network, two vectors are established between each pair of neighbors, pointing to each other. The initial batch of data follow such vectors, effectively flood through the network to explore all possible paths. Each node has a cache to detect loops and drop redundant packets (See Figure 6.3 for an example).

After receiving the initial batch of data reports along multiple paths, a sink uses a mechanism called *reinforcement* to select a path with the best forwarding quality. It sends a positive reinforcement message to the neighbor from

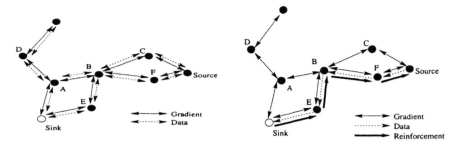

Figure 6.3. Interest flooding and multipath data forwarding

Figure 6.4. Reinforcement

which it receives data with the minimum delay (or other metrics, depending on applications' needs). It may also send negative reinforcement messages to other neighbors to de-activate low-quality paths. Such positive and negative reinforcements are repeated upstream towards data sources. Gradually, data reports flow only along the best path (Figure 6.4 shows an example where the data forwarding path F→B→E is reinforced).

To handle network dynamics such as node failures, each source floods reports periodically at a rate much lower than normal data, to maintain alternative paths. So nodes can reinforce other paths to route around failed nodes. This enhanced robustness is achieved at the cost of additional energy and bandwidth consumption due to flooding. In-network processing can be integrated into Diffusion. Each intermediate node actively aggregates queries carrying the same interest and reports containing correlated data. We will further elaborate in-network processing later.

Resilient multipath Directed Diffusion [3] enables quick path repairing in case of node failures by maintaining multiple backup paths together with the primary data forwarding path. These backup paths are used to quickly route around failed nodes. One method is to construct *braided* multipath. Every other node along the primary data forwarding path sends an *alternate path reinforcement* message to its next-preferred neighbor. Then the neighbor sends out a primary path reinforcement message to its most-preferred upstream neighbor. A node that is already on the primary forwarding path silently drops the alternate path reinforcement message. The final effect is that a node can immediately start pulling data from an alternative upstream neighbor if the current one fails, without waiting till the next data flooding.

6.2.3 SUMMARY

Reverse path based forwarding approaches naturally integrate in-network processing. Whenever multiple reports from different reverse paths merge at a node, the node can process and aggregate data based on their semantics. This in-network processing saves energy and bandwidth resources. Both DRP and Diffusion adopt this feature to achieve better efficiency.

Since the vector state kept in nodes depends on the local connectivity, overhead has to be paid to maintain states in the presence of node failures. DRP uses periodic beacon exchanges among neighbors to maintain state, whereas Diffusion uses reinforcement and periodic low-rate data flooding to adapt. These complexities and overhead are inevitable in reverse path approaches because the vector state has to be updated with the changes in connectivity.

Sink mobility is not well addressed in this approach. In DRP, the reverse path forms a tree rooted at a sink. Any change in the sink's location requires changes of the vector state in nodes. The sink needs to re-flood its queries to update the vector states in nodes to receive data reports when it roams. The same is true for Directed Diffusion.

6.3 COST-FIELD BASED FORWARDING

Cost-field based forwarding offers an alternative approach to data dissemination. In this approach, the forwarding states stored in an intermediate node take a form other than vector: a *scalar* denoting the node's "distance" to a sink. The scalar value denotes the total cost, which can be measured by the hop count, energy consumption or physical distance, to deliver a report from this node to the sink. The cost scalar of all the nodes forms a *cost-field*, where nodes closer to the sink having smaller costs and those farther away having higher costs. The cost value at each node is directionless, different from the vectors used in reverse-path forwarding. However, the cost field has implicit directions. A data report can flow from a node with a higher cost to a neighbor with a smaller cost, similar to water flowing from a hill to a valley, eventually reaching the sink (See Figure 6.5 for an illustration). A number of protocols, including Gradient Routing (GRAd) [13], GRAdient Broadcast (GRAB) [15, 16], ReInForM [2], Energy Aware Routing (EAR) [14] fall into this general approach.

6.3.1 FUNDAMENTALS OF COST-FIELD APPROACH

Cost-field approach has two main steps: setting up the cost field for a sink and how a node makes forwarding decisions. A sink sets up its cost field by flooding a query packet (ADV) similar to distance-vector routing. The packet carries a cost value, initially set to 0 by the sink. A node N_i receiving an ADV packet from node N_j calculates its cost as

$$C_i^{new} = C_j + C_{i,j}, \qquad (6.1)$$

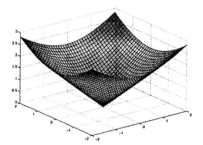
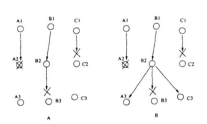

Figure 6.5. The "shape" of the cost field is like a funnel, with the sink sitting at the bottom. Data follow the decreasing cost direction to reach the bottom of the cost field, which is the sink.

Figure 6.6. A: any single node failure or packet loss ruins a single path; B: interleaving paths can recover each other from failures or packet losses

where C_j is node N_j's cost carried in its ADV packet, $C_{i,j}$ is the link cost to send a packet from N_i to N_j. If C_i^{new} is less than N_i's current cost C_i, it sets C_i to C_i^{new}; otherwise it does not change C_i. Then N_i broadcasts an ADV carrying its own cost C_i. N_i keeps the smallest one if it receives multiple ADVs carrying different costs from different neighbors during the flooding,

The cost at a node provides a sense of its "distance" to the sink. Nodes "closer" to the sink have smaller costs; the cost-decreasing direction points towards the sink. As long as a data report follows such directions, at each step it comes closer to the sink, and will not come back to the same node. Thus no loop can form. The vector in reserve path forwarding, however, does not provide any sense of distance. Vectors at different nodes may form loops, which have to be detected explicitly. The cost field is per-sink based. Each sink has to set up its own cost field to receive reports. A node keeps a cost for each sink. In the reverse-path-based approach, multiple sinks may share the same segments of data delivery paths. A node does not necessarily need to keep per-sink state.

In the cost-field-based approach, there are two ways a node makes forwarding decisions: *receiver-decided* and *sender-appointed.* In the former method, a sender includes its cost in a report and broadcasts the report, without addressing to any specific neighbor. Each receiver decides whether it wants to forward by comparing its own cost to that of the sender. Only receivers with costs less than that of the sender may forward the report. In the latter method, a sender decides which neighbor forwards the report and unicasts the report.

The receiver-decided method enables robust data forwarding with unreliable wireless channels and fallible nodes. Because typically several neighbors have smaller costs than the sender, at each hop multiple nodes can receive and forward different copies of a report in parallel. As long as there exists one node that receives the report successfully, packet losses or node failure of other nodes on this hop do not disrupt data delivery. Moreover, a single upstream forwarding node may spawn several copies to multiple downstream nodes successfully, thus recovering the parallelism in forwarding. The actual paths different copies take are not disjoint, but rather, they *interleave* and form a mesh (See Figure 6.6 for an example). This "self-recovery" mesh proves highly robust against packet losses and node failures, as demonstrated by GRAB.

The sender-appointed method may also form such a interleaving mesh, but the sender has to keep states regarding which neighbors are alive, and which links still function. In a harsh environment where both node topology and wireless connectivity change dynamically, such states have to be maintained with added complexity and may not be updated as quickly as changes happen.

GRAd, GRAB and ReInForm are receiver-decided. They achieve robustness in data delivery, but differ in whether and how they control the redundancy of the forwarding mesh. EAR is sender-appointed; it focuses on load balancing among different nodes.

6.3.2 RECEIVER-DECIDED PROTOCOLS

GRAd uses hop count as the cost. A sink builds its cost-field by flooding a REQUEST message. A source broadcasts a report; all neighbors with smaller costs forward the report. This process is repeated, and multiple copies of the report travel within a forwarding mesh of intermediate nodes towards the sink.

GRAB is designed for a harsh environment with high channel error rate and frequent node failures. The cost each node keeps is the total energy needed to send a packet to the sink. It assumes that each node knows the energy needed to transmit a report to a neighbor. GRAB further enables flexible trade-off between robustness and energy consumption. It controls the "width" of the forwarding mesh using a packet-carried state called *credit*. When a source sends a report, it assigns a credit to the report. The sum of the source's cost and the credit is the total budget that can be consumed to send the report downstream to the sink. The report can take any route that has a total cost less than or equal to the budget, but not those of costs exceeding the budget. A larger credit allows for a wider forwarding mesh; if the credit is "0", the report can only be forwarded over the minimum-cost route from the source. By controlling the amount of credit, the source effectively controls the degree of redundancy in the forwarding mesh.

ReInForm controls the redundancy differently. It calculates the number of parallel forwarding nodes needed at each hop to achieve a specified end-to-end delivery success probability r_s. It takes hop count as the cost. Each node keeps a measurement of local channel error rate e, and divides its neighbors into three groups H^-, H^0, H^+, i.e., neighbors with one less, the same, one more hop count. A source with a cost of h_s calculates P, the number of forwarding neighbors needed, as a function of e, r_s, h_s. If P exceeds the number of neighbors in H^-, some nodes in H^0 have to forward; if it exceeds the total number in H^- and H^0, some nodes in H^+ are needed. In such cases the source calculates three values P^-, P^0, P^+, representing the number of paths each node in the corresponding group needs to create. It then broadcasts the report, including the state P^-, P^0, P^+, e, h_s. A receiver knows its group and checks the corresponding P: it forwards the report if $P \geq 1$, or with probability P if $P < 1$. Each receiver that decides to forward repeats the same procedure as the source. It recalculates the P values for its three local neighbor groups using its own cost and channel error, then broadcasts the report.

6.3.3 SENDER-APPOINTED PROTOCOL

EAR achieves load balancing among multiple paths to extend the system lifetime. A node makes forwarding decision probabilistically among neighbors. A node N_a keeps one energy cost $C_{a,i}$ for each of the neighbors N_i from which an ADV is received (as in Equation 6.1). It assigns a probability P_i inverse proportional to $C_{a,i}$ of each of these neighbors. When sending a report, it selects one of these neighbors using the probability P_i's. The cost included in its ADV packets is $\sum_i P_i C_{a,i}$, a weighted average of the $C_{a,i}$'s.

Different from the other three cost-field protocols, EAR is single path forwarding. The sender chooses only one neighbor for each report. Its main goal is not to ensure robustness, but to statistically distribute the amount of consumed energy among different nodes. A path consuming more energy is used inversely proportionally less frequent. Thus forwarding nodes consume energy at about the same rate over time.

6.3.4 SUMMARY

Compared with the reverse-path-based approach, the cost-field approach enables mesh forwarding and receiver-decided delivery. By comparing its cost to that of a sender, multiple receivers can each decide whether it continues forwarding a packet or not. Loops cannot form when packets follow the direction of decreasing cost. Mesh forwarding greatly enhances robustness in the presence of unpredictable node failures and harsh channel conditions. The receiver-decided forwarding eliminates the complexity at each sender to maintain explicit path states.

Similar to reverse path based approach, cost-field base approach does not address mobile sinks well. Since the cost depends on the "distance" to the sink, it has to be updated when a sink moves. Cost-field based approach does not scale well with the number of sinks, either. Each sink needs a separate cost at every node, thus the amount of state increases proportionally to the number of sinks. In the reverse-path-based approach, the vector does not need to be associated with a particular sink (as in Diffusion). Thus multiple sinks can share the same segments of paths, at the cost of maintaining a cache at each node to detect loops.

6.4 ROUTING WITH VIRTUAL HIERARCHY

So far we have presented forwarding protocols that take the "flat" network model, in which every sensor node has the same functionality and no network hierarchy is formed during data dissemination. In this section, we present two designs that build a virtual hierarchy to address sink mobility and achieve energy efficiency.

6.4.1 TWO-TIER DATA DISSEMINATION

Sink mobility brings new challenges to large-scale sensor networking. Both reverse-path and cost-field based approaches require mobile sinks to continuously propagate queries to update sensor state throughout the network, so that all sensor nodes get updated with the direction of sending future data reports. However, frequent query dissemination from mobile sinks can lead to both increased collisions in wireless transmissions and rapid power consumption of the sensor's limited battery supply.

Two-tier Data Dissemination (TTDD) [12] aims at efficient and scalable sink mobility support. It considers a sensor network of homogeneous sensor nodes, but enforces a virtual grid infrastructure for query and data forwarding. In TTDD, the data source *proactively* builds a grid structure throughout the sensor field. In a two-dimensional sensor field, a source divides the plane into a *grid* of cells. Each cell is an $\alpha \times \alpha$ square. A source itself is at one crossing point of the grid. It propagates data announcements to reach all other crossings, called *dissemination points*, on the grid. The sensor closest to a dissemination point serves as a dissemination node. A dissemination node stores a few pieces of information for the grid structure: the data announcement message, the dissemination point L_p it is serving and the upstream dissemination node's location. It then further propagates the message to its neighboring dissemination points on the grid except the upstream one from which it receives the announcement. The data announcement message is *recursively* propagated through the whole sensor field so that each dissemination point on the grid is served by a dissemination node. Figure 6.7 shows a grid for a source B and its virtual grid.

With this grid structure in place, a query from a sink traverses two tiers to reach a source. The sink firstly floods its query within a cell. When the nearest dissemination node for the requested data receives the query, it forwards the query to its upstream dissemination node toward the source, which in turns

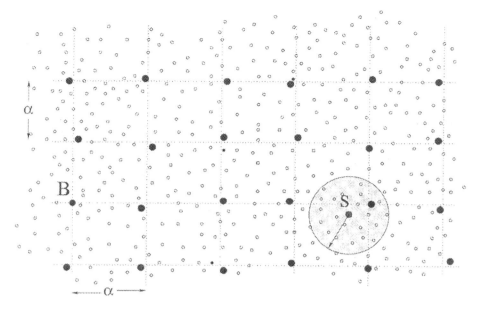

Figure 6.7. One source B and one sink S.

further forwards the query, until it reaches either the source or a dissemination node that is already receiving data from the source (e.g. upon requests from other sinks). This query forwarding process lays information of the path to the sink, to enable data from the source to traverse the same two tiers as the query but in the reverse order (see Figure 6.8 for an illustration).

TTDD also makes explicit effort to maintain its grid. The grid maintenance is triggered on-demand by on-going queries or upstream updates. Compared with periodic grid refreshing, it trades computational complexity for less consumption of energy, which is a more critical resource in the context of large-scale sensor networks.

6.4.2 LOW-ENERGY ADAPTIVE CLUSTERING HIERARCHY

Low-energy adaptive clustering hierarchy (LEACH) [6] is designed towards energy-optimal network organization. It assumes that each sensor node be able to adjust its wireless transceiver's power consumption, thereby controlling the network topology. At a sensor node's maximum transceiver's power, it is able to communicate directly with the sink, which is fixed in certain location of the sensor field.

LEACH is motivated by the comparison on energy consumption between two forwarding algorithms. One is direct transmission between each sen-

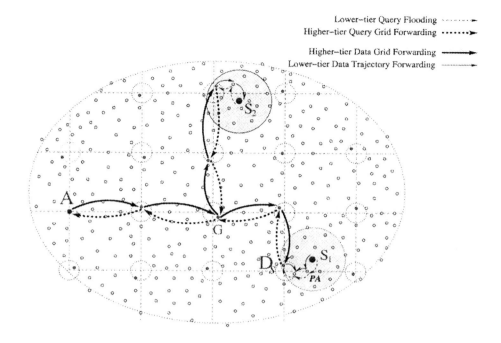

Figure 6.8. Two-tier query and data forwarding.

sor node and the sink. For distant sensor nodes, direct transmission to the sink requires a large amount of transmit power. However, direct transmission has the minimum energy consumption on reception, assuming the sink is not power-constrained. The other is minimum-transmission-energy (MTE). In MTE, rather than use one high-energy transmission, data messages go through n low-energy transmits and traverses n intermediate sensor nodes to reach the sink. MTE minimizes the energy consumption on transmission, however, the sensor network has to invest energy on reception. This amount of energy consumption is comparable to that on transmission given LEACH's radio model.

The analysis based on a certain network size and a first-order radio model results in LEACH's two-hop data dissemination, as a tradeoff between the transmission and receiving energy consumptions. Data messages from each sensor node are first transmitted to a local cluster head, and then forwarded to the base station using potentially long range radio. Periodically the system runs a randomized distributed algorithm to elect a number (determined a priori) of cluster heads. These cluster heads then advertise themselves to guide the clustering of the sensor networks. That is, a sensor node joins the cluster from which the strongest advertisement signal is received. Each sensor node reports itself to its cluster head. Based on cluster population, the cluster head creates a TDMA schedule telling each cluster member when it can transmit for intra-cluster transmission. LEACH also proposes to use CDMA spreading codes to decrease the inter-cluster interferences.

6.4.3 SUMMARY

Compared with flat forwarding approaches based on reverse path and cost-field, nodes in hierarchy forwarding carry different functions. Some special nodes form a structure that helps to address new issues or achieve better performance. TTDD's design exploits the fact that sensor nodes are both stationary and location-aware, and use simple greedy geographical forwarding to construct and maintain the grid structure with low overhead. In TTDD, the virtual grid confines the impact of sink mobility. A sink does not need to flood a query across the entire network when it moves. Instead, it only needs to disseminate queries within a local cell to find a new data dissemination node. Only data forwarding within the sink's local cell is affected; the higher-tier grid forwarding changes only incrementally and remain the same most of the time. In LEACH, the presence of a cluster head node enables data aggregation within the cluster, in order to save energy consumed to transmit data to the base station.

Hierarchy forwarding addresses the mentioned issues at a cost: the complexities needed to maintain the structure. In TTDD, the grid has to be built and tore down; it needs to be repaired when a dissemination node fails. In LEACH, nodes need to agree on their cluster header and take turns to be the header. These complexities not only increase the overhead, but may lead to less robust operations.

6.5 ADDITIONAL APPROACHES

In this section we describe a number of proposals that present other approaches to sensor network data forwarding. We first explain how two unique features of sensor networks, data-centric naming and location-awareness, can be exploited to improve forwarding efficiency. We then summarize protocols that seek to provide timely delivery.

6.5.1 EFFICIENCY BY IN-NETWORK PROCESSING

The data-centric nature of a sensor network allows for semantics-based in-network processing. [5] advocates that low-level naming in sensor networks should be based on attributes related to the application, rather than network topology. The application-specific paradigm of sensor networks is in contrast to the Internet, where routers' processing capability is the main constraint and the network is shared by many applications. [5] further proposes a Diffusion-based architecture utilizing application-specific code (called "filter") that processes data in the network. Semantics-based matching rules are used to identify which sensing data should be processed by a given filter. Two examples of in-network data aggregation and nested query illustrate the benefit achieved by such an architecture.

SPIN [7] exploits data-naming to reduce the overhead of disseminating data from one node to the rest of the network. A node first sends data descriptors in the form of meta-data to its neighbors. A receiver that needs those data sends back a request, upon which the sender sends the requested data. SPIN reduces the redundancy in receiving data by this semantics-based negotiation.

6.5.2 GEOGRAPHICAL FORWARDING

If each node knows its geographical location, this feature can be exploited to forward data to a given sink. The simplest method is greedy forwarding: each node always sends the packet to the neighbor that is closest to the destination. The problem with this simple approach is "void area," where the sender itself is closer to the destination than all its neighbors and the packet has to be forwarded backward temporarily to reach the destination. GPSR [9] solves this problem by forwarding a packet along the perimeter of a void area. It computes a planar topology where no two edges cross each other requiring only one-hop neighbor locations at each node. A packet follows the right-hand rule to circumvent the void area.

GEAR [17] exploits location information to address another issue of sending a query to all nodes within a target region. It first sends the query towards the centroid of the region; once the query reaches a node in the region, multiple copies are forwarded recursively to the sub-regions, until every node receives one copy. Different from pure geographical forwarding, GEAR uses a weighted sum of physical distance and residual energy when selecting its next-hop neighbor, thus achieving load balancing among nodes. It circumvents holes using a heuristic similar to the Learning Real Time A* algorithm in [10].

6.5.3 REAL-TIME DELIVERY

RAP [11] and SPEED [4] are two proposals providing real-time data delivery services in sensor networks. They are based on an idealized model that takes into account both the remaining lifetime before a data report expires and the remaining distance the report has to travel. Given the physical distance d to the destination and the lifetime t of a packet, a desired velocity $v = \frac{d}{t}$ can be calculated. If the packet travels along a straight line towards the destination at speed v, it will not miss the deadline. They both use greedy geographical forwarding, but differ in the exact mechanism used.

In RAP, the required velocity is updated at each hop to reflect the urgency of the report. A node uses multiple FIFO queues with different priorities to schedule reports of varying degrees of urgency. Each queue accepts reports of velocities within a certain range. RAP further adjusts the waiting and backoff times at MAC layer based on the priority of the report being transmitted, so that one with higher priority has greater probability of accessing the channel.

In SPEED, a node actively controls the data rate to avoid congestion and maintain a relatively stable relay-speed to each neighbor. The node measures the delay of sending a report to each neighbor using exponential weighted moving average. Given a report with velocity v, it computes the speed v_i of the report if neighbor N_i is chosen as the next hop. The sender chooses one neighbors with $v_i > v$ to forward the report. If no such a neighbor exists, the report is forwarded at a probability, which is the output of a proportional controller with the delivery miss ratios of neighbors as the input. This effectively reduces the traffic load on neighbors so that the desired relay speed can be achieved. Nodes in a congested region also feedback "back-pressure" messages upstream, so that data are detoured to bypass the region.

6.6 CONCLUSION

Wireless sensor networks hold great promises in providing a new networking paradigm which allows for interacting with the physical world. Though each individual sensor may have severe resource constraints in terms of energy, memory, communication and computation capabilities, thousands or even hundreds of thousands of them may collectively monitor the physical world, disseminate information upon critical environmental events, and process the information on the fly. They enable potentially numerous, new applications such as forest fire monitoring, chemical and biological agent detection, soil pollution measurement and autonomous ocean ecology sampling. Toward this end, efficient data dissemination protocol serves as a critical component to realize a large-scale sensor network.

In this chapter, we have provided a survey on the state-of-the-art dissemination protocols for sensor networks and discussed some future directions. Our writeup intends to provide a brief overview of this active research area and serves as a tutorial to attract more researchers to further refine this critically important protocol of a sensor network.

REFERENCES

[1] D. Coffin, D. V. Hook, S. McGarry, and S. Kolek. Declarative ad-hoc sensor networking. In *SPIE Integrated Command Environments*, 2000.

[2] B. Deb, S. Bhatnagar, and B. Nath. ReInForM: Reliable Information Forwarding Using Multiple Paths in Sensor Networks. In *28th Annual IEEE Conference on Local Computer Networks*, 2003.

[3] D. Ganesan, R. Govindan, S. Shenker, and D. Estrin. Highly Resilient, Energy Efficient Multipath Routing in Wireless Sensor Networks. In *ACM MOBIHOC*, 2001.

[4] T. He, J. A. Stankovic, C. Lu, and T. Abdelzaher. SPEED: A Stateless Protocol for Real-time Communication in Sensor Networks. In *ICDCS*, 2003.

[5] J. Heidemann, F. Silva, C. Intanagonwiwat, R. Govindan, D. Estrin, and D. Ganesan. Building Efficient Wireless Sensor Networks with Low-level Naming. In *ACM Symposium on Operating System Principles (SOSP'01)*, 2001.

[6] W. Heinzelman, A. Chandrakasan, and H. Balakrishnan. Energy-Efficient Communication Protocols for Wireless Microsensor Networks. In *Hawaaian Int'l Conf. on Systems Science*, 2000.

[7] W. Heinzelman, J. Kulik, and H. Balakrishnan. Adaptive Protocols for Information Dissemination in Wireless Sensor Networks. In *ACM MOBICOM*, 1999.

[8] C. Intanagonwiwat, R. Govindan, and D. Estrin. Directed Diffusion: A Scalable and Robust Communication Paradigm for Sensor Networks. In *ACM MOBICOM*, 2000.

[9] B. Karp and H. T. Kung. Gpsr: Greedy perimeter stateless routing for wireless networks. In *ACM MOBICOM*, 2000.

[10] R. Korf. Real-time Heuristic Search. *Artificial Intelligence*, 42(2-3), March 1990.

[11] C. Lu, B. M. Blum, T. F. Abdelzaher, J. A. Stankovic, and T. He. RAP: A Real-Time Communication Architecture for Large-Scale Wireless Sensor Networks. In *Eighth IEEE Real-Time and Embedded Technology and Applications Symposium*, 2002.

[12] H. Luo, F. Ye, J. Cheng, S. Lu, and L. Zhang. TTDD: A Two-tier Data Dissemination Model for Large-scale Wireless Sensor Networks. *ACM/Kluwer Journal of Mobile Networks and Applications (MONET)*, 2003.

[13] R. Poor. Gradient Routing in Ad Hoc Networks. http://www.media.mit.edu/pia/Research/ESP/texts/poorieeepaper.pdf.

[14] R. C. Shah and J. Rabaey. Energy Aware Routing for Low Energy Ad Hoc Sensor Networks. In *WCNC*, 2002.

[15] F. Ye, S. Lu, and L. Zhang. GRAdient Broadcast: A Robust, Long-lived Large Sensor Network. http://irl.cs.ucla.edu/papers/grab-tech-report.ps, 2001.

[16] F. Ye, G. Zhong, S. Lu, and L. Zhang. GRAdient Broadcast: A Robust Data Delivery Protocol for Large Scale Sensor Networks. *ACM Wireless Networks (WINET)*, 11(2), March 2005.

[17] Y. Yu, R. Govindan, and D. Estrin. Geographical and Energy Aware Routing: A Recursive Data Dissemination Protocol for Wireless Sensor Networks. Technical Report UCLA/CSD-TR-01-0023, UCLA Computer Science Dept., May 2001.

Chapter 7

ROUTING ON A CURVE

Dragos Niculescu and Badri Nath
Division of Computer and Information Sciences
Rutgers, the State University of New Jersey
110 Frelinghuysen Road
Piscataway, NJ 08854-8019
[dnicules,badri]@cs.rutgers.edu

Abstract Relentless progress in hardware sensor technology, and wireless networking have made it feasible to deploy large scale, dense ad-hoc networks. These networks, together with sensor technology can be considered as the enablers of emerging models of computing such as embedded computing, ubiquitous computing, or pervasive computing. A new position centric paradigm, called trajectory based forwarding (or TBF), is a generalization of source based routing and Cartesian forwarding. We argue that TBF is an ideal technique for routing in dense ad-hoc networks. Trajectories are a natural namespace for describing route paths when the topology of the network matches the topography of the physical surroundings in which it is deployed which by very definition is embedded computing. Simple trajectories can be used to implement important networking protocols such as flooding, discovery, and network management. TBF is very effective in implementing many networking functions in a quick and approximate way, as it needs very few support services. We discuss several research challenges in the design of network protocols that use specific trajectories for forwarding packets.

Keywords: ad hoc networks, trajectory based forwarding, routing, broadcasting, positioning

7.1 INTRODUCTION

Routing packets along a specified curve is a new approach to forwarding packets in large scale dense ad-hoc networks. The fundamental aspect of considering route paths as continuous functions is the decoupling of the path name from the path itself. The transition from a *discrete view of route path to a continuous view of route path* is important as we move from dealing with sparse networks to dealing with dense networks. The basic idea of routing on curve

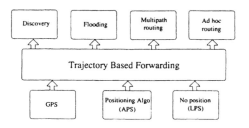

Figure 7.1. TBF layer

is to embed the trajectory of the route path in the packet and then let the intermediate nodes forward packets to those nodes that lie more or less on the trajectory. Representing route paths as trajectories is an efficient scalable encoding technique in dense networks. Since a trajectory does not explicitly encode the nodes in the path, it is impervious to changes in specific nodes that make up the topology. We believe that trajectories are a natural namespace to describe route paths when the topology of the network matches the topography of the physical surroundings in which it is deployed which by very definition is embedded computing. Forwarding packets along trajectories can be very effective in implementing many networking functions when standard bootstrapping or configuration services are not available, as will be the case in disposable networks where nodes are thrown or dropped to form a one-time use network.

This technique of forwarding packets along a curve or a trajectory is a generalization of source based routing [1] and Cartesian routing [2, 3]. As in source based routing, trajectories are specified by the source but do not explicitly encode the path on a hop-by-hop basis. As in Cartesian routing, the nodes are selected based on proximity but the proximity is to the trajectory instead of the final destination. TBF combines the best of the two methods: packets follow a trajectory established at the source, but each forwarding node makes a greedy decision to infer the next hop based on local position information, while the overhead of representing the trajectory does not depend on path length.

Trajectory based forwarding (TBF) has a number of unique features that makes it a candidate for a primitive in dense ad-hoc networks. Here are some distinguishing features of trajectory based forwarding:

- Forwarding based on trajectories decouples the path name from the path itself. Since a trajectory can be specified independent of the nodes that make up the trajectory, routing can be very effective even when the intermediate nodes that make up the path fail, go into doze mode, or move. Reactive routing algorithms (LSR [4], DSR [1]) explicitly encode the path. Proactive route maintenance algorithms (AODV [5]) need to update topology changes. Thus, communication costs increase in proportion to the topology changes induced by mobility. In DSR, any changes to the path results in route discovery, while in AODV, results in propagation of route updates. In TBF, route maintenance is virtually free and unaffected by mobility rate, as the path specification is independent of the names of the nodes involved in forwarding.

- The specification of the trajectory can be independent of the ultimate destination(s) of the packets. In fact, the trajectory specified in the packets need not have a final destination. A packet can be let go along a line or a curve for a desired length. This has implications for implementing networking functions such as flooding, discovery and multicast.

- For packets to follow the trajectory closely, nodes need to have the capability to position themselves in a coordinate system. A system such as GPS would be sufficient for this purpose, but TBF can be made to work even without GPS, as a coordinate system relative to the source can be constructed by positioning only the nodes in the neighborhood of the trajectory.

- Multipath routing has several advantages, such as increased reliability and capacity, but is seldom used because of the increased cost of maintenance. Using TBF, an alternate path is just another trajectory requiring no more maintenance than the main trajectory.

Integration in a layered architecture is essential in order to provide the advantages of TBF at both service and application layers. We envision a large array of services and applications including ad hoc routing, discovery, flooding, and multipath routing, as shown in Figure 7.1. Trajectory based forwarding (TBF) requires that nodes be positioned relative to a global coordinate system or a relative coordinate system. The strength of TBF lies in the flexibility of being able to work over a wide variety of positioning systems. In fact, TBF can be seen as a middle layer between global [6], ad hoc [7–10], and local [11] position providing services, and many network management services or applications.

7.2 FORWARDING ON A TRAJECTORY

The trajectory is usually decided by the source and can be conveniently expressed in parametric form $X(t), Y(t)$. The meaning of parameter t is also decided by the source, as well as the resolution at which the curve will be evaluated, indicated by dt. For the simplicity of the explanation, assume that t indicates the distance along the curve. The neighborhood of a node N_0(Figure 7.2) is defined as the portion of the curve and the nodes that are within a certain distance from N_0, indicated by a dashed circle in the figure. In the simplest case, the neighborhood could be the smallest rectangle enclosing all N_0's one hop neighbors.

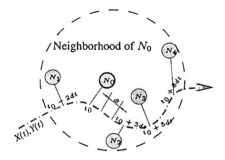

Figure 7.2. Forwarding on a curve

In a network in which node positions are known, the main question is how to choose a next hop that best approximates the trajectory. Assume node N_0 receives a packet with the trajectory indicated by the curve $X(t), Y(t)$ and the value t_0 that corresponds to the point on the curve that is closest to N_0. Using sampling of the curve at dt spaced intervals, indicated by dots in the dashed trajectory curve, N_0 can compute all the points of the curve that reside inside its neighborhood. For all neighbors $N_1..N_4$, their corresponding closest points on the curve are $t_0 - 2dt, t_0 + 3dt, t_0 + 5dt, t_0 + 8dt$. When referring to curve fitting, these values are called perpendicular residuals. In fact, the mentioned method computes an estimation of the residuals instead of the true ones, which would require either infinite precision, or usage of higher derivatives of $X(t)$ and $Y(t)$. Since choosing a next hop for the packet should be towards advancement on the trajectory, only portion of the curve with $t > t_0$ is considered.

Several policies of choosing a next hop are possible:

- *mindev* - node closest to the curve, with the minimum residual. This policy would favor node N_2 and would tend to produce a lower deviation from the ideal trajectory. This strategy should be chosen when it is important for the packets to follow the trajectory closely to possibly determine the state around the trajectory. Since the packet stays close to the trajectory, there is less likelihood for a packet to wander away from the intended trajectory due to errors in localization.

- *maxfwd* - most advancement on t, choosing N_4, first presented in [12], under the name of MFR (most forwarding within radius). This policy should also be controlled by a threshold of a maximum acceptable residual in order to limit the drifting of the achieved trajectory. It would produce paths with fewer hops than the previous policy, but with higher deviation from the ideal trajectory. This strategy should be favored when delay is an important metric and the packet needs to reach the destination or the perimeter of the network in minimum number of hops.

- centroid of the feasible set, choosing N_3. The path traversed will be more coarse compared to the choice of choosing the node closest to the trajectory and will cover the center of activity of the region without having too many small hops. The centroid is a way to uniformly designate clusters along the trajectory, and a quick way to determine the state of the network along the trajectory.

- randomly choose among the best nodes obtained from above three choices. This is useful when node positions are imperfect, or when it may be necessary to route around unknown obstacles. The randomness in the choice of the next hop can help mitigate the effects of small obstacles.

- in mobile networks a forwarding policy that might provide better results would be to choose the next hop which promises to advance along the trajectory, or one that is expected to have the least mobility in the future. Another interesting choice would be to choose a node that is moving towards the trajectory, rather than the one that is moving away. The previous strategies are expected to work better when trajectories and neighborhood information are cached, and for a given trajectory, all packets are forwarded to the same node. For mobility, either a periodic evaluation of the position of neighbors relative to the trajectory, or cache invalidation are necessary.

7.3 BASIC TRAJECTORIES AND APPLICATIONS

Having discussed the implementation details of TBF, we shall now investigate some of the applications that would benefit from an implementation under the TBF framework. There is a wide variety of trajectory shapes that can be used in applications, but a broad classification of trajectories may be into simple or composite. Simple trajectories describe a single continuous curve, and, in the context of routing, are used for unicast. Composite trajectories describe several, spatially different curves. They may also be used for unicast in an anchor based fashion, when a complicated trajectory is described as a list of simpler trajectories. Composite trajectories have a more obvious use in broadcast and multicast, where a unique curve is less appropriate.

A sampling of TBF based applications is shown in Figure 7.3. A naïve broadcasting scheme based on trajectories uses a number of radial outgoing lines that are reasonably close to each other to achieve a similar effect without all the communication overhead involved by receiving duplicates in classical flooding (Figure 7.3a). More generally, a source would indicate the directions and the lengths of the lines that would achieve a satisfactory coverage (Fig-

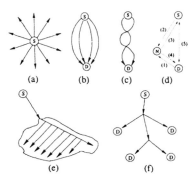

Figure 7.3. Examples of trajectory routing and forwarding

ure 7.3e). Coverage relies on the typical broadcast property of the wireless medium, in which several nodes overhear the packet being forwarded. Recovery from failure often involves multipath routing from a source to a destination. In a sensor network, both disjoint (Figure 7.3b) and braided (Figure 7.3c) paths are useful in providing resilience [13]. A simple five step discovery scheme (Figure 7.3d) based on linear trajectories may be used to replace traditional broadcast based discovery. If unicast communication is modeled by a simple curve, multicast is modeled by a tree in which each portion might be a curve, or a simple line. Distribution trees are used for either broadcasting (Figure 7.3e), or multicast routing (Figure 7.3f). A source knowing the area to be flooded can generate a tree describing all the lines to be followed by packets in order to achieve complete distribution with minimal broadcast communication overlap. A multicast source knowing positions for all members of a group may generate a spanning tree built of linear trajectories to be followed by packets. There is an overhead to be paid in describing the tree in each packet, but the solution saves in route maintenance.

7.3.1 UNICAST ROUTING

The prime application of forwarding is routing. The difference between the two is that in forwarding, a packet need not have a particular destination. Routing involves not only delivery to the destination, but the entire process that supports delivery. This includes forwarding, and also building or updating routing tables. In order to route, the position of a given destination node is needed, as provided by a location service [14, 15], to enable node centric applications run on top of position centric routing. The other central problem is how to determine the actual trajectory. So far we experimented with simple trajectories, such as lines and sine waves, but the more general question is how to determine the trajectory when the position of the destination is known. If the topology is uniform in terms of density and capacity, it is likely that simple lines and parabolas for alternate paths would suffice. If the network exhibits more variation, in general shape of policy defined islands (of security, or capabilities for example), determination of the trajectory cannot be done with localized decisions. For these cases, we intend to explore the creation of a service responsible for trajectory mapping, that would take into consideration all the aspects in the creation of routes.

More advantages are brought by TBF for multipath routing, which may be employed by a source to increase bandwidth, or resilience of communication. The key feature here is the **cheap path diversity**. Using TBF, the source may generate either disjoint paths as disjoint curves, or braided paths as two intersecting sine waves. In networks with low duty cycles, such as sensor networks, longer alternate paths might actually be more desirable in order to increase the resilience of the transmitted messages (concept similar to Fourier decomposition), or to distribute the load onto the batteries. Since there is essentially no route maintenance, each packet can take a different trajectory, depending on its resilience requirements (similar to different FEC requirements).

7.3.2   MOBILITY

Mobile networks are a case in which TBF provides a desirable solution due to its **decoupling of path name from the path itself**. In a mobile ad hoc network, route maintenance for trajectory based routing comes for free since all that is needed is the position of the destination. This is especially true when only the intermediate nodes or the source are moving, and the destination of the packet remains fixed. When the destination is moving, a location service [14] may be used, or the source may quantify its uncertainty about the destination by using a localized flooding around the destination (Figure 7.3e). If the destination is mobile and its path is known, it is possible to route towards this path. TBF can naturally use information about trajectories of destinations rather than locations of destinations.

7.3.3   DISCOVERY

One of the areas in which TBF is particularly appropriate is **quick and dirty implementation of services** without the support of preset infrastructure. Such is the case of discovery - of topology, or of some resource. Many algorithms use initial discovery phases based on flooding [1, 16] in order to find a resource or a destination. Generalizing an idea presented in [17], a replacement scheme using trajectories is as follows: possible destinations (servers S) advertise their position along arbitrary lines and clients C will replace their flooding phase with a query along another arbitrary line which will eventually intersect the desired destination's line. The intersection node then notifies the client about

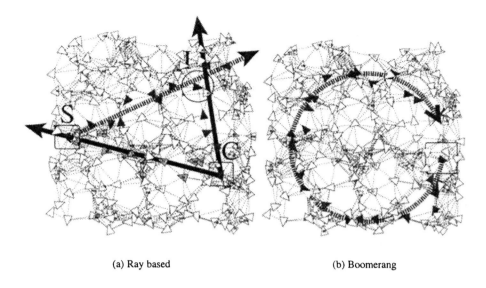

(a) Ray based   (b) Boomerang

Figure 7.4. TBF based discovery

the angle correction needed to contact the server directly (figures 7.3d and 7.4a). In order to guarantee that the server and client lines intersect inside the circle with diameter CS, it is in fact necessary for the nodes each to send in four cardinal directions.

Another interesting trajectory is a **boomerang,** where the packet takes a certain route and returns to the sender (Figure 7.4b). A circle is a simple example of this trajectory and can be used to secure the perimeter of the network by sending a challenge response token along the trajectory. All nodes who respond properly can be considered to be authenticated. A packet that returns to the source after being routed along a circle or a boomerang path is in effect implementing a self ARQ.

7.3.4   BROADCASTING

Broadcasting is one of the most used primitives in any network, used for tasks ranging from route discovery at the network layer, to querying and resource discovery at the application layer. Its most frequent implementation is under the form of suppressed flooding, which entails each node of the network broadcasting the message exactly once. It is a **stateful** method, since it requires bits to mark the status of a node - covered or uncovered. The problem with the marking bits is that they have to be provisioned on a per broadcast basis, if several broadcasts are to be supported simultaneously. If only $O(1)$ marking

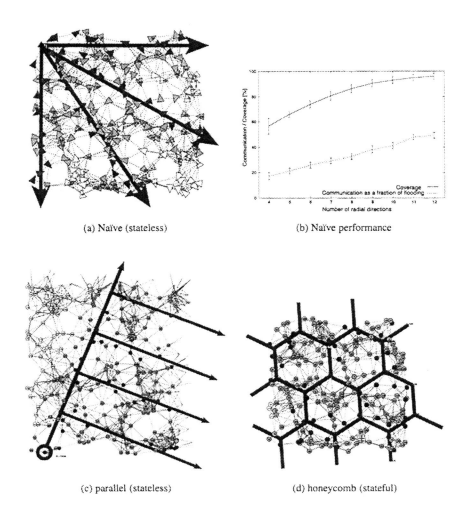

(a) Naïve (stateless)

(b) Naïve performance

(c) parallel (stateless)

(d) honeycomb (stateful)

Figure 7.5. TBF based broadcast

bits are used, some global serialization is necessary. For example if one bit is used, one broadcast is supported in the network, and after the settling time (time at which last copy of a message is broadcast), the bit has to be cleared to allow for another broadcast. Suppressed flooding also incurs several other problems: increased communication [18], increased settling time, poor scalability and delivery ratio in congested networks [19]. Probabilistic flooding [20] addresses some of these problems by flipping a coin each time a node has to re-broadcast the message. This reduces the number of duplicates a node receives, but the method exhibits a bimodal behavior, meaning that either the broadcast is successful in covering most of the network, or it dies very early, covering only a small portion around the source. While broadcasting is not the main application of TBF, we can provide solutions that address most shortcoming of traditional flooding and of probabilistic flooding. The broadcast achieved by TBF also has an approximate nature, just like probabilistic flooding, meaning that there may be nodes which do not receive the message even under ideal collision free conditions.

We discuss here two classes of broadcast approximation: stateless and stateful. Stateful broadcasting (Figure 7.5d) is similar in nature to classical flooding in that it requires per source marking bits in order to suppress duplicates. Stateless broadcasting (figures 7.5a,c) has the property that no memory is used at nodes in order to mark the visited/non visited status. The broadcasting scheme is designed to not overlap itself indefinitely, although it may intersect itself. Settling time is therefore less of a factor, since different broadcasts need not choose between serial execution and the use of marking bits. They can overlap without interference problems at the algorithmic level. As an example of naïve stateless broadcasting, in Figure 7.5a, the node in the upper left corner initiates radial spokes that cover directly the nodes involved in forwarding, and indirectly nodes which are at most one hop away from the trajectory. The performance curves in figure (7.5b) are for an initiator node in the middle of the network. As the number of spokes used increases, the coverage increases, but also does the communication spent. When seen as a fraction of communication spent by classical suppressed flooding, the naïve method performs remarkably well achieving 95% coverage with less than half the communication. However, for larger networks, spokes will diverge, leaving large areas uncovered.

In Figure 7.5(c,d) we investigate alternative broadcasting schemes. The parallel structure of the spokes proves to be quite economical in terms of overhead because spokes spaced at about twice the communication range tend to reduce the number of duplicate packets, while providing high coverage. Both stateless schemes - the radial and the parallel spokes - have a lower predictability since large portions of the network may remain uncovered due to a break in the trajectory.

For stateful schemes, we experimented with three plane filling patterns - triangles, squares and hexagons (Figure 7.5d, the honeycomb structure is also explored in [21], under the name of "optimal flooding"). These scheme resemble classical suppressed flooding with respect to coverage and predictability, but still have the property of limiting the amount of duplicates specific to all TBF based methods. They presented similar performance under ideal conditions, although a square shape seems to be a little better, however not as good as the parallel spoke stateless scheme.

Simulation results showed that broadcasting along predefined plane filling lines performs very well with increasing density in a fixed area, because the number of transmissions only depends on the fixed area to be covered, and not on the actual number of nodes receiving the packet. The main advantage stems from the fact that a node receives a packet only about three or four times, instead of receiving it from all of its neighbors, as in classical suppressed flooding.

It is worth noting that plane filling patterns also have applications in topology discovery: if a node wants a low resolution image of the network, without the burden of querying for the entire detailed topology, it may employ a pattern like the honeycomb, but with a larger size. This will have the flowing property of flooding to wrap around obstacles and go through bottlenecks in the topology, but will touch only a fraction of the nodes. It is true that obstacles and bottlenecks have to be larger than the pattern size in order to achieve full coverage, but the result is a finely tunable tradeoff between communication spent and resolution obtained.

7.3.5 MULTICAST

Another example of quick and dirty implementation of a network service is multicast. Traditional multicast implementation is overkill in many sensor or ad hoc networks applications, because of group setup and tree maintenance. There are situations when only one shot multicast is necessary, to send a query or a notification, and a multicast tree setup would be unjustified. Even when the membership has a longer duration, mobility could render the connecting tree useless. For such cases, the source node, assuming it has the positions of all the receivers, can determine an approximate Euclidean Steiner tree to be used as a distribution tree, without using other communication except position notification from the receivers.

7.4 TRAJECTORY SPECIFICATION
7.4.1 ENCODING

There are a number of choices in representing a trajectory: functional, equation, or a parametric representation. Functional representation(e.g. $y(x) = ax + b$) cannot be used to specify all types of curves, such as vertical lines. Equation representation (e.g. $x^2 + y^2 = R^2$) requires explicit solution to determine the points on the curve and cannot easily handle the notion of progress. Parametric representation is ideally suited for the purpose of forwarding. The parameter of the curve is a natural metric to measure the forward progress along the path and can be considered a proxy for the hop count. Given the choice of a parametric form, the next issue is how the trajectories should be interpreted. Trajectories can have several parameters and each node needs to correctly interpret these parameters. One approach is to have a convention where the nodes know how to interpret the fields given a well known set of trajectories.

A trajectory can also be composed of several simple trajectories. These simple trajectories can be viewed as segments and can be specified along with the appropriate intervals of the parameter over which it is valid. To obtain a continuous trajectory, these intervals need to overlap. A trajectory may involve a complex shape that may not have a simple parametric representation. However, in many cases, this shape can be represented by a simpler set of Fourier components. The more Fourier components in the specification of the trajectory, the better is the reproduction of the trajectory. There is an interesting trade-off between the accurate reproduction of the trajectory and the overhead of specifying the components and interpreting them. Other flexible and compact ways to encode a complex trajectory are fractal encoding and compiled form encoding. The latter approach, also used in active networking [22], sends in each packet the binary code needed for the parametric evaluation of the curve. Another possibility for encoding the parametric curve is the reverse polish notation (ready to be interpreted), which we used in our current implementation on Mica motes, for the increased flexibility.

The complex trajectories required for multicast and broadcast may increase the overhead in the packet, depending on the type of trajectory. The trajectory in Figure 7.5d is compactly represented because of being self similar. The encoding only specifies a straight line and two recursive calls at $\pi/3$ left and right of the line. An arbitrary multicast tree however must be described in its entirety in the packet if it has an irregular shape. Even in this case, the tree representation might be pruned as the packet descends towards the leaves, so that only the portion representing the subtree is sent.

7.4.2 DETERMINATION

Another important question is how the trajectories are determined. In our initial work on TBF, we assumed that the source knows the trajectories a priori, that is being derived from the application at hand. This may be the case in many applications of embedded computing where the topology of the network closely mirrors the underlying physical infrastructure in which the network is embedded. As was shown with applications such as flooding and discovery, simple lines or rays are sufficient. However, in other circumstances, the source or the intermediate nodes may have to determine the actual trajectory given the destination or group of destinations. Our initial thinking is to rely on a trajectory mapping service similar to DNS where, given a destination, or a group of destinations, the mapping service returns a trajectory for the source to use. How such a service can be built is a topic for further research.

7.4.3 MODIFICATION

Once the source has determined the trajectory to use for routing, it may have to be modified due to network conditions such as obstacles, failures, or mobility. The question is how to detect that a modification is needed and who should modify the trajectory. A route failure detection mechanism or an ACK/NACK scheme can be used for detecting the need for a change in the trajectory. A source can then choose a new trajectory for forwarding subsequent packets. Here, the onus of successful delivery is completely on the source. Another choice would be for some authorized intermediate nodes to detect the need for modification and provide a "patch" to the trajectory so that packets can be forwarded around obstacles, failures or local routability conditions such as network congestion. Whatever the reason for providing these patches, the patches should only serve as local detours where the packets eventually get back on track and are forwarded along the original trajectory.

7.5 TBF IN SPARSE NETWORKS

Dealing with the sparseness of the network addresses in fact a larger class of problems: obstacles, low connectivity areas, dead or sleeping nodes, policy defined islands that are to be avoided – islands of low energy or vulnerability. In TBF, a trajectory is most of the time advanced by greedy decisions (section 7.2) taken during forwarding. To find a path around the obstacle for cartesian forwarding [2], localized methods explored in the literature include flooding and depth first search. These however require maintenance of per destination states in the forwarding nodes, and extra communication, posing scalability problems. The most relevant work related to avoiding obstacles in routing is the FACE algorithm [23](see also [24]), which is stateless and localized. It operates on planar graphs only, so the first step involves localized elimination of the edges that are not part of a known planar graph, such as relative neighborhood graph, Gabriel graph, or Delaunay graph. Routing is then done in one of the two modes: greedy, choosing the node closest to the destination ($maxfwd$, or MFR [12]), or FACE mode. The latter uses the "right hand rule", which involves following one side of the polygon, until finding a edge that intersects the source destination line, at which point greedy forwarding mode is resumed. While the online nature of this algorithm is desirable, as it allows for localized decisions, the algorithm cannot be trivially adapted to work with the general trajectories. An intuitive argument is that inside the obstacle the curve may oscillate back and forth, regardless of the advance made by FACE packet on the border, making it harder to detect the exit point. Detecting the exit point for

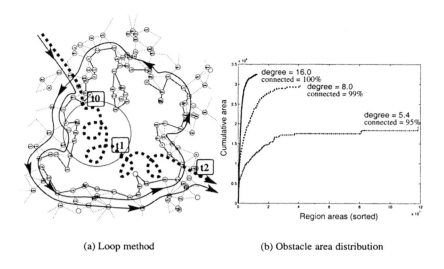

Figure 7.6. Obstacle avoidance

a straight line trajectory involves just the intersection of a line and a segment, whereas for arbitrary trajectories it requires equation solving using numerical methods.

A simple solution would be to premap all the polygons determined by the planarization, and store each polygon in all the border nodes it is made of. When a trajectory arrives at a node where greedy TBF is not possible, the node would have all its adjacent polygons stored locally, thus being able to plot a path around the offending polygon, using the FACE strategy. This preprocessing method would only be appropriate for static networks and when polygon sizes are small. For example, nodes on the outer face would have to store the entire perimeter of the network, which may be prohibitive in scale.

What we propose eliminates the need of preprocessing, but may incur either extra computation or extra communication. It requires the sender of the trajectory to have a rough estimate of the size of the obstacles, and of the possible behavior of the curve inside such sized obstacles. Associated with the trajectory, the sender attaches Δ – an estimated diameter of the obstacles in terms of t. In Figure 7.6a, the dashed arrow shows the desired trajectory, while the continuous one shows the trajectory achieved by the obstacle avoidance method. When greedy mode is no longer possible at t_0 on the curve, the responsible node switches to FACE mode and continues forwarding along the border using the "right hand rule", shown in the picture with a continuous arrow. The problem is that each side of the polygon must be tested for a possible exit point of the curve. Each node forwarding in FACE mode evaluates the curve from the point t_0 to the point $t_1 = t_0 + \Delta$, and switches back to greedy mode if it is the case. If Δ is underestimated, as in Figure 7.6a, FACE mode will loop around the obstacle reaching point t_0 without finding an exit point. In a second tour of the polygon, all nodes evaluate the curve between $t_1 = t_0 + \Delta$ and $t_2 = t_0 + 2\Delta$, with a node being successful in finding the exit point. A large underestimation of Δ leads to wasted communication, as in the example here. Overestimation leads to high computational strain on the border nodes, as they have to compute a large portion of the curve that is outside the obstacle they are a part of. A way to limit the amount of curve evaluations is to send along with the FACE packet a circle enclosing the evaluated $[t_0 + i\Delta, t_0 + (i+1)\Delta]$ portion of the curve. This circle is computed by the node at t_0 and enlarged slightly to ensure that segments on the obstacle perimeter having both ends outside the circle need not evaluate the Δ portion of the curve.

The method is a good engineering compromise for obstacle avoidance if the approximate diameter of the obstacle is known statistically. Its advantages are that it requires no state in the border nodes, and no preprocessing, thus being appropriate for mobile scenarios. In networks with a uniform distribution of nodes, it turns out that when there is enough density to guarantee connectivity, the uncovered areas (obstacles) are fairly small and with predictable maximum size. In Figure 7.6b, 1000 unit disk graph networks with 1000 nodes each are considered at three different densities by using different sizes of the disks. The graphs are planarized by retaining only the edges in Gabriel graph, and the subdivisions resulted are sorted by area. For low densities, there are two noticeable aspects: the cumulative of all subdivisions barely covers half of the network, and there is a large number of large area regions. Higher densities show a distribution of obstacle size that is more friendly to TBF: most area is covered by small regions, and there may be little need for using FACE mode. This means that in uniform density networks, we can have a reasonable estimate for Δ, that would avoid excessive looping or excessive curve evaluations.

7.6 TBF USING APPROXIMATE POSITIONING

In many networks GPS is not available, but TBF can make use of alternative positioning methods based on heterogeneous capabilities, such as angle measurement, range measurement and compasses. One option is to run a network wide ad hoc positioning algorithm, such as APS or AhLOS [7, 9, 10]. Another option is to use multimodal sensing to determine local ranges and angles, and eventually to set up local coordinate systems which allow handling of trajectories. This, however, involves successive coordinate systems alignment and is more sensitive to the measurement errors than the first option. For all the mentioned positioning options, the performance of TBF (accuracy, drift) depends on the quality of the positions obtained at each node and therefore, ultimately, on the quality of the measurements.

7.6.1 ALTERNATIVES TO POSITIONING

Trajectory based forwarding is simpler when all nodes positions are known relative to a reference coordinate system by use of a global capability such as GPS, or running a self positioning algorithm such as [7, 9, 25, 26]. In many cases however, running a network wide positioning algorithms is expensive and unnecessary. In this section, we shortly review LPS [11], a method that enables TBF even when such a global or ad hoc capability is not available due to place of deployment (e.g. indoors), high communication costs [7, 9, 10], or additional infrastructure requirements [6, 27, 25]. With this method, nodes use some other capabilities (ranging, AOA, compasses) to establish local coordinate systems in which all immediate neighbors are placed. It is then possible to register all these coordinate systems with the coordinate system of the source of the packet. Local Positioning System (LPS) is a method to achieve positioning **only for the nodes along the trajectory**, with no increase in communication cost, as if all node positions were known. Each node touched by the trajectory will spend some computation to position itself in the coordinate system of the source of the packet, trading off some accuracy of the trajectory.

The localized capabilities that can be used by LPS to replace positioning are ranging, AOA, compasses, and accelerometers. Range based estimation of distance between two nodes has been previously used for positioning purposes [7, 9, 26], even with the high measurement error involved. In most implementations, ranging is achieved either by using an estimate from the strength of the signal (unreliable), or using time difference of arrival (TDOA). In AOA approach, each node in the network is assumed to have one main axis against which all angles are reported and the capacity to estimate with a given precision the direction from which a neighbor is sending data. When interacting with two neighbors, a node can find out the angle between its own axis and the direction the signal comes from. A small form factor node that satisfies conditions outlined has been developed at MIT by the Cricket Compass project [28]. Its principle of operation is based on both time difference of arrival (TDOA) and phase difference of arrival. Other node capabilities that are available in small form factors include accelerometers and compasses. The accelerometer's main use is to indicate pose, while compass indicates the absolute orientation of each node.

7.6.2 LOCAL POSITIONING SYSTEM (LPS)

In Figure 7.7, if node A is able to measure distances to all its neighbors, via a ranging method, it can compute the sides and the angles for all triangles

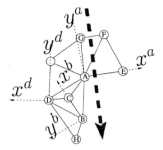

Figure 7.7. Local coordinate systems

created with all pairs of neighbors which are also neighbors of each other. This would enable A to place itself in $0,0$ of its local coordinate system and all its immediate neighbors at positions that satisfy all the range requirements known to A, ranges indicated by continuous lines in the figure. To establish its coordinate system, A randomly chooses E as an indicator for its x^a axis and F as an indicator for its y^a axis. All the other neighbors are then placed in this coordinate system such that all the range readings are respected (in a least square sense). Most nodes can be unambiguously positioned if they are neighbors with two previously positioned nodes. Node B sets up a similar local coordinate system by initiating its axes using neighbors C and D.

Registration between two local coordinate systems is the process that computes a transformation which will translate any point from one coordinate system to the other. The input to this process are points for which the coordinates are known in both coordinate systems with some accuracy. If perfect ranging were used in creating the local coordinate systems, the registration would produce a rigid transformation. In the example in Figure 7.7, a translation and a rotation are enough to overlap the local system of B over the local system of A. In the case of D, however, the first two neighbors randomly chosen as axes indicators produce a coordinate system that cannot be overlapped over A's using only translation and rotation. In this case, due to $D's$ original, localized and independent choice of axes, a mirroring transformation will also be produced by registration. The main cause for these mirroring transformations is the fact that ranging information does not provide a sense of direction (at least in the first phase, before the local coordinate systems are set).

It is possible to use both AOA and ranging in creating local coordinate systems, possibly enhanced with local compasses. The following table indicates all the possible combinations of node capabilities, and the transformations involved in the registration process (T=translation, R=rotation, S=scaling, M=mirroring).

Capability	Transformations
Range	T, R, M
AOA	T, R, S, (M)
AOA+Compass	T, S, (M)
AOA+Range	T, R, (M)
AOA+Range+Compass	T, (M)

7.6.3 FORWARDING USING LPS

The forwarding procedure works with a node selecting the next hop based on the proximity to the desired trajectory, or any of the other possible policies. In Figure 7.7, the ideal trajectory is shown as a thick dashed arrow. Assume that A knows the equation of the trajectory in its own coordinate system, which has been already registered to the coordinate system of the packet. If the next node to be selected along the trajectory is B, it will receive from A all the information needed to perform the registration. Node B is then able to select one of its own neighbors that is closer to the trajectory, in order to continue the process.

What is in fact achieved by LPS is the registration of all coordinate systems of visited nodes to the coordinate system of the initiating node, which achieves positioning of all these nodes in the coordinate system of the source. The method is localized to the nodes actually touched by the trajectory. Unlike a network wide positioning algorithm, such as [7, 9], which involves collaboration and coordination of a large number of nodes, LPS involves only the nodes "touched" by the desired trajectory.

According with the results presented in [11], LPS based forwarding maintains a low deviation, making it usable for discovery purposes, and reaches the destination with high probability, making it usable for routing. Although error compounds with distance, it can be countered efficiently by using additional hardware available in small form factors. An accelerometer present in each node detects node flipping, eliminating false mirroring, while a digital compass eliminates rotations from registrations process, when angular measurements are used.

7.6.4 IMPRECISE LOCATIONS

Errors in forwarding have two main causes: trajectory encoding and measurement error. Errors might be produced by encoding when a complex trajectory is not represented accurately, in order to reduce overhead. Measurement errors affect TBF when multimodal sensing is used for positioning (when GPS is not available). Depending on the hardware available on the nodes, errors accumulate along the trajectory and can be modeled as different random walk processes. If, for example, a compass is available in each node, the actual trajectory might drift left or right of a true linear trajectory, but will not change direction. If only ranges or angles are used for local positioning the trajectory may completely go off course. Characterizing these errors analytically would help in providing confidence areas for points on the achieved trajectory. Another research direction would be to investigate the possibility of correcting the trajectories on course, possibly with the use of a small fraction of GPS enabled nodes.

An implicit assumption we have made so far, and one that is made by most position centric schemes is that if node positions are available, they are perfect. This assumption is not always valid, and not only due to positioning device accuracy. In the event that GPS is not available throughout the network, it is possible to either agree on some relative coordinate system [29], or have a small fraction of GPS enabled nodes disseminate location to the rest of the network [7, 9, 10]. Most of these schemes employ ranging methods based on signal strength, ultrasound, angle of arrival, or just plain hop distance. All these provide estimates with some amount of error, which is finally translated in positioning error. In this section we explore the behavior of TBF under the assumption that node positions are approximative, as they are provided by a positioning algorithm. For this purpose we used DV-hop [7], a method that only uses hop distances to estimate positions. Its basic idea is that when a node knows shortest distances in hops to three different known landmarks, it can infer its position. This is done by approximating euclidean distance using hop distances, and assuming that the exact positions of the landmarks are known. DV-hop does not require additional capabilities, such as ranging, or AoA, but only produces acceptable positions in isotropic networks (having the same physical properties - connectivity, density, node degree - in all directions).

As we will show, when using approximate (warped) positions, TBF incurs only a small penalty in efficiency, thus being usable even when not all nodes are GPS equipped. The networks that we consider for this experiment are all isotropic, with uniform distribution of nodes, and with varying densities obtained by using various number of nodes in a fixed area, with a fixed communication radius. The position error, or distance from the true location, is on average one half the communication range, but may be larger for fringe nodes, due to the well known edge effects. The central aspect of this setup is that although positions are affected by errors, the resultant image of the network is coherent. In Figure 7.8, a network with 200 nodes 10 of which are landmarks, is shown with true locations on the left, and with warped locations obtained by DV-hop, on the right. The warped graph, although not a unit disk graph, like the original, has the same set of nodes and edges like the original. It can also

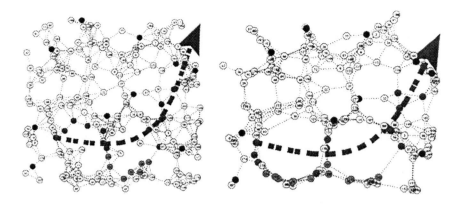

Figure 7.8. Warped network graph

also be planarized, for the purposes of guaranteed delivery [23]. The problem is that in the warped space two nodes can be very close and not have a link between them, or be far apart and have the link. This prevents the use of localized planarization procedures available for planar subgraphs such as Gabriel graph or relative neighborhood graph. However, for the rest of this section, we assume that the graph is planarized properly, in order to use the obstacle avoiding strategies outlined in section 7.5. Given that the trajectory is expressed in the warped coordinate system, on which all nodes implicitly agree, forwarding can then be performed just like in any planar subdivision.

While the actual forwarding path will always deviate from the ideal trajectory (when density is finite), even with perfect positions, we want to see how this deviation is affected by the warped positions provided by DV-hop. In order to quantify this drift, in networks with 1300-3000 nodes, we randomly chose 200 pairs of nodes and forwarded along linear trajectories in the two spaces. The paths had on average 15-25 hops, depending on the density. For TBF forwarding, two policies were considered - $maxfwd$ and $mindev$, explained in section 7.2. The main parameter of these simulations was density, as we wanted to capture the behavior in both avoiding obstacles (low densities), and greedy forwarding (higher densities).

In Figure 7.9a, average deviation from the ideal parametric trajectory for both the real and the warped networks is under one communication radius for all but the lowest density. The deviation is computed as the length of the perpendicular from the actual node used in forwarding to the ideal trajectory in the true space, whether in greedy or in FACE mode. As we expect, deviation decreases with growing density, and $mindev$ sticks closer to the trajectory than $maxfwd$. The unexpected result is the small difference in deviation between routing with true positions and with warped positions. In Figure 7.9b, we

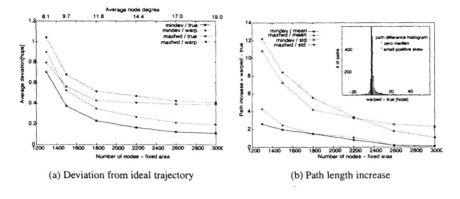

(a) Deviation from ideal trajectory (b) Path length increase

Figure 7.9. TBF on DV-hop

examine the difference in the length of the path in hops between the warped network and the real network. As the inset histogram shows, this difference is not always positive, that is, the path in the warped network is not always longer. In fact, the distribution of this difference has a zero median, and a small positive mean with a large standard deviation. Because obstacle avoidance decisions are localized, certain decisions in forwarding might incur or avoid large detours. The intuition behind this behavior is that certain obstacles may be avoided in the warped network, and not in the real one, or vice-versa.

Overall, the performance in terms of deviation and path length is only slightly deteriorated by the use of imprecise positions, which shows that TBF is robust in the face of realistic corruptions in node positioning. Investigation of alternative positioning schemes and efficient planarization of the warped graph remain interesting open issues.

7.7 CONCLUSIONS

We presented Trajectory Based Forwarding (TBF), a novel paradigm that makes the transition from a discrete view of the paths to a continuous view of the paths in future dense networks. The method is a hybrid between source based routing and cartesian forwarding in that the trajectory is set by the source, but the forwarding decision is local and greedy. Its main advantages are that it provides cheap path diversity, decouples path naming from the path itself, and trades off communication for computation. When GPS is not available, TBF may make use of alternate techniques, such as global and local positioning algorithms. It is robust in front of adverse conditions, such as sparse networks, and imprecise positioning. We believe that TBF should be used as an essential layer in position centric ad hoc networks, as a support for basic services: routing (unicast, multicast, multipath), broadcasting and discovery.

REFERENCES

[1] David B. Johnson and David A. Maltz. Dynamic source routing in ad hoc wireless networks. *Mobile Computing, Kluwer Academic Publishers*, 353, 1996.

[2] G. Finn. Routing and addressing problems in large metropolitan-scale internetworks. Technical Report ISI Research Report ISI/RR-87-180, University of Southern California, March 1987.

[3] J. C. Navas and Tomasz Imielinski. Geographic addressing and routing. In *ACM MOBICOM*, Budapest, Hungary, September 26-30 1997.

[4] D. B. Johnson. Mobile host internetworking using ip loose source routing. Technical Report CMU-CS-93-128, Carnegie Mellon University, February 1993.

[5] Charles E. Perkins and Elizabeth M. Royer. Ad hoc on-demand distance vector routing. In *2nd IEEE Workshop on Mobile Computing Systems and Applications*, pages 90–100, New Orleans, LA, February 1999.

[6] B.W. Parkinson and J.J. Spilker. *Global Positioning System: Theory and Application.* American Institute of Astronautics and Aeronautics, 1996.

[7] Dragoş Niculescu and Badri Nath. Ad hoc positioning system (APS). In *GLOBECOM*, San Antonio, November 2001.

[8] Chris Savarese, Jan Rabaey, and Koen Langendoen. Robust positioning algorithms for distributed ad-hoc wireless sensor networks. Technical report, Delft University of Technology, 2001.

[9] A. Savvides, C.-C. Han, and M. Srivastava. Dynamic fine-grained localization in ad-hoc networks of sensors. In *ACM MOBICOM*, Rome, Italy, 2001.

[10] Dragoş Niculescu and Badri Nath. Ad hoc positioning system (APS) using AoA. In *INFOCOM*, San Francisco, CA, April 2003.

[11] Dragoş Niculescu and Badri Nath. Localized positioning in ad hoc networks. In *Sensor Network Protocols and Applications*, Anchorage, Alaska, April 2003.

[12] Leonard Kleinrock and John Silvester. Optimum transmission radii for packet radio networks or why six is a magic number. In *IEEE National Telecommunications Conference*, pages 4.3.1–4.3.5, Birmingham, Alabama, 1978.

[13] Deepak Ganesan, Ramesh Govindan, Scott Shenker, and Deborah Estrin. Highly resilient, energy efficient multipath routing in wireless sensor networks. In *Mobile Computing and Communications Review (MC2R)*, volume 1, 2002.

[14] J. Li, J. Jannotti, D. De Couto, D. Karger, and R. Morris. A scalable location service for geographic ad hoc routing. In *ACM MOBICOM*, pages 120–130, Boston, MA, August 2000.

[15] P. H. Hsiao. Geographical region summary service for geographical routing. *Mobile Computing and Communication Review*, 5(4):25–39, January 2002.

[16] C. Intanagonwiwat, R. Govindan, and D. Estrin. Directed diffusion: a scalable and robust communication paradigm for sensor networks. In *ACM MOBICOM*, Boston, MA, August 2000.

[17] Ivan Stojmenović. A scalable quorum based location update scheme for routing in ad hoc wireless networks. Technical Report TR-99-09, SITE, University of Ottawa, September 1999.

[18] Sze-Yao Ni, Yu-Chee Tseng, Yuh-Shyan Chen, and Jang-Ping Sheu. The broadcast storm problem in a mobile ad hoc network. In *ACM MOBICOM*, pages 151–162. ACM Press, 1999.

[19] B. Williams and T. Camp. Comparison of broadcasting techniques for mobile ad hoc networks. In *Proceedings of the ACM International Symposium on Mobile Ad Hoc Networking and Computing (MOBIHOC)*, pages 194–205, 2002.

[20] Zygmunt Haas, Joseph Halpern, and Li Li. Gossip based ad hoc routing. In *INFOCOM*, New York, USA, June 2002.

[21] Vamsi S. Paruchuri, Arjan Durresi, Durga S. Dash, and Raj Jain. Optimal flooding protocol for routing in ad-hoc networks. Technical report, Ohio State University, CS Department, 2002.

[22] D. L. Tennenhouse and D. Wetherall. Towards an active network architecture. *Multimedia Computing and Networking*, January 1996.

[23] P. Bose, P. Morin, I. Stojmenović, and J. Urrutia. Routing with guaranteed delivery in ad hoc wireless networks. In *3rd International Workshop on Discrete Algorithms and methods for mobile computing and communications*, Seattle, WA, August 1999.

[24] B. Karp and H.T. Kung. GPSR: Greedy perimeter stateless routing for wireless networks. In *ACM MOBICOM*, Boston, MA, August 2000.

[25] N.B. Priyantha, A. Chakraborty, and H. Balakrishnan. The cricket location-support system. In *ACM MOBICOM*, Boston, MA, August 2000.

[26] Paramvir Bahl and Venkata N. Padmanabhan. RADAR: An in-building RF-based user location and tracking system. In *INFOCOM*, Tel Aviv, Israel, March 2000.

[27] Nirupama Bulusu, John Heidemann, and Deborah Estrin. GPS-less low cost outdoor localization for very small devices. In *IEEE Personal Communications Magazine*, Special Issue on Smart Spaces and Environments. October 2000.

[28] N.B Priyantha, A. Miu, H. Balakrishnan, and S. Teller. The cricket compass for context-aware mobile applications. In *ACM MOBICOM*, Rome, Italy, July 2001.

[29] S. Capkun, M. Hamdi, and J.-P. Hubaux. GPS-free positioning in mobile ad-hoc networks. In *Hawaii International Conference On System Sciences*, Outrigger Wailea Resort, January 3-6 2001. HICSS-34.

Chapter 8

RELIABLE TRANSPORT FOR SENSOR NETWORKS
PSFQ - Pump Slowly Fetch Quickly Paradigm

[+]Chieh-Yih Wan, [+]Andrew T. Campbell and [*]Lakshman Krishnamurthy
[+]*Columbia University,* [*]*Intel Lab*

Abstract: We propose PSFQ (Pump Slowly, Fetch Quickly), a reliable transport protocol suitable for a new class of reliable data applications emerging in wireless sensor networks. For example, currently sensor networks tend to be application specific and are typically hard-wired to perform a specific task efficiently at low cost; however, there is an emerging need to be able to re-task or reprogram groups of sensors in wireless sensor networks on the fly (e.g., during disaster recovery). Due to the application-specific nature of sensor networks, it is difficult to design a single monolithic transport system that can be optimized for every application. PSFQ takes a different approach and supports a simple, robust and scalable transport that is customizable to meet the needs of different reliable data applications. To our knowledge there has been little work on the design of an efficient reliable transport protocol for wireless sensor networks, even though some techniques found in IP networks have some relevance to the solution space, such as, the body of work on reliable multicast. We present the design and implementation of PSFQ, and evaluate the protocol using the ns-2 simulator and an experimental wireless sensor testbed based on Berkeley motes. We show through simulation and experimentation that PSFQ can out perform existing related techniques (e.g., an idealized SRM scheme) and is highly responsive to the various error conditions experienced in wireless sensor networks, respectively.

Keywords: Reliable transport protocols, wireless sensor networks.

8.1 INTRODUCTION

There is a considerable amount of research in the area of wireless sensor networks ranging from real-time tracking to ubiquitous computing where users interact with potentially large numbers of embedded devices. This paper addresses the design of system support for a new class of applications emerging in wireless sensor networks that require reliable data delivery. One such application that is driving our research is the reprogramming or "re-tasking" of groups of sensors. This is one new application in sensor networks that requires underlying transport protocol to support reliable data delivery. Today, sensor networks tend to be application specific, and are typically hard-wired to perform a specific task efficiently at low cost. We believe that as the number of sensor network applications grows, there will be a need to build more powerful general-purpose hardware and software environments capable of reprogramming or "re-tasking" sensors to do a variety of tasks. These general-purpose sensors would be capable of servicing new and evolving classes of applications. Such systems are beginning to emerge. For example, the Berkeley motes [1-2] are capable of receiving code segments from the network and assembling them into a completely new execution image in EEPROM secondary store before re-tasking a sensor.

Unlike traditional networks (e.g., IP networks), reliable data delivery is still an open research question in the context of wireless sensor networks. To our knowledge there has been little work on the design of reliable transport protocols for sensor networks. This is, as one would expect, since the vast majority of sensor network applications do not require reliable data delivery. For example, in applications such as temperature monitoring or animal location tracking, the occasional loss of sensor readings is tolerable, and therefore, the complex protocol machinery that would ensure the reliable delivery of data is not needed. Directed diffusion [3] is one of a representative class of data dissemination mechanisms, specifically designed for a general class of applications in sensor networks. Directed diffusion provides robust dissemination through the use of multi-path data forwarding, but the correct reception of all data messages is not assured. We observed that in the context of sensor networks, data that flows from sources to sinks is generally tolerable of loss. On the other hand, however, data that flows from sinks to sources for the purpose of control or management (e.g., re-tasking sensors) is sensitive to message loss. For example, disseminating a program image to sensor nodes is problematic. Loss of a single message associated with code segment or script would render the image useless and the re-tasking operation a failure.

There are a number of challenges associated with the development of a reliable transport protocol for sensor networks. For example, in the case of a re-tasking application there may be a need to reprogramming certain groups of sensors (e.g., within a disaster recovery area). This would require addressing groups of sensors, loading new binaries into them, and then, switching over to the new re-tasked application in a controlled manner. Another example of new reliable data requirements relates to simply injecting scripts into sensors to customize them rather than sending complete, and potentially bandwidth demanding, code segments. Such retasking becomes very challenging as the number of sensor nodes in the network grows. How can a transport offer suitable support for such a retasking application where possibly hundreds or thousands of nodes need to be reprogrammed in a controlled, reliable, robust and scalable manner?

Reliable point-to-point, or more appropriately, multicast transport mechanisms are well understood in conventional IP-style communication networks, where nodes are identified by their end-points. However, these schemes (e.g., TCP, XTP [4], SRM [5], RMP [19]) cannot be efficiently applied to sensor networks mainly because of the unique communication challenges presented by wireless sensor networks, including the need to support cluster-based communications, wireless multi-hop forwarding, application-specific operations, and lack of clean layering for the purposes of optimization, etc. There is a need for the development of a new reliable transport protocol, which can respond to the unique challenges posed by sensor networks. Such an approach must be lightweight enough to be realized even on low-end sensor nodes, such as, the Berkeley mote series of sensors. A reliable transport protocol must be capable of isolating applications from the unreliable nature of wireless sensor networks in an efficient and robust manner. The error rates experienced by these wireless networks can vary widely, and therefore, any reliable transport protocol must be capable of delivering reliable data to potentially large groups of sensors under such conditions. This is very challenging.

In this paper, we propose PSFQ (Pump Slowly, Fetch Quickly), a new reliable transport protocol for wireless sensor networks. Due to the application-specific nature of sensor networks, it is hard to generalize a specific scheme that can be optimized for every application. Rather, the focus of this paper is the design and evaluation of a new transport system that is simple, robust, scalable, and customizable to different applications' needs. PSFQ represents a simple approach with minimum requirements on the routing infrastructure (as opposed to IP multicast/unicast routing requirements), minimum signaling thereby reducing the communication cost for data reliability, and finally, responsive to high error rates allowing successful operation even under highly error-prone conditions.

The paper is organized as follows. Section 2 presents the PSFQ model and discusses the design choices. Section 3 presents the detail design of PSFQ's pump, fetch and report operations. Section 4 presents an evaluation of the protocol and comparison to Scalable Reliable Multicast (SRM) [5] using the ns-2 simulator.

Section 5 presents experimental results from the implementation of PSFQ in an experimental wireless sensor testbed based on Berkeley motes. Section 6 discusses related work on reliable transport for sensor networks. Finally, we present some concluding remarks in Section 7.

8.2 PROTOCOL DESIGN SPACE

The key idea that underpins the design of PSFQ is to distribute data from a source node by pacing data at a relatively slow speed ("pump slowly"), but allowing nodes that experience data loss to fetch (i.e., recover) any missing segments from immediate neighbors very aggressively (local recovery, "fetch quickly"). We assume that message loss in sensor networks occurs because of transmission errors due to the poor quality of wireless links rather than traffic congestion since most sensor network applications generate light traffic most of the time. Messages that are lost are detected when a higher sequence number than expected is received at a node triggering the fetch operation. Such a system is equivalent to a negative acknowledgement system. The motivation behind our simple model is to achieve loose delay bounds while minimizing the lost recovery cost by localized recovery of data among immediate neighbors. PSFQ is designed to achieve the following goals:
– to ensure that all data segments are delivered to all the intended receivers with minimum[1] support from the underlying transport infrastructure;
– to minimize the number of transmissions for lost detection and recovery operations with minimal signaling;
– to operate correctly even in an environment where the radio link quality is very poor; and
– to provide loose delay bounds for data delivery to all the intended receivers.

8.2.1 Hop-by-Hop Error Recovery

To achieve these goals we have taken a different approach in comparison to traditional end-to-end error recovery mechanisms in which only the final destination node is responsible for detecting loss and requesting retransmission. Despite the various differences in the communication and service model, the biggest problem with end-to-end recovery has to do with the physical characteristic of the transport medium: sensor networks usually

---

[1] PSFQ only requires a MAC that is capable of broadcasting operations (e.g., CSMA, TDMA).

operate in harsh radio environments, and rely on multi-hop forwarding techniques to exchange messages. Error accumulates exponentially over multi-hops. To simply illustrate this, assume that the packet error rate of a wireless channel is p then the chances of exchanging a message successfully across a single hop is (1-p). The probability that a message is successfully received across n hops decreases quickly to $(1-p)^n$. For a negative acknowledgement system, at least one message has to be received correctly at the destination after a loss has happened in order to detect the loss. Figure 8.1 illustrates this problem numerically. The success rate denotes the probability of a successful delivery of a message in end-to-end model, which is $(1-p)^n$. Figure 8.1 plots the success rate as a function of the network size (in terms of the number of hops) and shows that for larger network it is almost impossible to deliver a single message using an end-to-end approach in a lossy link environment when the error rate is larger than 20%.

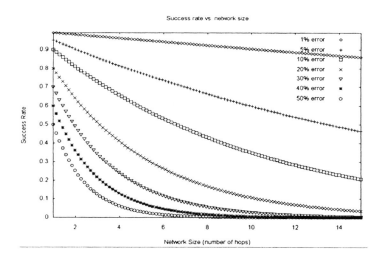

Figure 8.1. Probability of successful delivery of a message using an end-to-end model across a multi-hop network.

From Figure 8.1, we can see that end-to-end approach performs fine even across large numbers of hops in highly reliable link environments where the channel error rate is less than 1%, e.g., found in a wired network. Under such conditions the probability of a successful delivery is well above 90%. This requirement can be easily met in wired network and even in wireless LAN networks, such as IEEE 802.11. However, it is not the case in sensor networks. Due to the various resources and design constraints on a sensor node, sensor network operations require low-power RF communications, which cannot rely on using high power to boost the link reliability when operating under harsh radio conditions. In military applications or disaster recovery efforts, it is not unusual to have channel error rate that is in the range of 5% ~ 10% or even higher. This observation suggests that end-to-end error recovery is not a good candidate for reliable transport in wireless sensor networks, as indicated by the result shown in Figure 8.1.

We propose hop-by-hop error recovery in which intermediate nodes also take responsibility for loss detection and recovery so reliable data exchange is done on a hop-by-hop manner rather than an end-to-end one. Several observations support this choice. First, this approach essentially segments multihop forwarding operations into a series of single hop transmission processes that eliminate error accumulation. The chance of exchanging a message successfully across a single hop is (1-p). Therefore, the probability of detecting loss in a negative acknowledgement system is proportional to (1-p) in a hop-by-hop approach (independent of network size), rather than decreasing exponentially with growing network size as in the case of end-to-end approach. The hop-by-hop approach thus scales better and is more error tolerable. Second, the extra cost of involving intermediate nodes in the loss detection process (i.e., intermediate nodes must keep track of the data they forward, which involves allocating sufficient data cache space) can be justified in sensor networks. Typically, communication in wireless sensor networks is not individual-based but is group or cluster-based communications. Consider some of the example applications that require reliable data delivery (e.g., re-tasking the sensor nodes, or for control or management purposes), the intended receivers are often the whole group of sensor nodes in the vicinity of a source node (a user). In this case, intermediate nodes are also the intended receiver of data; therefore there is no extra cost in transiting data through nodes.

8.2.2 Fetch/Pump Relationship

For a negative acknowledgement system, the network latency would be dependent on the expected number of retransmissions for successful delivery. In order to achieve loose delay bound for the data delivery, it is essential to maximize the probability of successful delivery of a packet within a "controllable time frame". An intuitive approach to doing this would be to enable the possible multiple retransmissions of packet n (therefore increasing the chances of successful delivery) before the next packet n+1 arrives; in other words, clear the queue at a receiver (e.g., an intermediate sensor) before new packets arrive in order to keep the queue length small and hence reduce the delay. However, it is non-trivial to determine the optimal number of retransmissions that tradeoff the success rate, probability of successful delivery of a single message within a time frame, against wasting too much energy on retransmissions. In order to investigate and justify this design decision, we analyze a simple model, which approximates this mechanism. Let p be the packet loss rate of a wireless channel. Assume that p stays constant at least during the controllable time frame, it can be shown that in a negative acknowledgement system, the probability of a successful delivery of a packet between two nodes that allows n retransmissions can be expressed recursively as:

- $(1-p) + p \times \Omega(n)$ $\quad\quad\quad\quad\quad\quad\quad\quad\quad\quad\quad\quad\quad\quad\quad\quad (n \geq 1)$

- $\Omega(n) = \Phi(1) + \Phi(2) + \ldots + \Phi(n)$

- $\Phi(n) = (1-p)^2 \times [1 - p - \Phi(1) - \Phi(2) - \ldots - \Phi(n-1)]$ $\quad (\Phi(0) = 0)$

Where $\Omega(n)$ is the probability of a successful recovery of a missing segment within n retransmission, $\Phi(n)$ is the probability of a successful recovery of the missing segment at n^{th} retransmission.

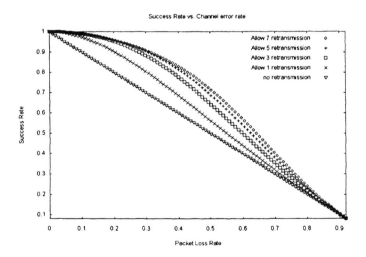

Figure 8.2 Probability of successful delivery of a message when the mechanism allows multiple retransmissions before the next packet arrival.

The above expressions are evaluated numerically against packet loss rate p, as shown in Figure 8.2. The straight line that denotes "no retransmission" is simply the probability of receiving an error free packet over the channel, which is 1-p; this line represents the case when no retransmission is attempted within a time frame before next segment is "pumped" into the channel. Figure 8.2 demonstrates the impact of increasing the number of retransmissions up to n equal to 7. We can see that substantial improvements in the success rate can be gained in the region where the channel error rate is between 0 and 60%. However, the additional benefit of allowing more retransmission diminishes quickly and becomes negligible when n is larger than 5. This simple analysis implies that the optimal ratio between the timers associated with the pump and fetch operations is approximately 5. This simple model also shows that at most ≈20% gain in the success rate can be achieved with this approach, as indicated from the result shown in Figure 8.2.

Multi-modal Operations

There are several important considerations associated with the pump operation's ability to localize loss events while maximizing the probability of in-sequence data delivery. As a result of these considerations the PSFQ pump operation is designed for multi-modal operations, providing a graceful tradeoff between the classic "packet forwarding" and "store-and-forward" communication paradigms depending on the wireless channel conditions experienced. In what follows, we discuss the reasoning behind this key design choice.

Figure 8.3 illustrates an example in which a local loss event propagates to downstream nodes. The propagation of a loss event could cause a serious waste of energy. A loss event will trigger error recovery operations that attempt to fetch the missing packet quickly from immediate neighbors by broadcasting a "Nack" message. However, for nodes B and C in Figure 8.3, none of their neighbors have the missing packet, therefore the loss cannot be recovered and the control messages associated with the fetch operation are wasted. As a result, it is necessary to make sure that intermediate nodes only relay messages with continuous sequence numbers. In other words, node A in Figure 8.3 should not relay message #4 until it successfully recovers message #3.

The use of a data cache is required to buffer both message #3 and #4 to ensure in-sequence data forwarding and ensure complete recovery for any fetch operations from downstream nodes. Note that cache size effect is not investigated here but in our reference application, the cache keeps all code segments. This pump mechanism not only prevents propagation of loss events and the triggering of unnecessary fetch operations from downstream nodes, but it also greatly contributes toward the error tolerance of the protocol against channel quality. By localizing loss events and not relaying any higher sequence number messages until recovery has taken place, this mechanism operates in a similar fashion to a store-and-forward approach where an intermediate node relays a file only after the node has received the complete file. The store-and-forward approach is effective in highly error-prone environments because it essentially segments the multi-hop forwarding operations into a series of single hop transmission processes (errors accumulate exponentially for multi-hop communication, as discuss in Section 2.1).

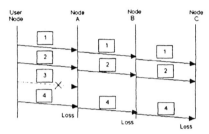

Figure 8.3. Propagation of a loss event. The packet with sequence number 3 sent by the user node to node A is lost or corrupted due to channel error. The subsequent packet with sequence number 4 received by node A triggers a loss event. If this packet is forwarded to node B, another loss event is triggered at node B. When this packet is forwarded from node B to node C, it will again trigger another loss event at node C. The loss event will keep on propagating in this manner until the TTL reaches 0 and packet is dropped.

However, the classic store-and-forward approach suffers from large delay even in error free environments. Therefore, store-and-forward is not a suitable choice in most cases although it could be the only choice in highly error-prone environments. PSFQ benefits from the following tradeoff between store-and-forward and multihop forwarding. The pump operation operates in a multihop packet forwarding mode during periods of low errors when lost packets can be recovered quickly, and behaves more like store-and-forwarding communications when the channel is highly error-prone. Therefore, as mentioned earlier, the PSFQ exhibits a novel multi-modal property that provides a graceful tradeoff between forwarding and store-and-forward paradigms, depending on the channel conditions encountered.

The observations presented in this section motivate our "pump slowly, fetch quickly" paradigm. The fetch operation should be fast relative to the pump operation as to allow a reasonable number of retransmissions in order to maximize the success rate of receiving a data segment within a controllable time frame. In addition, these insights suggest the need for in-sequence forwarding at intermediate nodes for the pump operation.

8.3 PROTOCOL DESCRIPTION

PSFQ comprises three functions: message relaying (pump operation), relay-initiated error recovery (fetch operation) and selective status reporting (report operation). A user (source) injects messages into the network and intermediate nodes buffer and relay messages with the proper schedule to achieve loose delay bounds. A relay node maintains a data cache and uses cached information to detect data loss, initiating error recovery operations if necessary. As in many negative acknowledgement systems, there is no way for the source to know when the receivers have received the data messages. This has several drawbacks. First, the data segments must be retained indefinitely at the source for possible retransmissions. Next, it is important for the user to obtain statistics about the dissemination status (e.g., the percentage of nodes that have obtained the complete execution image for a re-tasking application) in the network as a basis for subsequent decision-making, such as the correct time to switch over to the new task in the case of re-tasking. Therefore, it is necessary to incorporate a feedback and reporting mechanism into PSFQ that is flexible (i.e., adaptive to the environment) and scalable (i.e., minimize the overhead).

In what follows, we describe the main PSFQ operations (viz. pump, fetch and report) with specific reference to a re-tasking application -- one in which a user needs to re-task a set of sensor nodes in the vicinity of its location by distributing control scripts or binary code segments into the targeted sensor nodes. A number of concerns associated with this simple PSFQ model are related to the timer issues that control the loose service properties, such as, statistical delay bounds. Important protocol parameters include message pumping speed and loss recovery speed.

8.3.1 Pump Operation

Recall that PSFQ is not a routing solution but a transport scheme. In the case where a specific node (instead of a whole group) needs to be addressed, PSFQ can work on top of existing routing or data dissemination scheme, such as directed diffusion, DSDV [20], etc., to support reliable data transport. A user node uses TTL-based methods to control the scope of its re-tasking operation; note that, since the term "source" in sensor network usually denotes a sensor node which has sensed data to be sent, we use the term "user node" in this paper to refer to a node which distributes the code segments to avoid confusion. To enable local loss recovery and in-sequence data delivery, a data cache is created and maintained at intermediate nodes.

We define an "inject message" associate with the pump operation in PSFQ. The inject message has four fields in its header: *i)* file ID, *ii)* file length *iii)* sequence number, and *iv)* TTL[2]. The message payload carries the data fragment (code segment).

The pump operation is important in controlling four performance factors associated with our example re-tasking application. First, the timely dissemination of code segments to all target nodes used for re-tasking the sensor nodes. Second, to provide basic flow control so that the re-tasking operation does not overwhelm the regular operations of the sensor network, (e.g., monitoring environmental conditions). Next, for densely deployed sensor networks in which nodes are generally within transmission range of more than one neighboring node, we need to avoid redundant messaging to save power and to minimize contention/collision over the wireless channel. Finally, we want to localize loss, avoiding the propagation of loss events to downstream nodes. This requires mechanisms to ensure in-sequence data forwarding at intermediate nodes, as discussed in Section 2.3. The first two performance factors discussed above require proper scheduling for data forwarding. We adopt a simple scheduling scheme, which use two timers T_{min} and T_{max} for scheduling purposes.

I. Pump Timers

A user node broadcasts a packet to its neighbors every T_{min} until all the data fragments has been sent out. In the meantime, neighbors that receive this packet will check against their local data cache discarding any duplicates. If this is a new message, PSFQ will buffer the packet and decrease the TTL field in the header by 1. If the TTL value is not zero and there is no gap in the sequence number, then PSFQ sets a schedule to forward the message. The packet will be delayed for a random period between T_{min} and T_{max} and then relayed to its neighbors that are one or more hops away from the source. In this specific reference case, PSFQ simply rebroadcast the packet. Note that the data cache has several potential uses, one of which is loop prevention, i.e., if a received data message has a matching data cache entry then the data message is silently dropped. A packet propagates outward from the source node up to TTL hops away in this mode. The random delay before forwarding a message is necessary to avoid collisions because RTS/CTS dialogues are inappropriate in broadcasting operations when the timing of rebroadcasts among interfering nodes can be highly correlated.

---

[2] One bit of the TTL field in the inject message is used as the "report" bit in order to solicit a report message from target nodes. The use of this bit is discussed in Section 3.3.

T_{min} has several considerations. First, there is a need to provide a time-buffer for local packet recovery. One of the main motivations behind the PSFQ paradigm is to recover lost packets quickly among immediate neighboring nodes within a controllable time frame. T_{min} serves such a purpose in the sense that a node has an opportunity to recover any missing segment before the next segment come from its upstream neighbors, since a node must wait at least T_{min} before forwarding a packet in pump operation. Next, there is a need to reduce redundant broadcasts. In a densely deployed network, it is not unusual to have multiple immediate neighbors within radio transmission range. Since we use broadcast instead of unicast for data relaying in our reference application, too many data forwarding rebroadcasts are considered to be redundant if all its neighbors already have the message. In [7], the authors show that a rebroadcast system can provide only 0 ~ 61% additional coverage[3] over that already covered by the previous transmissions. Furthermore, it is shown that if a message has been heard more than 4 times, the additional coverage is below 0.05%. T_{min} associated with the pump operation provides an opportunity for a node to hear the same message from other rebroadcasting nodes before it would actually have started to transmit the message. A counter is used to keep track of the number of times the same broadcast message is heard. If the counter reaches 4 before the scheduled rebroadcast of a message then the transmission is cancelled and the node will not relay the specific message because the expected benefit (additional coverage) is very limited in comparison to the cost of transmission. T_{max} can be used to provide a loose statistical delay bound for the last hop to successfully receive the last segment of a complete file (e.g., a program image or script). Assuming that any missing data is recovered within one T_{max} interval using the aggressive fetch operation described in next section, then the relationship between delay bound D(n) and T_{max} is as follows:

$$D(n) = T_{max} \times n \times (\text{Number of hops}),$$

where n is the number of fragments of a file.

Fetch Operation

Since most sensor network applications generate light traffic most of the time, message loss in the sensor networks usually occurs because of transmission errors due to poor quality wireless links and not because of traffic congestion. This is not to say that congestion cannot occur but that the

---

[3] Corresponds to the number of additional nodes that can be reached by a rebroadcast.

vast majority of loss in these networks is associated with errors. This is especially true considering the environment in which sensor networks operate in is highly unpredictable, and therefore, the quality of the communication links can vary considerably due to obstructions or hostile ambient conditions.

A node goes into fetch mode once a sequence number gap in a file fragments is detected. A fetch operation is the proactive act of requesting a retransmission from neighboring nodes once loss is detected at a receiving node. PSFQ uses the concept of "loss aggregation" whenever loss is detected; that is, it attempts to batch up all message losses in a single fetch operation whenever possible.

Loss Aggregation

There are several considerations associated with loss aggregation. The first consideration relates to bursty loss. Data loss is often correlated in time because of fading conditions and other channel impairments. As a result loss usually occurs in batches. Therefore, it is possible that more than one packet is lost before a node can detect loss by receiving a packet with higher sequence numbers than expected. PSFQ aggregates loss such that the fetch operation deals with a "window" of lost packets instead of a single packet loss. Next, in a dense network where a node usually has more than one neighbor, it is possible that each of its neighbors only obtains or retains part of the missing segments in the loss window. PSFQ allows different segments of the loss window to be recovered from different neighbors. In order to reduce redundant retransmissions of the same segment, each neighbor waits for a random time before transmitting segments, i.e., sets a retransmission timer to a random value, and sends the packet only when the timer goes off. Other nodes that have the data and scheduled retransmissions will cancel their timers if they hear the same "repair" (i.e., retransmission of a packet loss) from a neighboring node. Third, in poor radio environments successive loss could occur including loss of retransmissions and fetch control messages. Therefore, it is not unusual to have multiple gaps in sequence number of messages received by a node after several such failures. Aggregating multiple loss windows in the fetch operation increases the likelihood of successful recovery in the sense that as long as one fetch control message is heard by one neighbor all the missing segments could be resent by this neighbor.

8.3.3 Nack Messaging

We define a NACK message associate with the fetch operation as the control message that requests a retransmission from neighboring nodes. The NACK message has at least three header fields (this could be more) with no payload: *i)* file ID, *ii)* file length, and *iii)* loss window. The loss window represents a pair of sequence numbers that denote the left and right edge of a loss window (see example below). When there is more than one sequence number gap, each gap corresponds to a loss window and will be appended after the first three header fields in the NACK message. For example, if a node receives messages with sequence number (3,5,6,9,11), then computes 3 gaps and hence 3 loss windows that are (4,4), (7,8) and (10,10), respectively.

II. Fetch Timer

In fetch mode, a node aggressively sends out NACK messages to its immediate neighbors to request missing segments. If no reply is heard or only a partial set of missing segments are recovered within a period T_r ($T_r < T_{max}$, this timer defines the ratio between pump and fetch) then the node will resend the NACK every T_r interval (with slight randomization to avoid synchronization between neighbors) until all the missing segments are recovered or the number of retries exceed a preset threshold thereby ending the fetch operation. The first NACK is scheduled to be sent out within a short delay that is randomly computed between 0 and Δ ($<<T_r$). The first NACK is cancelled (to keep the number of duplicates low) in the case where a NACK for the same missing segments is overheard from another node before the NACK is sent. Since Δ is small, the chance of this happening is relatively small. In general, retransmissions in response to a NACK coming from other nodes are not guaranteed to be overheard by the node that cancelled its first NACK. In [7] the authors show that at most there is a 40% chance that the canceling node receives the retransmitted data under such conditions. Note, however that a node that cancel its NACK will eventually resend a NACK within T_r if the missing segments are not recovered, therefore, such an approach is safe and beneficial given the tradeoffs.

To avoid the message implosion problem, NACK messages never propagate; that is, neighbors do not relay NACK messages unless the number of times the same NACK is heard exceeds a predefined threshold while the missing segments requested by the NACK message are no longer retained in a node's data cache. In this case, the NACK is relayed once, which in effect broadens the NACK scope to one more hop to increase the chances of recovery. Such a situation should be a rare occurrence, since loss is triggered when a packet with a higher sequence number than expected is received. The upstream node that sent this packet maintains a data cache and must have obtained all the preceding segments prior to sending this higher sequence number packet, which in this scenario, failed to reach the "fetching" node. The probability that all neighbors do not have the missing segments is very low. In our reference application, since all nodes must keep the code for re-tasking purposes, all segments that have been received correctly can be pulled out of cache or external storage. Therefore, NACK messages never need to be propagated in this case.

Each neighbor that receives a NACK message checks the loss window field. If the missing segment is found in its data cache, the neighboring node schedules a reply event (sending the missing segment) at a random time between ($\frac{1}{4}T_r$, $\frac{1}{2} T_r$). Neighbors will cancel this event whenever a reply to the same NACK for the same segment (same file ID and sequence number) is overheard. In the case where the loss window in a NACK message contains more than one segment to be resent, or more than one loss window exists in the NACK message, then neighboring nodes that are capable of recovering missing segments will schedule their reply events such that packets are sent (in-sequence) at a speed that is not faster than once every ¼ T_r.

III. Proactive Fetch

As in many negative acknowledgement systems, the fetch operation described above is a reactive loss recovery scheme in the sense that a loss is detected only when a packet with higher sequence number is received. This could cause problems on rare occasions; for example, if the last segment of a file is lost there is no way for the receiving node to detect this loss[4] since no packet with higher sequence number will be sent. In addition, if the file to be injected into the network is small (e.g., a script instead of binary code), it is not unusual to lose all subsequent segments up to the last segment following a bursty loss. In this case, the loss is also undetectable and thus non-recoverable with such a reactive loss detection scheme. In order to cope with these problems, PSFQ supports a "proactive fetch" operation such that a node can also enter fetch mode proactively and send a NACK message for the next segment or the remaining segments if the last segment has not been received and no new packet is delivered after a period of time T_{pro}. When a proactive fetch operation is triggered, a node will manually create a loss event and send out a NACK control message with the desired loss window.

---

[4] A node knows that it has not received the last segment, but it does not know whether the last segment is lost or will be relayed at some point in the future.

The proactive fetch mechanism is designed to autonomously trigger the fetch mode at the proper time. If fetch mode is triggered too early, then the extra control messaging might be wasted since upstream nodes may still be relaying messages or they may not have received the necessary segments. In contrast, if fetch mode is triggered too late, then the target node might waste too much time waiting for the last segment of a file, significantly increasing the overall delivery latency of a file transfer. The correct choice of T_{pro}, must consider two issues. First, in our reference application, where each segment of a file needs to be kept in a data cache or external storage for the re-tasking operation, the proactive fetch mechanism will "Nack" for all the remaining segments up to the last segment if the last segment has not been received and no new packet arrives after a period of time T_{pro}. T_{pro} should be proportional to the difference between last highest sequence number (S_{last}) packet received and the largest sequence number (S_{max}) of the file (the difference is equal to the number of remaining segments associated with the file), i.e., $T_{pro} = \alpha * (S_{max} - S_{last}) * T_{max}$ ($\alpha \geq 1$). S_{max} is the file length found in the header field, T_{max} is the timer defined in the previous section, α is a scaling factor to adjust the delay in triggering proactive fetch and should be set to 1 for most operational cases.

This definition of T_{pro} guarantees that a node will wait long enough until all upstream nodes have received all segments before a node moves into the proactive fetch mode. In addition, this enables a node to start proactive fetch earlier when it is closer to the end of a file, and wait longer when it is further from completion. Such an approach adapts nicely to the quality of the radio environment. If the channel is in a good condition, then it is unlikely to experience successive packet loss; therefore, the reason for the reception of no new messages prior to the anticipated last segment is most likely due to the loss of the last segment, hence, it is wise to start the proactive fetch promptly. In contrast, a node is likely to suffer from successive packet loss when the channel is error-prone; therefore, it makes sense to wait longer before pumping more control messages into the channel. If the sensor network is known to be deployed in a harsh radio environment then α should be set larger than 1 so that a node waits longer before starting the proactive fetch operation. Finally, a node that starts proactive fetch will create a loss window with the left edge equal to ($S_{last}+1$) and right edge equal to S_{max} before sending a NACK message. The rest of the actions taken in response to a NACK message are exactly the same as normal fetch operations including the retransmission of NACKs and the handling of loss windows, as discussed earlier.

In other applications where the data cache size is small and nodes only can keep a portion of the segments that have been received, the proactive fetch mechanism will "Nack" for the same amount of segments (or less) that the data cache can maintain. In this case, T_{pro} should be proportional to the size of the data cache. If the data cache keeps n segments, then $T_{pro} = \alpha * n * T_{max}$ ($\alpha \geq 1$). As in discussed previously, α should be set to 1 in low error environments and to a larger value in harsher radio environments. This approach keeps the sequence number gap at any node smaller than n, i.e., it makes sure that a node will fetch proactively after n successive missing segments. Recall that a node waits at most T_{max} before relaying a message in the pump operation so that the probability of finding missing segments in the data cache of upstream nodes is maximized.

Report Operation

In addition to the pump and fetch operations described above, PSFQ supports a report operation designed specifically to feedback data delivery status information to users in a simple and scalable manner. In wireless communication, it is well known that the communication cost of sending a long message is less than sending the same amount of data using many shorter messages [14]. Given the potential large number of target nodes in a sensor network in addition to potential long paths (i.e., longer paths through multi-hops greatly increase the delivery cost of data), the network would become overwhelmed if each node sends feedback in the form of report messages. Therefore, there is a need to minimize the number of messages used for feedback purposes. PSFQ's report message and feedback mechanisms are designed to address these issues. The report message is designed to travel from the furthest target node back to the user on a hop-by-hop basis. Each node en route toward the user is capable of piggybacking their report message in an aggregated manner. Nodes can add/append their own feedback information to the original report message sent by the most distant target node as it propagates back toward the user that initially requested the report.

Report Message

The report message has only one field in its header representing the destination node ID of the node that should relay this report. The payload is a chain of node IDs and sequence number pairs that feedback the current status of each node along the path from the last hop toward the source user node.

Report Timers

A node enters the report mode when it receives an inject data message with the "report bit" set in the TTL field. The user node sets the report bit in the inject message whenever it needs to know the latest status of the surrounding nodes. To reduce the number of report messages and to avoid report implosion, only the last hop[5] nodes will respond immediately by initiating a report message sending it to its parent[6] node at a random time between (0, Δ). Each node along the path toward the source node will piggyback their report message by adding their own node ID and sequence number pair into the report, and then propagate the aggregated report toward the user node. Each node will ignore the report if it found its own ID in the report to avoid looping. Nodes that are not last hop nodes but are in report mode will wait for a period of time ($T_{report} = T_{max} \times TTL + \Delta$) sufficient to receive a report message from a last hop node, enabling it to piggyback its state information. A node that has not received a report message after T_{report} in report mode will initiate its own report message and send it to its parent node. If the network is very large then it is possible for a node to receive a report message that has no space to append more state information. In this case a node will create a new report message and send it prior to relaying the previously received report that had no space remaining to piggybacking data. This ensures that other nodes en route toward the user node will use the newer report message rather than creating new reports because they themselves receive the original report with no space for piggybacking additional status.

8.3.5 Single-packet message consideration

Loss is detected by observing sequence number gap or timeout in the PSFQ paradigm. For short messages that can fit into a single packet (smaller than the network MTU), delivery failure is undetectable in a NACK-based protocol without explicit signaling. One possible solution is to make use of the Report primitive described above to acquire application-specific feedback at the sink. For example, a sink would set the report bit (to solicit a report message from receivers in an aggregated manner) in every single-packet message that required a delivery guarantee. Based on the feedback status, a sink would resend the packet until all receivers confirm reception. This essentially turns PSFQ into a positive aggregated-ACK protocol used in

---

[5] The last hop can be identified from the "TTL" field of the inject message, i.e., nodes that receive an inject message with TTL=1.
[6] The node where the previous segment came from.

an on-demand basis by the sink. This also illustrates one of the flexible usages of the Report primitive in PSFQ.

8.4 PERFORMANCE EVALUATION

We use packet-level simulation to study the performance of PSFQ in relation to several evaluation metrics and discuss the benefits of some of our design choices. Simulation results indicate that PSFQ is capable of delivering reliable data in wireless sensor networks even under highly error prone conditions, whereas, in contrast, other relevant approaches "retooled" to operate under such conditions cannot.

8.4.1 Simulation Approach

We implemented PSFQ as part of our reference re-tasking application using the ns-2 network simulator [13]. In order to highlight the different design choices made we compare the performance of PSFQ to an idealized version of Scalable Reliable Multicast (SRM) [5], which has some similar properties to PSFQ, but is designed to support reliable multicast services in IP networks. While there is growing body of work in multicast [8-9] in mobile ad hoc networks and some initial work on reliable multicast support [11-12], we have chosen SRM as the best possible candidate that is well understood in the literature. SRM supports reliable multicast on top of IP and uses three control messages for reliable delivery, including session, request and repair messaging. Briefly, session messages are sent by each node in a multicast group to inform members of the last data message received by a node. Session messages are time-stamped and their exchange is also used to calculate the delay between pairs of nodes. Request messages are multicast by a node when it discovers that a data message is missing. Repair message, on the other hand, responds with the missing data requested in the request message. Repair messages may be sent not only by the original source but also by any other node able to retransmit the missing data. SRM represents a scheme that uses explicit signaling for reliable data delivery while PSFQ is a more minimalist transport that can be unicast (on top of routing) or broadcast and does not require periodic signaling.

We compare PSFQ with the loss detection/recovery approach of SRM but extract out the IP multicast substrate and replace it with an idealized omniscient multicast routing scheme. In this sense, we present SRM in the best possible light. There are several considerations for doing this. First, SRM is based heavily on the group delivery model and Application Level Framing [10], which make it a good match for cluster-based communications and the application-specific nature of sensor networks. On the other hand, SRM relies heavily on an IP multicasting mechanism for data routing. It is, however, unrealistic to assume an IP substrate in the context of sensor networks, as discussed previously. In addition, PSFQ is solely a reliable data transport scheme in our reference application, it does not provide a general routing solution as in the case of SRM, and therefore it is only fair to isolate SRM from the routing cost incurred by an IP multicast substrate for the purpose of our evaluation. We therefore only compare the reliable delivery portions of the SRM and PSFQ protocols. Since PSFQ uses a simple broadcast mechanism as a mean for routing in our reference application, it makes sense to layer SRM over an ideal omniscient multicast routing layer for simulation purposes. Using omniscient multicast, the source transmits its data along the shortest-path multicast tree to all intended receivers in which the shortest path computation and the tree construction to every destination is free in term of communication cost.

The major purpose of our comparison is to highlight the impact of different design choices made. SRM represents a traditional receiver-based reliable transport solution and is designed to be highly scalable for Internet applications. SRM's service model has the closest resemblance to our reference application in sensor networks. However, SRM is designed to operate in the wired Internet in which the transport medium is highly reliable and does not suffer from the unique problems found in wireless sensor networks, such as, hidden terminal and interference. To make a fair comparison, we try to idealize the lower layer to minimize the differences of the transport medium (which SRM is designed for) for simulation purposes, and, solely focus on the reliable data delivery mechanism – we term this idealized SRM scheme as *SRM-I*.

The goal of our evaluation is also to justify the design choices of PSFQ. We choose three metrics, which underpin the major motivations behind PSFQ:

- *average delivery ratio,* which measures the ratio of number of message a target node received, to the number of message a user node injects into the network. This metric indicates the error tolerance of the scheme at the point where a scheme fails to deliver 100% of the messages injected by a user node within certain time limits.
- *average latency,* which measures the average time elapsed from the transmission of the first data packet from the user node until the reception of the last packet by the last target node in the sensor network. This metric examines the delay bound performance of a scheme.
- *average delivery overhead,* which measures the total number of messages sent per data message received by a target node. This metric

examines the communication cost to achieve reliable delivery over the network.

We study these metrics as a function of channel error rate as well as the network size.

To evaluate PSFQ in a realistic scenario, we simulate the re-tasking of a simple sensor network in a disaster recovery scenario within a building. Typically, sensor nodes in a building are deployed along the hallway on each floor. Figure 8.4 depicts such a simple sensor network in a space of dimensions 100m x 100m. Each sensor node is located 20 meters from each other. Nodes use radios with 2 Mbps bandwidth and 25 meters nominal radio range. The channel access is the simple CSMA/CA and we used a uniformly distributed channel error model. A user node at location 0 attempts to inject a program image file with size equal to 2.5KB into every node on the floor for the purposes of re-tasking. The typical packet size used by the sensors is used in this evaluation, and for which the radio is designed, is 50 bytes. This is equivalent to injecting 50 packets into the sensor network from the user node. Packets are generated from the user node and transmitted at a rate of one packet per 10ms. For PSFQ, the timer parameters were set conservatively to follow PSFQ paradigm: T_{max} is 100ms, T_{min} is 50ms and T_r is 20ms. Therefore, the fetch operation can be 5 times faster than pump operation. Each experiment is run 10 times and the results shown are an average of these runs.

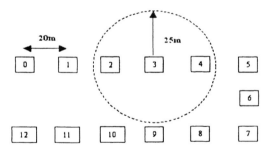

Figure 8.4. Sensor network in a building. A user node at location 0 injects 50 packets into the network within 0.5 seconds.

8.4.2 Simulation Results

One of the major goals of PSFQ is to be able to work correctly under a wide variety of wireless channel conditions. The first experiment examines the "error tolerance" of PSFQ and SRM-I, and compares their results. Following the consideration of optimizing lower layer support for SRM, SRM-I is given extra benefit in channel access by using CSMA/CA with RTS/CTS and ACK support while PSFQ only uses CSMA broadcasting. The use of omniscient multicast along with the RTS/CTS channel access greatly contributes to the error tolerance of SRM-I in two respects. First, RTS/CTS eliminates the hidden terminal problems and reduces possible interference between nodes. Second, ACK support in our simulation allows up to 4 link-layer retransmissions (note that from the simple analysis shown in Figure 8.2, this could provide an improvement up to 20%) after an RTS-CTS exchange, this essentially incorporates the loss recovery mechanism into lower layer (i.e., MAC-level ARQ) in addition to that offered by SRM.

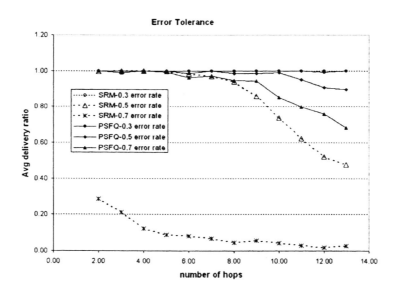

Figure 8.5. Error tolerance comparison - average delivery ratio as a function of the number of hops under various channel condition for different packet error rate.

In Figure 8.5, we present the results for PSFQ and SRM-I under various channel error conditions as we increase the number of hops in the network. As one might expect, the average delivery ratio of both schemes decrease as channel error rate increases. In addition, for larger error rates, the delivery ratio decreases rapidly when the number of hops increases. Notice that the user node starts sending data packets into the network at a constant rate of one packet per 10ms at 2 seconds into the simulation trace and finishes sending all 50 packets within 0.5 seconds. The simulation ran for 100 seconds after the user node stopped sending data packets. This arbitrary cutoff point was chosen as the time after which the delivery of data would be meaningless for a time critical re-tasking operation. Of course, this time limit is very much application-specific; in this case, consider that 100 seconds is 200 times the amount of time required by the user node to inject the entire program image file into the network. Observe from Figure 8.5, SRM-I (dotted line) can achieve 100% delivery at up to 13 hops away from the source node only when the channel error rate is smaller than 30%. For 50% error rate, the 100% delivery point decreases to within 5 hops; and for larger error rates, SRM-I is only able to deliver a portion of the file two hops away from the user node. In contrast, PSFQ (solid line) can achieve a much higher delivery ratio for all cases under consideration for a wide range of channel error conditions. PSFQ achieves 100% delivery up to 10 hops away from the user node even at 50% error rate and delivers more than 90% of the packet up to 13 hops away. Even under extremely error-prone channel error rate of 70%, PSFQ is still able to deliver 100% data up to 4 hops away and 70% of the packets up to 13 hops, while SRM-I can only deliver less than 30% of data even within 2 hops.

The better error tolerance exhibited by PSFQ in comparison to SRM-I justifies the design paradigm of pump slowly and fetch quickly for wireless sensor networks. The in-sequence data pump operation prevents the propagation of loss events, as discuss in Section 2.3. While SRM-I does not attempt to provide ordered delivery of data and loss events are propagated along the multicast tree. PSFQ's aggressive fetch operation and loss aggregation techniques support multiple loss windows in a single control message. In contrast SRM-I is conservative in loss recovery operations. This is because SRM is intended for applications without fixed deadlines, it also does not support the aggregation of multiple loss windows in a single control message.

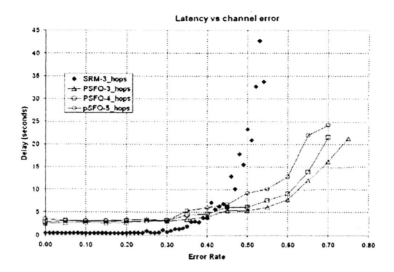

Figure 8.6. Comparison of average latency as a function of channel error rate.

Our second experiment examines the data delivery latency of both schemes under various channel conditions. The results are shown in Figure 8.6. Delivery latency is defined only when all intended target nodes receive all data packets before the simulation terminates. For SRM-I, we know that 100% delivery can be achieved only within a limited number of hops when the error rate is high. In this experiment, we compare the two schemes using a 3-hop network and investigate PSFQ's performance with a larger number of hops since PSFQ has better error tolerance. Figure 8.6 shows that SRM-I has a smaller delay than PSFQ when the error rate is smaller than 40%, but its delay grows exponentially as the error rate increases, while PSFQ grows more slowly until it hits its error tolerance barrier at 70% error rate. The reason that SRM-I performs better than PSFQ in terms of delay in the small error region is due to the "pump slowly" mechanism, in which each node delays a random period of time between T_{min} and T_{max} before forwarding packets. Despite this small penalty in the smaller error region, the coupling of this mechanism with the "fetch quickly" operation proves to be very effective. As shown in Figure 8.6, PSFQ can provide delay assurances even at very high error rates. Figure 8.6 also shows that as the number of hops increases, the delay in PSFQ increases rapidly in the higher error rate region, but it is still within the anticipated delay bound.

In the next experiment, we study the communication cost for reliability in both schemes under various channel conditions using a 3-hop network.

Communication cost is measured as the average number of transmissions per data packet (i.e., average delivery overhead). For SRM-I, we separate the communication cost of the SRM-specific loss recovery mechanisms from the total communication cost, which includes the cost associated with the link-layer loss recovery mechanisms (RTS/CTS/ACK). Figure 8.7 shows that the cost for PSFQ is consistently smaller than SRM-I by an order of magnitude even after excluding the link-layer cost of SRM-I. Figure 8.7 also illustrates the 100% delivery barrier of both schemes (the two vertical lines). The 52% error rate mark shows the limit for SRM-I while the 70% error rate mark shows the operation boundary for PSFQ. The different performance observed under simulation is rooted in the distinct design choices made for each protocol. PSFQ utilize a passive, on-demand loss recovery mechanism, whereas SRM employ periodic exchange of session messages for loss detection/recovery.

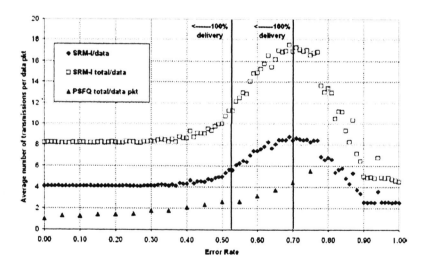

Figure 8.7. Average delivery overhead as a function of channel error rate.

If we consider the results for SRM-I in Figure 8.8, we can observe four distinct portions of the curve: *i)* from 0 to 30% error rate, the curve is linear where the link-layer loss recovery mechanisms are able to recover all packet losses and there is no need for the SRM (REQ, REP) mechanisms to be used; *ii)* from 30% to 50% there is a rapid increase in messages as the packet loss overwhelms the link-layer mechanisms, and the SRM reliable delivery mechanisms must be used to maintain 100% delivery; *iii)* after that point, message growth continues, but despite SRM, not all packets can be delivered to all nodes; and *iv)* the curve declines rapidly when the channel error rate increases beyond 70%, at which point the error rate is so high that hardly any transmission can take place since the RTS-CTS exchange is rarely completed.

8.5 WIRELESS SENSOR TESTBED

In what follows, we discuss some early experiences implementing PSFQ in an experimental wireless sensor testbed. We implemented PSFQ using the TinyOS platform [1, 6] on RENE mote [1]. The sensor device has an ATMEL 4MHz, low power, 8-bit micro-controller with 8K bytes of program memory and 512 bytes of data memory; 16KB EEPROM serves as secondary storage. The radio is a single channel RF transceiver operating at 916MHz and capable of transmitting at 10kbps using on-off-keying encoding. The radio performs transmission and bit sampling in software (TinyOS). TinyOS [6] is an event-based operating system, employing a CSMA MAC and performs encoding and decoding of the byte stream using Manchester encoding with a 16-bit CRC. The packet size is 30 bytes. With a link speed of 10kbps the channel capacity can delivers at most 20 packets per second. Tuning the transmission power can change the radio transmission range of the motes.

We implemented the PSFQ pump, fetch and report operations as a component of TinyOS that interfaces with the lower layer radio components. The component code size for PSFQ is 2KB. In the implementation, every data fragment that is received correctly is stored into the external EEPROM at a predefined location based on its sequence number. The sequence number is used as an index to locate and retrieve data segments when a node receives a NACK from its neighbors.

We conduct several experiments using a simple scenario for preliminary evaluation of PSFQ. In order to emulate a wireless channel with different packet error rates, we manipulate the radio transmission power of the motes before starting the experiments, and measure the packet loss rate between every pair of motes that are separated by a fixed distance by calculating the number of missing packet for the transmission of 100 packets between the node pairs. Due to the irregular fading conditions in the laboratory, it is difficult to obtain accurate and fine-grained channel error rates. We are only able to obtain four different error rates that are relatively consistent over the period of our experiments. These simple experiments measure the delivery latency of sending 30 packets from a source node to multiple target nodes hops away. In order to monitor the completion time in which a mote received all the packets, additional motes connected to a laptop computer are placed close to the target motes to snoop the reports sent by each node.

Figure 8.8 shows the result of our experiments. Every data point in the Figure 8.8 is an average of 5 independent experiments. The timer parameters are set as follows: $T_{max} = 0.3s$ and $T_r = 0.1s$. In the Figure 8.8, the delay for a 5-hop network increases rapidly for an error rate of 15%, whereas in smaller-hop networks, the delay increases more slowly even at higher error rates. The results observed from the testbed show poorer performance in comparison to the simulation results where a rapid increase occurs at much higher error rates. This discrepancy is expected because of several factors. First, channel conditions of the wireless link in the real world are highly irregular especially for mote that uses a very simple radio. The error rates presented in Figure 8.8 are not accurate enough. Second, the computational overhead cannot be captured in simulation. Since the processor has to process every bit received off the link, intensive computation could overwhelm the processor forcing packets to be missed. Therefore, the actual loss rate during the experiment could be higher. Finally, our implementation on the mote uses a very simple random number generator (16-bit LFSR) therefore the likelihood of collision in the testbed is higher than under simulation conditions.

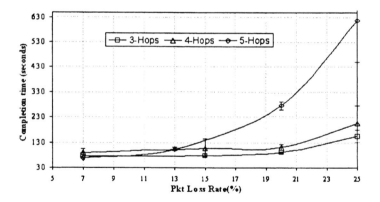

Figure 8.8. Delay experiments for mote testbed.

8.6 RELATED WORK

RMST (Reliable Multi-Segment Transport) [15] is a transport layer paradigm for sensor networks that is closest to our work in PSFQ. RMST is designed to complement Directed Diffusion [3] by adding on reliable data transport service on top of it. RMST is a NACK-based protocol, which has primarily timer driven loss detection and repair mechanisms. The authors analyze the tradeoff between hop-by-hop repair vs. end-to-end repair and conclude the importance of hop-by-hop recovery that is consistent with our analysis and simulation results. Unlike the PSFQ paradigm which provides reliability purely at the transport layer, in [15] the authors propose to involve both the transport and MAC layers, i.e., employ a NACK-based transport layer running over a selective-ARQ MAC layer assuming that it is possible to provide an energy efficient MAC layer with ARQ such as S-MAC [16].

In ESRT [17] the authors propose using an event-to-sink reliability model to provide reliable event detection that embeds a congestion control component. Unlike PSFQ, ESRT does not deal with data flows that require strict delivery guarantee; instead the authors define *"desired event reliability"* as the number of data packets required for reliable event detection that is determined by the application. In [17] the transport problem is therefore to configure the report rate of source nodes so as to achieve the required event detection reliability at the sink.

A Sink-to-Sensors reliability solution is presented in [18] that focus on communication reliability from the sink to the sensors in a static network. The authors propose using a two-radio approach where each node is equipped with a low frequency "busy-tone" radio in addition to the default radio that is used for data transmission and reception. The "busy-tone" radio is used to ensure delivery of single-packet messages or the first packet of a longer message. A NACK-based recovery core is constructed from the minimum dominating set (MDS) of the underlying graph; the recovery core would then assume the responsibility of recovering all losses and retransmit those packets to their non-MDS neighbors.

SRM [5] is representative of a traditional wire-line approach with explicit signaling as described in section 4. RMP [19] is an early proposal for many-to-many reliable communication protocols in wire-line networks that is based upon IP multicasting. RMP is a modified centralized scheme in which the receivers take turns to serve as a token site to assume the responsibility for all acknowledgments and retransmissions. While reliable multicast in wired networks is of interest they do not address the same problems as sensor networks.

8.7 CONCLUSION

In this paper, we have presented PSFQ, a reliable transport protocol specifically designed for wireless sensor networks. PSFQ is a lightweight, simple mechanism that is scalable and robust making minimum assumptions about the underlying transport infrastructure. We have discussed the need for reliable data delivery in sensor network, especially the application where a user node needs to re-task a group of sensor nodes in its vicinity by injecting program images into target nodes. Base on this reference application, we have described the design of PSFQ to achieve several goals, including operation under high error rate conditions and support for loose delay bounds for data delivery. We evaluated PSFQ and compared its performance to an idealized SRM implementation under simulation. We found that PSFQ outperforms SRM-I in terms of error tolerance, communication overhead, and delivery latency. We also presented some initial results from an experimental wireless sensor network that supports PSFQ using the TinyOS platform on top of the RENE motes. Results show a basic proof-of-concept indicating that the approach looks very promising in an actual wireless testbed. Future work includes more experimentation with larger numbers of sensors. We also plan to explore other transport issues in sensor networks utilizing PSFQ paradigm, for example to look into different variants of PSFQ that are optimized for different metrics such as delay sensitive reliable delivery. The driving force behind our work remains transport and system support for programming wireless sensor networks. Along those lines we intend to study the impact of re-tasking motes on-the-fly using PSFQ. Results from this phase of our work will be the subject of a future publication.

ACKNOWLEDGEMENTS

Many thanks to Shane Eisenman for helping in data measurement, and we would also like to thank the editors for their efforts and insightful comments.

REFERENCES

1. Jason Hill, Robert Szewczyk, Alec Woo, Seth Hollar, David Culler and Kristofer Pister, "System architecture directions for network sensors", *Proc. of the 9th International Conf. on Architectural Support for Programming Languages and Operating Systems*, Nov 2000, pp. 93-104.

2. Cots Dust, Large Scale Models for Smart Dust. http://www-bsac.eecs.berkeley.edu/~shollar/macro_motes/macromotes.html.
3. C. Intanagonwiwat, RC. Govindan and D. Estrin, "Directed Diffusion: A Scalable and Robust Communication Paradigm for Sensor Networks", *Proc. of the sixth annual international conf. on Mobile computing and networking*, Aug. 2000, pp. 56-67.
4. J. Atwood, O. Catrina, J. Fenton and W. Strayer, "Reliable Multicasting in the Xpress Transport Protocol", *Proc. of 21st conference on Local Computer Networks*, Oct 1996, pp. 202-211.
5. S. Floyd, V. Jacobson, C. Liu, S. Macanne and L. Zhang. "A Reliable Multicast Framework for Lightweight Session and Application Layer Framing". *IEEE/ACM Transactions on Networking*, vol. 5, no. 6, pp. 784-803, Dec. 1997.
6. TinyOS Homepage. http://webs.cs.berkeley.edu/tos/.
7. S.-Y. Ni, Y.-C. Tseng, Y.-S. Chen and J.-P. Sheu, "The broadcast storm problem in a mobile adhoc network", *Proc. of the fifth annual ACM/IEEE international conference on Mobile computing and networking*, Aug. 1999, pp. 151-162.
8. J.J. Garcia-Luna-Aceves and E. L. Madruga, "The core assisted mesh protocol", *IEEE Journal on Selected Areas in Communications*, Aug. 1999, vol. 17, no. 8, pp. 1380-94.
9. S.-J. Lee, M. Gerla and C.-C. Chiang, "On-demand multicast routing protocol", *Proc. IEEE Wireless Communications and Networking Conf.*, Sept. 21-25 1999, pp. 1298-1304.
10. D. Clark and D. Tennenhouse, "Architectural Considerations for a New Generation of Protocols", *Proceedings of ACM SIGCOMM '90*, Sept. 1990, pp. 201-208.
11. C. Ho, K. Obraczka, G. Tsudik and K. Viswanath, "Flooding for Reliable Multicast in Multi-Hop Ad Hoc Networks", *Mobicom Workshop on Discrete Algorithms & Methods for Mobility* (DialM'99), Aug, 1999.
12. E. Pagani and G. Rossi, "Reliable Broadcast in Mobile Multihop Packet Networks", *Proc. of the third annual ACM/IEEE international conference on Mobile computing and networking*, Sept. 1997, pp. 34-42.
13. The Network Simulator - ns-2. http://www.isi.edu/nsnam/ns/.
14. D. A. Maltz. "On-Demand Routing in Multi-hop Wireless Mobile Ad Hoc Networks", PhD thesis, Carnegie Mellon University, 2001.
15. Red Stann and John Heidemann, "RMST: Reliable Data Transport in Sensor Networks", Appearing in *1st IEEE International Workshop on Sensor Net Protocols and Applications (SNPA)*, Anchorage, Alaska, USA, May 2003.
16. W. Ye, J. Heidemann and D. Estrin. "An Energy Efficient MAC Protocol for Wireless Sensor Networks", In *Proc. of the 21st Intl. Annual Joint Conf. of the IEEE Comp. & Comm. Soc. (INFOCOM 2002)*, New York, NY, USA, June 2002.
17. Y. Sankarasubramaniam, O.B. Akan, and I.F. Akyildiz, "ESRT: Event-to-Sink Reliable Transport in Wireless Sensor Networks", To appear in *Proceedings of ACM MobiHoc*, Annapolis, MD, USA, June 2003.
18. S-J. Park and R. Sivakumar, "Sink-to-Sensors Reliability in Sensor Networks", Extended Abstract to appear in *Proceedings of ACM MobiHoc*, Annapolis, MD, June 2003.
19. Nicholas F. Maxemchuk, "Reliable Multicast with Delay Guarantees", *IEEE Communications Magazine*, pp. 96-102, September 2002.
20. C.E. Perkins and P. Bhagwat. "Highly dynamic Destination- Sequenced Distance-Vector routing (DSDV) for Mobile Computers", in *SIGCOMM Symposium on Communications Architectures and Protocols*, pp. 212-225, September 1994.

III

DATA STORAGE AND MANIPULATION

Chapter 9

DATA-CENTRIC ROUTING AND STORAGE IN SENSOR NETWORKS

Ramesh Govindan
Department of Computer Science
University of Southern California
Los Angeles, CA 90089.
ramesh@usc.edu

Abstract Early sensor network data access systems were built upon the the idea of *data-centric routing*. Data-centricity allows robust and energy-efficient communication by abstracting the physical location of data. In this chapter, we introduce a complementary idea called *data-centric storage*. Data-centric storage provides efficient rendezvous between the producer of a data item, and its consumer. The primitives underlying data-centric storage can be used to construct different distributed data structures for efficiently answering a variety of queries.

Keywords: Sensor networks, distributed hash tables, database indexing, geographic routing.

9.1 INTRODUCTION

Wireless sensor networks promise the ability to observe natural and man-made phenomena at finer spatial scales than is currently possible. Sensor networks have both commercial and scientific applicability, a potential that arises from the untethered nature of the devices that constitute these networks—individual sensor nodes are battery-operated and use wireless technologies for communication. These features enable quick, inexpensive, and relatively non-intrusive deployments of large numbers of nodes within a given physical environment. But their untethered nature also poses interesting design challenges: in order to achieve reasonable lifetimes, these networks will need to conserve energy, and an important component of energy usage in these systems is wireless communication [10].

This energy constraint crucially impacts those subsystems of sensor networks that provide access to the data gathered by individual devices. Researchers have recognized that *data-centric* communication paradigms enable energy-efficient designs for large-scale data access from these systems [10]. Unlike traditional network communication where individual nodes are named ("give me data from node 57"), data-centric communication relies on naming the data ("I'm interested in the average temperature of from geographic region X"). Naming the data enables nodes within the network to store or cache the data transparently, as well as process (*e.g.* compress or aggregate) the data. These capabilities can reduce communication cost, thereby increasing network lifetimes [9].

The earliest examples of data-centric communication in sensor networks are Directed Diffusion [10] and TAG [14]. These systems implement *data-centric routing*: "interests" or "queries" are routed to nodes that might contain matching data, and responses are routed back to the querying nodes. Because data-centricity abstracts the identity of nodes producing the data, such systems usually require *flooding* the interest or query (either globally or scoped geographically) to discover nodes that contain matching data. These systems are appropriate for long-lived queries initiated by users from outside the network. Examples of such queries include tracking an object, or continuously computing aggregates over a sensor field.

Sensor networks will need support for more flexible and efficient ways of accessing the data, however, than that provided by data-centric routing systems. An obvious requirement is that of a "one-shot" query: that which computes, say, the maximum temperature observed in a sensor field in the last 5 minutes. Implementing such queries using data-centric routing may be inefficient, for two reasons: the cost of flooding may dominate the overall cost of the query in large networks; in smaller networks, such queries will be frequently issued from within the network (for example, by individual nodes that perform some action based on the results of the query such as turning on a camera) and the cost of flooding each query from internal nodes will adversely impact the lifetime of the overall system.

In this chapter, we discuss a complementary class of systems called *data-centric storage* systems [18]. In such systems, data that is generated at a node is stored at another node determined by the name of the data. Such an approach enables efficient *rendezvous*—given the name of a data item, one can retrieve the data item without flooding. More generally, data-centric storage enables *distributed data structures* (such as indices) for sensor network applications. Such data structures will provide efficient access to data in large-scale sensor networks, or where queries are frequently initiated from within the network. In a deployed sensor network, data-centric storage systems may co-exist with data-centric routing; applications will pick the appropriate mode of data access for their needs.

Data-centric storage is the subject of active research. In this chapter, we will attempt to summarize the state of this research. Section 9.2 introduces data-centric storage and places it in the context of other ways of accessing data from within a sensor network. The subsequent sections describe different distributed data structures that can be built using data-centric routing: geographic hash tables (Section 9.3), hierarchical search structures (Section 9.4), and indices on single (Section 9.5) or multiple keys (Section 9.6).

9.2 DATA-CENTRIC STORAGE

Traditional data communication networks have built communication mechanisms that rely on addressing individual nodes. This is appropriate when each end system is usually associated with a distinct user and most traffic is point-to-point. By contrast, in a sensor network the individual identity of nodes is almost always unimportant. Moreover, there are usually a small number of users of a sensor network, and traffic patterns can be varied (ranging from multipoint-to-multipoint for local collaboration between nodes, to multipoint-to-point for communicating data from within the network to external *sinks*). For these reasons, sensor network data access systems will be based upon *data-centric abstractions* which abstract node identity.

9.2.1 DATA-CENTRIC ROUTING AND STORAGE

Fundamental to data-centricity is the ability to *name* data. There are many approaches to data naming, and the naming scheme sometimes depends on the application. For example, Internet information access systems such as the Intentional Naming System [2], use *hierarchical naming*. In such systems, data generated by a camera sensor in a parking lot may be named:

```
US/Universities/USC/CS/camera1
```

On the other hand, systems such as Directed Diffusion use an attribute-value naming scheme. In such systems, the data generated by the camera sensor in the previous example would be named:

```
type = camera
value = image.jpg
location = ``CS Dept, Univ.of Southern California''
```

These naming schemes implicitly define a set of ways in which the data may be accessed (*e.g.*all cameras within US universities, cameras in computer science departments *etc.*for our hierarchical naming example). For our purposes, the particular naming scheme that an application chooses is unimportant; our schemes merely rely on data being named.

In data-centric routing, routing decisions are based on the name(s) associated with data. For example, in Directed Diffusion [10], each node matches a data item with forwarding state created by *interest* messages before deciding which neighbor to send the data packet to. In Directed Diffusion, data-centric routing is implemented at lower levels of the communication stack, and can lead to efficient communication [9]. Other systems (such as INS) are *application-level* data-centric routing systems—they are built upon a lower-level primitive that supports node-to-node communication.

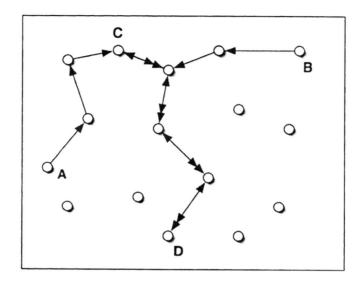

Figure 9.1. This figure illustrates the key ideas behind data-centric storage. Assume that nodes A and B want to insert an event named "bird sighting". Assume further that the name of the event maps to a node C. These events can then be routed to C. Knowing the name, another node D can retrieve the events by directly routing a message to C. In such a system, no message need be flooded through the network.

In data-centric storage, on the other hand, data can be stored and retrieved by name. Generally speaking, a data-centric storage system provides primitives of the form put(data) and data = get(name). Data-centric storage systems are not new; distributed hash table (DHT) systems [19, 23, 16, 15] are instances of such systems. This chapter explores the application of data-centric storage ideas to sensor networks.

There are, of course, many ways to implement data-centric storage. Efficient implementations of data-centric storage usually involve an algorithmic mapping between a data name and a "location" (a location can be a network node, a precise geographic location or region, or an identifier in a key space). Such an algorithmic mapping provides efficient *rendezvous*; nodes can retrieve data efficiently knowing only the name of the data (or part thereof). More precisely, unlike data-centric routing systems, retrieving a data item does not require (geographically-scoped) flooding of queries. As we show below, this property can be used to design a class of systems that are complementary to data-centric routing systems.

Thus far, in describing data-centric storage for sensor networks, we have avoided precisely specifying what data gets stored in a data-centric storage system. To understand this, it helps to introduce some terminology. Individual sensor nodes generate low-level *observations* [18]. Examples of such observations include a sequence of temperature readings at a node, or the readings from an accelerometer. Observations can be combined or aggregated to produce either *events* or *features*. An event is a semantically-rich description of a physical phenomenon. Examples of events include "a tank went through this intersection", or a "bird alighted upon this nest". By contrast, a feature is a geometric or statistical pattern in the observations generated by one or more sensor nodes, perhaps across time. Examples of features include "the average temperature reading in the last 10 minutes", or "a geometric curve enclosing all sensor nodes with comparable temperature readings".

Given the potential volume of observations, and the inefficiency of transporting these across the network, we expect that data-centric storage systems will be used to store (and retrieve) events and features.

9.2.2 PLACING DATA-CENTRIC STORAGE IN CONTEXT

Before we discuss the mechanistic aspects of data-centric storage, we describe how it would fit into a sensor network software framework. We organize this discussion in the form of layers purely for the purposes of exposition. We emphasize that the overall software structure of sensor networks is still the subject of active research and experimentation.

At the lowest layer of the sensor network software framework are systems that provide device-level access and spatio-temporal context: radio medium-access [21] localization [17] and time-synchronization [5]. Above this is a layer that provides packet-routing primitives. Flooding a packet to the entire network or flooding a packet within some scope (number of hops) are primitives that will be available at this layer. Also relevant to this layer are *geographic* routing primitives; those, like GPSR [12], which allow a packet to be sent to a specified location, and those like GEAR [22] which allow a packet to be delivered to all nodes within a specified geographic region. We expect that local *collaborative information processing* systems [20] will be built upon these packet routing primitives. These systems can be used to generate events and features.

Data-centric routing and storage will belong to the wide-area information dissemination layer, which sits above these layers. Data-centric routing is largely complementary to data-centric storage; below, we discuss and quantify where each technique might be applicable.

Finally, complex query processing systems can be built on top of data-centric routing and storage. For example, TinyDB [14] is a distributed sensor network database that is layered on top of a data-centric routing primitive. Later in this chapter, we describe distributed data structures that can be built upon data-centric storage. Eventually, mechanisms like these will be integrated into a comprehensive large-scale query processing system for sensor networks.

9.2.3 THE PERFORMANCE OF DATA-CENTRIC STORAGE SYSTEMS

In the preceding paragraphs, we have placed data-centric storage in the context of other sensor network components. We conclude this introduction to data-centric storage by comparing its performance against other data dissemination techniques. Even though we have not introduced specific mechanisms for data-centric storage, we can use asymptotic arguments to get a qualitative sense of the performance regimes where data-centric storage is the preferred dissemination alternative.

In our comparisons, we consider two other alternatives: an *External Storage* scheme in which all events (in what follows, we use the term *event* to encompass events and features) are stored at a node outside the network; and a *Local Storage* where each event is stored at the node at which it is generated.

It is easy to see the performance tradeoffs exhibited by these different alternatives. For external storage, the cost of accessing events is zero, since all events are available at one node. However, the cost of conveying data to this external node is non-trivial; there is an energy cost in communicating events to this node, and significant energy is expended at nodes near the external node in receiving all these events (these nodes become hotspots). If events are accessed far more frequently than they are generated, external storage might be an acceptable alternative (assuming the event generation rate does not deplete the bottleneck). At the other end of the spectrum, local storage incurs zero communication cost in storing the data, but incurs a large communication cost—a network flood—in accessing the data. Local storage may therefore be feasible when events are accessed less frequently than they are generated. Data-centric storage lies somewhere in between, since it incurs a non-zero cost both in storing events and retrieving them.

It is possible to make these intuitions more precise using some simple algebra. Consider a network of n nodes, in which the cost of sending messages to all nodes (*e.g.* a flood) is $O(n)$ and the cost of sending a message to a designated node is $O(\sqrt{n})$. Let us denote by D_e the total number of events detected, Q the number of queries, and D_q the number of events which are returned as

Canonical Method	Total	Hotspot
External	$D_e \sqrt{n}$	D_e
Local	$Qn + D_q \sqrt{n}$	$Q + D_q$
Data-centric	$Q\sqrt{n} + D_e\sqrt{n} + D_q\sqrt{n}$ (list) $Q\sqrt{n} + D_{events}\sqrt{n} + Q\sqrt{n}$ (summary)	$Q + D_q$ (list) $2Q$ (summary)

Figure 9.2. Analysis of asymptotic message transmission costs for three storage approaches.

answers for the Q queries. Furthermore, we assume that the data that is stored can either be a *listing* of events of a given type, or a *summary* of such events. Figure 9.2 shows the network-wide communication cost, as well as the hotspot energy usage for our three schemes. This analysis shows that data-centric storage becomes more preferable as the size of the network increases, or when many more events are generated than can be usefully queried. This performance advantage increases when summaries are returned as answers. Thus, data-centric storage is an attractive alternative as sensor networks scale.

9.3 MECHANISMS FOR DATA-CENTRIC STORAGE: GEOGRAPHIC HASH-TABLES

Our discussion has introduced data-centric storage as a concept, without describing the specifics of a particular instance of a data-centric storage system. In this section, we describe a system called a Geographic Hash table (GHT) and focus on its mechanistic underpinnings. GHT's mechanisms are powerful in that they can be re-used to design a variety of other distributed data structures in sensor networks.

9.3.1 GEOGRAPHIC HASH TABLES: AN OVERVIEW

We have said that the essence of a data-centric storage system is captured in its interface, which supports a put() operation that stores data by name within the network, and a get() operation that retrieves data by name. The simplest instance of such a system is one that provides hash-table semantics: event names are mapped to a key k (this mapping can be many-to-one) and the system allows applications to $put(event, k)$ (*i.e.* store an event based on its key), and $eventset = get(k)$ (*i.e.* retrieve all events matching a given key k).

Geographic Hash Tables (GHTs) represent one implementation of this interface for sensor networks. In a GHT, event names are *randomly* hashed to a geographic location (*i.e.* an x, y coordinate). For this hashing to work well, GHT assumes that all nodes know the geographic boundary of the sensor network. The boundary does not have to be precisely known (it can, for example, be a bounding rectangle encompassing all nodes) but all nodes must have a consistent picture of the boundary. Then, a put() operation stores an event at the node nearest the hashed location, and a get() operation retrieves one or more events from that node.

Implicitly, a GHT specifies a set of *queries* that it allows. The simplest kind of query that it allows is an enumeration query, a listing of all events matching a key k. It can also trivially support any statistic (*e.g.*count, sum, average) defined on that enumeration. Applications can determine which parts of an event name are used to compute the geographic hash. This choice can crucially affect performance of the overall system, as well as the set of supported queries, of course. For example, an application can hash events by sensor type so that all temperature sensor readings are stored at one node. Alternatively, an application may choose to hash sensor *values* to different locations, so that temperature readings with a value of 10° can be stored at a different node than, say readings of 20°.

GHTs derive inspiration from the many distributed hash table systems that have been recently proposed [19, 16, 23, 15]. There is one major difference between these systems and GHT—they all devise novel and efficient routing algorithms for locating the node at which a key can be stored. GHTs can leverage geographic routing protocols such as GPSR [12] directly, as we describe next.

9.3.2 GPSR: AN OVERVIEW

GPSR is a geographic routing protocol that was originally designed for mobile ad-hoc networks. Given the coordinates of a node, GPSR routes a packet to that node using location information alone. To do this, GPSR contains two different algorithms: *greedy* forwarding and perimeter forwarding.

Greedy forwarding is conceptually very simple. Assume each node in a network knows its own location, and that of its neighbors. When a node receives a message destined to location D, it sends the message to that neighbor N which closer to D than to itself. Such a neighbor might not always exist; in this case, GPSR invokes *perimeter routing* at that node. We now describe perimeter routing.

When a packet finds itself at a node which has no neighbors closer to the destination than itself, we say that the packet has encountered a *void*. Figure 9.3 depicts a situation where a packet destined for node D arrives at a node A. No neighbor of A is closer to D than A itself, indicating the existence of a void. Voids can result from irregular deployment of nodes, as well as from radio-opaque obstacles.

A natural technique to route around a void is the right-hand rule (Figure 9.3). Such a traversal walks around the *perimeter* of the void, hence the term *perimeter forwarding*. However, it can be shown that using the right hand rule on the graph formed by mere connectivity between nodes may not work in many

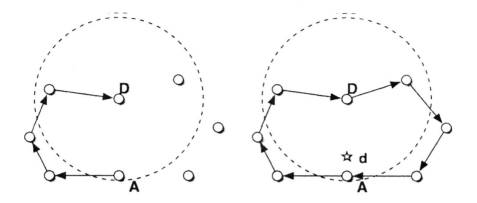

Figure 9.3. Using the right-hand rule to traverse a void.

Figure 9.4. How GHT uses perimeter traversal to find the home node.

cases [11]. To circumvent this, GPSR computes a planar subgraph of this graph in a distributed manner, and applies the right-hand rule to this subgraph. When this traversal reaches a node that is closer to D than A, greedy forwarding resumes.

9.3.3 HOW GHT USES GPSR: THE HOME NODE

From our discussions so far, two facts are evident. First, GPSR provides the functionality to deliver a packet to a node at a specified location. Second, GHT requires the ability to store an event at the node *closest* to a specified location (called the *home node* for an event). (More precisely, GHT maps an event's name to a geographic location, and one way of implementing a GHT would be to store the event at the node closest to that location). Thus, there is a subtle difference between what GPSR provides, and what GHTs require. In particular, GPSR would drop a packet if there exists no node at the destination location specified in the packet header.

To get around this, we can leverage an interesting property of GPSR. Assume that GHT hashes an event to a destination location d, and, without loss of generality, that no node exists at that location (Figure 9.4). When GPSR routes the packet containing the event, it will enter perimeter mode at the home node (by definition, the home node will have no neighbor closer to the destination than itself). Furthermore, the packet will remain in perimeter mode and return to the home node in that mode (again, by definition, no other node on the perimeter can be closer to the destination than the home node). GHT uses this property to detect and store events at the home node. When a packet returns in perimeter mode to the node that originated the perimeter traversal, the corresponding event is stored at that node.

This alone is not sufficient, however, since events can be lost if the home node fails. Furthermore, if new nodes are deployed, the definition of a home node may change. We now discuss a simple extension to the GHT design that maintains home node associations in the face of node dynamics.

9.3.4 GHT ROBUSTNESS: PERIMETER REFRESH

GHTs use a simple perimeter-refresh protocol to maintain home node associations. Periodically, for a given event, a home node will send a message destined to the corresponding location. This *refresh* message will traverse the perimeter around the specified location. As this message traverses the perimeter, each node on the perimeter stores a local copy of the event. Each node also associates a timer with the event. This timer is reset whenever a subsequent refresh is received. If the timer expires, nodes use this as an indication of home node failure, and initiate a refresh message themselves. In this manner, home node failures are detected, and the correct new home node is discovered.

As the refresh message traverses the perimeter, it may encounter a new node that is closer to the specified location. By GPSR's forwarding rules, this node will initiate a new perimeter traversal that will pass through the previous home node. The previous home node will detect this condition and remove its association as a home node for that event (the node will, of course, cache a replica of that event to enable failure detection of the new home node). When the perimeter traversal returns to the new home node, the association between the event and that node will have been completed. In this manner, GHTs are able to detect and adjust for the arrival of new nodes into the system.

The perimeter traversal technique trades off increased communication cost for robustness to node dynamics. This tradeoff is reasonable, since in a dense sensor network, most perimeters will be about three hops.

9.3.5 GHT SCALING: STRUCTURED REPLICATION

The final piece in the design of GHTs is called *structured replication*. This is motivated by the observation that if many events hash to the same location (for example, a certain type of event occurs frequently), the home node can become a *hotspot*.

To avoid this, structured replication hierarchically decomposes the geographical region enclosing the sensor network in a manner shown in Figure 9.5. In structured replication, the rectangular or square boundary encompassing the sensor network is split into 2^d equally sized sub-regions, where d is the depth of the replication. Consider an event whose home node is x; x is called the root node for the event. Clearly x is located in one of the sub-regions defined by the spatial decomposition. In each of the remaining sub-regions, one can then compute the location of a *mirror* of x; this mirror has the same coordinates in its sub-region as x has in its own.

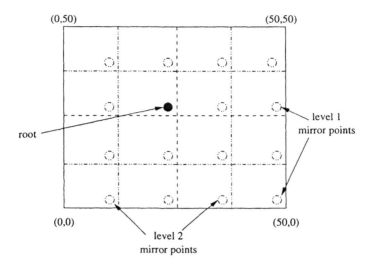

Figure 9.5. Structured Replication.

This spatial decomposition can be used to define a hierarchy of mirrors, as shown in Figure 9.5. The level-1 mirrors of the root are those that would have been selected if d had been 1; these level-1 mirrors are children of the root in this hierarchy. A similar recursive definition can be applied to levels greater than 1. Notice that different events will have different hierarchies since the spatial decomposition is defined for a given event name.

A node that generates an event would, instead of storing the event at the root (or home) node, store it in the nearest mirror. This mirror is computed using simple geometric operations, knowing d and the boundary of the sensor network. Thus, the root node is no longer a hotspot since events are distributed across the network. However, queries for a particular event now have to be directed to all the mirrors. The mirror hierarchy is used for this in a straightforward way: the query is sent directly to the root, which then forwards the query to its children, and so on. Replies make their way up the hierarchy in the reverse direction, and may get aggregated along the way.

Contrary to what the name might suggest, structured replication does not replicate data at multiple nodes; rather, in some sense, the home node for an event is now replicated in several sub-regions to alleviate hotspots. Structured replication allows an application designer to trade off between storage cost and query cost, by adjusting d. Looked at this way, it spans the solution space between GHT ($d = 0$) and local storage (large d). Ghose et al. [1] describe an extension to GHTs which improves resilience by allowing events to be replicated across a few mirrors.

9.3.6 GENERALIZING FROM GHTS

GHTs represent one instance of data-centric storage. The design of GHTs relies upon two key ideas: *hashing* an event name to a geographic location, and *spatial decomposition*. In GHTs, geographic routing is used to map the hashed location to nearby nodes, and spatial decomposition is used to replicate the home node for scalability.

Viewed another way, GHTs can be described as a distributed data structure (a hash table) which supports a well-defined set of queries (enumeration and aggregate queries on events matching a given name). In the rest of this chapter, we will consider other distributed data structures on sensor networks that support the following kinds of queries:

- Spatio-temporal aggregates on observations.
- Range queries on a single key.
- Multi-dimensional range queries.

We show that these data structures can be designed using the same principles (event hashing and spatial decomposition) as GHTs.

9.4 HIERARCHICAL SEARCH STRUCTURES

In Section 9.2, we used the term *feature* to describe a spatial or statistical pattern in the data generated by nodes. Consider a system that allows users to efficiently *search* for such patterns by, for example, asking a sequence of questions such as "What was the average temperature in area X an hour ago?" and then "What was the average temperature in sub-area A of area X in the last 10 minutes of that hour?". Such a system can be useful in sensor networks where users do not *a priori* know what to look for, but can use low-overhead queries like these to drill-down to phenomena of interest, such as regions where the temperature is/was unusually high. An important leverage point for the design of such systems is that the answers to such questions can be *approximate*.

The DIMENSIONS system [7, 6] is a data-centric storage system designed to handle such queries. It builds upon prior work that describes approximate querying of large datasets using *wavelet coefficients* of the data [4]. While an extended treatment of wavelets is beyond the scope of this chapter, we briefly introduce wavelet encoding of data sets to give the reader an intuitive understanding of what is possible with the DIMENSIONS system. Our description follows the example in [4]. Consider a vector $V = [5, 6, 4, 4]$. The simplest wavelet transform of this vector takes elements pairwise and computes averages recursively. The table below shows this process, where each level of averaging constitutes a different *resolution*. Clearly, each step of the averaging loses some information; the lost information is captured by the *detail coefficients* shown in this table. Each detail coefficient is the difference between the computed average and the second of the pair of numbers from which that number was computed. The overall average, together with an ordered (starting from the lowest resolution) list of the detail coefficients, constitutes the *wavelet coefficients* of V.

Resolution	Averages	Detail Coefficients
2	[5, 6, 4, 4]	–
1	[5.5, 4]	[−0.5, 0]
0	[4.75]	[0.75]

There are several important properties of this coding technique. First, we can reconstruct the original data given all the wavelet coefficients. As a corollary, we can also efficiently compute averages of arbitrary subsequences of the original data vector V by using the coefficients at the appropriate levels of resolution. Second, we can compute wavelet coefficients at level i merely from the wavelet coefficients at level $i + 1$. Finally, if at some scale (*i.e.* at higher resolutions) the detail coefficients are small, they can be neglected with some loss of accuracy in computed averages. In this sense, wavelet coding is *multi-resolution*.

DIMENSIONS leverages these properties, together with the fact that this kind of wavelet coding generalizes to higher dimensions, to support spatio-temporal queries. Consider a sensor network where each sensor node periodically generates temperature samples. Viewed across the entire network, the collection of temperature samples can be represented by a 2-dimensional array of temperature values indexed by location and time. If all these values could be collected in a central location, then it is clear that one could compute wavelet coefficients on the data to efficiently answer spatio-temporal queries of the form described above.

However, the design of the DIMENSIONS system makes the observation that these wavelet coefficients can be computed and stored far more efficiently in a *distributed* fashion. This capability rests on the two properties: resolution in the spatial dimension corresponds to differently sized regions of the sensor field, and that lower resolution wavelet coefficients can be computed from higher-resolution coefficients. DIMENSIONS constructs a hierarchy using data-centric storage concepts, and stores successively lower resolution coefficients in the higher levels of the hierarchy. Given a number of levels of resolution d, the system effectively tesselates the geographic region occupied by the sensors into 2^d tiles. To construct the hierarchy (Figure 9.1), the system first hashes a unique name corresponding to this dataset to a sensor network location. The home node corresponding to this location is the root of the hierarchy, called the *apex*. At the next level of the hierarchy are clusterheads, one in each quadrant of the sensor network's geographic region. For a given quadrant, the clusterhead is defined as follows: take the centroid of the quadrant, and compute the location at which the line between the centroid and the apex intersects the quadrant. The home node corresponding to this location is the clusterhead for the quadrant. Clusterheads at each lower-level in the hierarchy are computed recursively. Note that given the name of the dataset and the boundary of the network, any node can compute the apex and the clusterheads entirely locally.

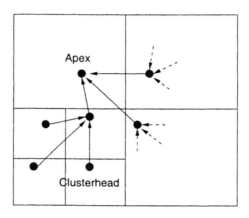

Figure 9.6. The DIMENSIONS hierarchy.

On this hierarchy, wavelets corresponding to the dataset are propagated upward. The clusterhead of each finest-grain tile stores the highest resolution wavelet coefficients, based on readings gathered from all sensors within that tile. The clusterhead of the next larger tile obtains the wavelet coefficients from its children and computes the next lower resolution wavelet coefficients, and so on up the hierarchy. Computing the spatio-temporal averages is then a matter of selectively traversing the hierarchy.

There are many interesting practical challenges in designing this system. One such challenge is in determining the tradeoff between accuracy and communication or storage overhead. Specifically, a clusterhead can choose to omit some wavelet coefficients from those it transmits up the hierarchy, at the expense of accuracy. Another is in designing load balancing schemes such that the root of the hierarchy is not a bottleneck. Finally, when local storage is limited, we need *aging* strategies that gracefully degrade query quality for older data, while providing high query quality for more recent data. Some of these issues are pertinent to DIMENSIONS, but others such as aging are applicable to all data-centric storage systems.

9.5 DISTRIBUTED INDICES ON A SINGLE KEY

Besides spatio-temporal aggregates and exact match queries, sensor network applications will need *range* queries of the form: "List all events for which the temperature values were between 50° and 60°". Equally useful are *geographically constrained* range queries: "List all events that occured in region A of the sensor network for which the temperature values were between 50° and 60°". This section describes the design of a distributed data structure that enables such range queries on a specified attribute of an event name.

There are several simple implementations of such queries. Assume that the event attribute for which we are constructing this data structure is integer-valued and bounded; this is true of many sensor readings in practice. We will use the temperature attribute, ranging between 0 and 100, as a running example. A simple GHT-based implementation of range queries would hash the event by the *value* of the temperature attribute. A range query for, say, all events with temperature between x_1 and x_2 would then translate to a GHT get() operation for each integer value in that range. It is possible to be slightly more efficient by using structured replication—all temperature events are stored in local mirrors, the range query is propagated down the hierarchy, and the range predicate is applied at each mirror to select the matching events. Using this latter approach, it is also trivial to implement geographically constrained range queries.

The DIFS [8] system takes a slightly different approach that increases the efficiency of range queries. It has two salient features. First, it builds a geographically-aligned multi-rooted hierarchy designed to avoid the root as a bottleneck in a simple tree-based hierarchy. As we show below, the construction of this hierarchy illustrates a novel application of hashing and spatial decomposition (the two key mechanisms in GHT design). Second, it efficiently propagates information summaries up the hierarchy that can be used to prune traversal at higher levels of the hierarchy.

DIFS is best described with reference to Figure 9.7. Assume that the geographic region defined by the sensor field is tesselated into 2^d sub-regions. As with DIMENSIONS, this tesselation implicitly defines a hierarchy of sub-regions. Now, suppose that an event with a temperature value 26 is generated at some node within the network. That node stores the event at the *local leaf* node of the DIFS hierarchy: this local leaf is obtained by hashing a unique name for the index (say "temperature") together with a textual representation of the *range* of values ("0:100" in our example; the reason for this will become clear). The event is stored at this local leaf node; thus, each local leaf is responsible for storing all events detected in its sub-region.

Each local leaf has b parents, where b is a tunable system parameter. Each leaf node computes a histogram of the event values it has; it then splits up the

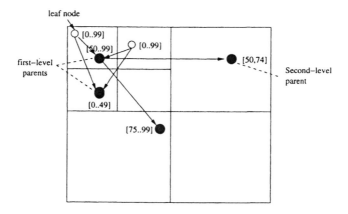

Figure 9.7. The DIFS multi-rooted hierarchy.

histogram into b equal parts, and passes these parts to its b parents. Thus, in Figure 9.7 where b is 2, the leaf node would pass up the histogram for temperature values between 0 and 50 to one parent, and the histogram for values 51 to 100 to another parent. (These histograms help prune the range query, as discussed below). How does the leaf node locate its parents? Again, through hashing the name of the index ("temperature") together with the appropriate range ("0:50" for the first parent and "51:100" for the second). This hash function is however, computed on the geographic region *one level up in the spatial hierarchy*. In this way, each parent node is responsible for $\frac{1}{b}$ of the value range, but each parent node serves 4 times the geographic area covered by the child. Higher-levels of the hierarchy are built up recursively. The top-level of this hierarchy can contain many roots, each of which is responsible for a small part of the value range, but which covers the entire geographic region. While conceptually similar to the DIMENSIONS hierarchy, DIFS uses the hash function in different ways.

Given a range query for events within a range say $[50, 60]$, the originating node first picks the set of root nodes whose individual ranges cover the query range, and starts traversal. At each node, the histograms obtained from the children can be used to determine if matching events exist, thereby pruning the descent. For geographically constrained range queries (*e.g.* all events within the range $[50, 60]$ and in region A), the querying node can pick intermediate nodes in the hierarchy (those that cover part of the query range and the querying region A) and initiate traversal from there.

In summary, DIFS creatively uses geographic hashing and spatial decomposition to answer range queries in a load-balanced manner.

9.6 DISTRIBUTED MULTI-DIMENSIONAL INDICES

While we have presented DIFS as an index over a single attribute, it is more appropriate to think of DIFS as a data structure that allows range queries over two keys or attributes where one of the attributes is geographical location. In this section we consider the design of an index that permits range queries over multi-dimensional data in general, where the dimensions are either not constrained to be of a particular type, or there is no limit to the number of dimensions. As an example of the need for such queries, consider scientists analyzing the growth of marine micro-organisms. They might be interested in events that occurred within certain temperature and light conditions because that helps them understand the correlations between a phenomenon and ambient conditions: "List all events that have temperatures between 50°F and 60°F, and light levels between 10 and 20".

Such range queries can be used in two distinct ways. They can help users efficiently drill-down their search for events of interest. The query described above illustrates this, where the scientist is presumably interested in discovering, and perhaps mapping, the combined effect of temperature and light on the growth of marine micro-organisms. More importantly, they can be used by *application software* running within a sensor network for *correlating* events and triggering actions. For example, if in a habitat monitoring application, a bird alighting on its nest is indicated by a certain range of thermopile sensor readings, and a certain range of microphone readings, a multi-dimensional range query on those attributes enables higher confidence detection of the arrival of a flock of birds, and can trigger a system of cameras.

In traditional database systems, such range queries are supported using pre-computed indices. Indices trade-off some initial pre-computation cost to achieve a significantly more efficient querying capability. We discuss the design of a distributed data structure called DIM [13] for supporting multi-dimensional queries in sensor networks. DIMs are inspired by classical database indices, and are essentially embeddings of such indices within the sensor network.

The key to resolving range queries efficiently is *data locality*: *i.e.* events with comparable attribute values are stored nearby. The basic insight underlying DIM is that data locality can be obtained by a *locality-preserving geographic hash* function. Our geographic hash function finds a locality-preserving mapping from the multi-dimensional space (described by the set of attributes) to a 2-d geographic space; this mapping is inspired by *k*-d trees [3] and is described later. Moreover, each node in the network self-organizes to claim part of the attribute space for itself (we say that each node owns a *zone*), so events falling into that space are routed to and stored at that node. Intuitively, a zone is a sub-division of the geographic extent of a sensor field.

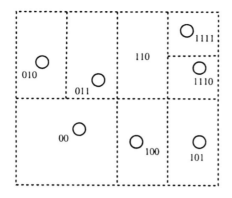
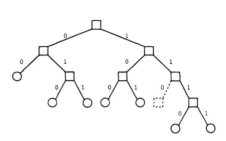

Figure 9.8. The Zone Boundaries.   *Figure 9.9.* The Corresponding Zone Tree.

A zone is defined by the following constructive procedure. Consider a rectangle R on the x-y plane. Intuitively, R is the bounding rectangle that contains all sensors withing the network. We call a sub-rectangle Z of R a *zone*, if Z is obtained by dividing R k times, $k \geq 0$, using a procedure that satisfies the following property:

> After the i-th division, $0 \leq i \leq k$, R is partitioned into 2^i equal sized rectangles. If i is an odd number, the i-th division is parallel to the y-axis; otherwise the division is parallel to the x-axis.

That is, the bounding rectangle R is first sub-divided into two zones at level 0 by a vertical line that splits R into two equal pieces, each of these sub-zones can be split into two zones at level 1 by a horizontal line, and so on. We call the non-negative integer k the level of zone Z, i.e. $level(Z) = k$. As a result of this division process, each can assign each zone a bit string, called its *zone code* that reflects its position in the splitting process. The splitting process can itself be represented as a binary tree, called the *zone tree*.

In DIM, each network node is mapped to a unique zone. If the sensor network were deployed in a grid-like (*i.e.* very regular) fashion, then it is easy to see that there exists a k such that each node maps into a distinct level-k zone. In general, however, the node placements within a sensor field are likely to be less regular than the grid. For a given k, some zones may be empty and other zones might have more than one node situated within them. In DIM, nodes can have differently-sized zones (as Figure 9.8 illustrates), and each node automatically discovers its zone codes. The zone tree corresponding to Figure 9.8 is shown in Figure 9.9.

To complete the picture, we need to describe how events are hashed to zones. Consider a DIM that aims to support m distinct attributes. Let us denote these attributes $A_1 \ldots A_m$. For ease of exposition, assume that the depth of every zone in the network is k, that m is a sub-multiple of k, and that this value of k

is known to every node. In the design of DIM, however, these assumptions can be relaxed. Furthermore, assume that all attribute values have been normalized to be between 0 and 1. The DIM hashing scheme assigns a k bit zone code to an event as follows. For i between 1 and m, if $A_i < 0.5$, the i-th bit of the zone code is assigned 0, else 1. For i between $m + 1$ and $2m$, if $A_{i-m} < 0.25$ or $A_{i-m} \in [0.5, 0.75)$, the i-th bit of the zone is assigned 0, else 1. This procedure repeats until all k bits have been assigned. As an example, consider event $E = \langle 0.3, 0.8 \rangle$. For this event, the 5-bit zone code is $code(Z_E) = 01110$.

At least conceptually, the rest of the DIM system is relatively easy to describe. To insert an event E, DIM computes the zone code of the event. GPSR is then used to route the event to the zone whose zone code is the longest-matching prefix of the event's zone code. The basic idea is straightforward, since the event's zone code can be mapped to a geographic location, and the target zone will be somewhere "near" that location. Finally, notice that a two-dimensional range query for all events in $[0.3 - 0.5, 0.4 - 0.8]$ intersects with a set of zones and the query can be resolved by routing the request to the nodes that own those zones. The exact technique for doing this also leverages GPSR, together with some optimizations that *split* the query carefully as it is routed within the network.

9.7 CONCLUSIONS

In this chapter, we have seen how some elegant distributed data structures can be defined using two basic principles: data hashing by name, and spatial decomposition. Because these systems share underlying primitives and other systems concerns (robustness to node dynamics, finite node storage *etc.*), it is reasonable to believe that they can all be collectively implemented in a comprehensive sensor network storage system (a *senstore*, as it were) that will enable application designers to pick and tailor data structures according to their needs. The design of a senstore is the subject of ongoing research.

ACKNOWLEDGEMENTS

This chapter describes the work of several people: Deborah Estrin, Deepak Ganesan, Ben Greenstein, John Heidemann, Wei Hong, Brad Karp, Young Kim, Xin Li, Sylvia Ratnasamy, and Scott Shenker.

REFERENCES

[1] Abhishek Ghose, Jens Grossklags, John Chuang (2003). Resilient data-centric storage in wireless ad-hoc sensor networks. In *Mobile Data Management - MDM 2003*, pages 45–62.

[2] Adjie-Winoto, W., Schwartz, E., Balakrishnan, H., and Lilley, J. (1999). The Design and Implementation of an Intentional Naming System. In *Proceedings of the Symposium on Operating Systems Principles*, Kiawah Island, SC.

[3] Bentley, J. L. (1975). Multidimensional Binary Search Trees Used for Associative Searching. *Communications of the ACM*, 18(9):475–484.

[4] Chakrabarti, Kaushik, Garofalakis, Minos, Rastogi, Rajeev, and Shim, Kyuseok (2001). Approximate query processing using wavelets. *VLDB Journal: Very Large Data Bases*, 10(2–3):199–223.

[5] Elson, Jeremy, Girod, Lewis, and Estrin, Deborah (2002). Fine-grained network time synchronization using reference broadcasts. In *Proceedings of the Fifth USENIX Symposium on Operating Systems Design and Implementation*, Boston, MA, USA. USENIX.

[6] Ganesan, D., Greenstein, B., Perelyubskiy, D., Estrin, D., and Heidemann, J. (2003). An Evaluation of Multi-Resolution Search and Storage in Resource-Constrained Sensor Networks. In *Proceedings of the ACM Sensys*, Los Angeles, CA.

[7] Ganesan, Deepak, Estrin, Deborah, and Heidemann, John (2002). DIMENSIONS: Why do we need a new data handling architecture for sensor networks? In *Proceedings of the ACM Workshop on Hot Topics in Networks*, Princeton, NJ, USA. ACM.

[8] Greenstein, B., Estrin, D., Govindan, R., Ratnasamy, S., and Shenker, S. (2003). DIFS: A Distributed Index for Features In Sensor Networks. In *Proceedings of the IEEE ICC Workshop on Sensor Network Protocols and Applications*, Anchorage, AK.

[9] Heidemann, John, Silva, Fabio, Intanagonwiwat, Chalermek, Govindan, Ramesh, Estrin, Deborah, and Ganesan, Deepak (2001). Building efficient wireless sensor networks with low-level naming. In *Proceedings of the Symposium on Operating Systems Principles*, pages 146–159, Chateau Lake Louise, Banff, Alberta, Canada. ACM.

[10] Intanagonwiwat, Chalermek, Govindan, Ramesh, Estrin, Deborah, Heidemann, John, and Silva, Fabio (2002). Directed diffusion for wireless sensor networking. *ACM/IEEE Transactions on Networking*.

[11] Karp, B. (2000). *Geographic Routing for Wireless Networks*. PhD thesis, Harvard University.

[12] Karp, Brad and Kung, H. T. (2000). GPSR: Greedy perimeter stateless routing for wireless networks. In *Proceedings of the ACM/IEEE International Conference on Mobile Computing and Networking*, pages 243–254, Boston, Mass., USA. ACM.

[13] Li, X., Kim, Y. J., Govindan, R., and Hong, W. (2003). Multi-Dimensional Range Queries in Sensor Networks. In *Proceedings of the ACM Sensys*, Los Angeles, CA.

[14] Madden, Samuel, Franklin, Michael J., Hellerstein, Joseph, and Hong, Wei (2002). TAG: Tiny AGgregate queries in ad-hoc sensor networks. In *Proceedings of the USENIX Symposium on Operating Systems Design and Implementation*, page to appear, Boston, Massachusetts, USA. USENIX.

[15] Ratnasamy, Sylvia, Francis, Paul, Handley, Mark, Karp, Richard, and Shenker, Scott (2001). A scalable content-addressable network. In *Proceedings of the ACM SIGCOMM Conference*, pages 13–25, San Diego, CA, USA. ACM.

[16] Rowstron, A. and Druschel, P. (2001). Pastry: Scalable, Distributed Object Location and Routing for Large-Scale Peer-to-peer Systems. In *Proc. 18th IFIP/ACM Conference on Distributed Systems Platforms*, Heidelberg, Germany.

[17] Savvides, Andreas, Han, Chih-Chien, and Srivastava, Mani (2001). Dynamic fine-grained localization in ad-hoc networks of sensors. In *Proceedings of the ACM/IEEE International Conference on Mobile Computing and Networking*, page to appear, Rome, Italy. ACM.

[18] Shenker, Scott, Ratnasamy, Sylvia, Karp, Brad, Govindan, Ramesh, and Estrin, Deborah (2002). Data-Centric Storage in Sensornets. In *Proc. ACM SIGCOMM Workshop on Hot Topics In Networks*, Princeton, NJ.

[19] Stoica, Ion, Morris, Robert, Karger, David, Kaashoek, M. Frans, and Balakrishnan, Hari (2000). Chord: A scalable peer-to-peer lookup service for internet applications. In *Proceedings of the ACM SIGCOMM Conference*, Stockholm, Sweden. ACM.

[20] Yao, Kung, Hudson, Ralph E., Reed, Chris W., Chen, Daching, and Lorenzelli, Flavio (1998). Blind beamforming on a randomly distributed sensor array system. *IEEE Journal of Selected Areas in Communication*, 16(8):1555–1567.

[21] Ye, Wei, Heidemann, John, and Estrin, Deborah (2003). Medium access control with coordinated, adaptive sleeping for wireless sensor networks. Technical Report ISI-TR-567, USC/Information Sciences Institute.

[22] Yu, Yan, Govindan, Ramesh, and Estrin, Deborah (2001). Geographical and energy aware routing: A recursive data dissemination protocol for wireless sensor networks. Technical Report TR-01-0023, University of California, Los Angeles, Computer Science Department.

[23] Zhao, B. Y., Kubatowicz, J., and Joseph, A. (2001). Tapestry: An Infrastructure for Fault-Tolerant Wide-Area Location and Routing. Technical Report UCB/CSD-01-1141, University of California Berkeley.

Chapter 10

COMPRESSION TECHNIQUES FOR WIRELESS SENSOR NETWORKS

Caimu Tang [1] and Cauligi S. Raghavendra [2]

[1] *Department of Computer Science*
University of Southern California
Los Angeles, CA 90089
caimut@cs.usc.edu

[2] *Departments of EE-Systems and Computer Science*
University of Southern California
Los Angeles, CA 90089
raghu@usc.edu

Abstract In wireless sensor network applications, an event is observed by a group of spatially distributed sensors which collaborate to make decisions. Since there is limited bandwidth in wireless sensor networks, it is important to reduce data bits communicated among sensor nodes to meet the application performance requirements. It also saves node energy since less bits are communicated between nodes. An approach is to compress sensor data before transmissions to reduce energy as some loss is acceptable without affecting the results of applications. Data collected by sensors that are in close proximity exhibit spatial correlation, further, since the samples collected over time are from the same source(s) also show temporal correlation. We shall first review state-of-the-art results on compression and issues in deploying these techniques in wireless sensor networks. We then present a new data compression scheme, called ESPIHT, that exploits this spatio-temporal correlation present in sensor networks to reduce the amount of data bits transmitted in a collaborative signal processing application. The proposed ESPIHT seamlessly embeds a distributed source coding (DSC) scheme with a SPIHT based iterative set partitioning scheme to exploit both spatial and temporal correlation. Instead of being a generic compression scheme, ESPIHT is coupled with application on which it adapts to application fidelity requirement so that application level false alarm rate is kept low. We evaluated the proposed scheme using dataset from an acoustic sensor based ATR application on a network of iPAQs with Wireless LAN connection. Our results show that ESPIHT reduces data rate by a factor of 8 and maintains Signal-to-Noise Ratio (SNR) gain of 20dB or better. The coding/decoding processing is simple and takes on the order of 10 msec on iPAQs. It is superior to known schemes in terms of SNR gain as shown in the experimental study based on field data when the sampling rate is relatively high and the network is dense.

Keywords: Distributed Source Coding (DSC), Discerete Wavelet Transform, Wireless Sensor Network, Spatio-Temporal Compression.

10.1 INTRODUCTION

Inexpensive sensors capable of significant computations and wireless communication are increasingly becoming available [1, 2]. A typical sensor node will have one or more types of sensors to collect data, an embedded processor with limited memory and operating with battery power, and the ability to communicate wirelessly with other sensor nodes and devices. Sensor nodes can collect audio, seismic, pressure, temperature, and other types of data. A variety of modern data sensors are being built including Sensoria [3], Smart Dust Mote, Rockwell Hidra [4], Dust Motes [5], to name a few. A network consisting of sensor arrays can be deployed to collect useful information from the field in a variety of scenarios including military surveillance, building security, in harsh physical environments, for scientific investigations on other planets, etc. [6, 7].

In general, in sensor network applications, sensor nodes collaborate and share their sensed data to achieve robust results. Sensors typically perform signal processing tasks on the collected data and make decisions collaboratively on the events happening in the sensor field. In their collaborations, usually sensors require two types of communications: (1) local communications in the neighborhood and (2) global communications to distant sensor nodes and/or to basestations. Local communications in the neighborhood, such as in a sensor array, is required in many applications. Long distance communications are required when a result from the sensor network need to be communicated to other nodes and/or basestations, for example, chemical contamination detection, and some type of beamforming algorithms [8].

Nodes in wireless sensor networks communicate via radio. Radios in sensor nodes typically operate at low power with short ranges, and usually have limited bandwidth for communication. Typical data rates in modern sensor node radio range from few tens of Kbps to few hundred Kbps. It is therefore, critical to ensure that a sensor network meets its application data communication requirements. It will be useful to compress data communicated among sensor nodes to reduce the communication bandwidth required. Since sensors operate with batteries, an added benefit from data compression is reduction in energy consumption, as the energy for 1-bit communication is at least an order of magnitude higher than one computation operation [9]. Therefore, data compression in sensor networks will be useful in reducing the data rate requirement to meet application performance constraints as well as to reduce the overall energy consumption for an application.

Compression has been widely used in many applications, including storage, communication, multimedia. In previous applications, compression has been used to reduce memory and/or reduce communication delay of some multimedia contents. There are many algorithms to meet various application requirements, i.e. SNR requirement (lossy or lossless). Compression in sensor network provides another benefit. Due to the reduced number of bits to be transmitted, energy can be saved provided that the encoding and decoding computation cost is low. For lossy compression algorithms, some transform is usually performed on the source bits in order to compact the signal energy so that a better SNR can be achieved. There are various kinds of transforms for different applications. Discrete wavelet transform (DWT) is one of these types of transforms. The compression algorithm in general compresses the wavelet coefficients. Good compression ratio and SNR can be achieved via allocating big portion of bit budgets to big coefficients while discarding insignificant coefficients.

An application of sensor network is detecting a target using acoustic signal, e.g. ATR. In this type of application, sensors collaboratively process signal data to detect and classify targets. In one scenario, a network consists of a number of sensor arrays, with each array having a cluster of homogeneous sensors performing the target detection. Sensors in an array are geometrically close to each other in the range of 100 meters while inter-array distances could be much farther (Figure 10.1 shows one sensor array with 7 sensors). One sensor is elected as the head which aggregates data from its members, e.g, node 1 in Fig. 10.1, and the head sensor node also observes the same event. In this scenario, clearly there is both spatial and temporal correlation in sensed data.

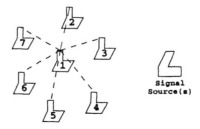

Figure 10.1. A Sensor Array with 7 Sensors

10.1.1 RELATED WORK AND OUR CONTRIBUTIONS

Distributed source coding has been studied in [10, 11, 13]. In [13], a two-stage iterative approach is devised followed by an index-reuse which is aimed to exploit spatial correlation. More recently, a more concise and rigorous treatment on this topic including estimation, source coding and channel coding, called DISCUS, has been presented in [12]. A data compression technique that exploits spatial correlation in wireless sensor networks is presented in [14]. The idea is to send only few coded bits and the recipient will infer the correct data through transmitted syndromes. There, trellis code is employed at the decoder to achieve data compression. DISCUS has been shown effective when sensors are densely deployed [14], however this technique only exploits spatial correlation. Additional gains can be achieved by exploiting both spatial and temporal correlation which is the main focus in this work and it is especially useful when the sampling frequency is high as it is in many target tracking applications. An interesting observation from the experimental study is that spatio-temporal, spatial only and temporal only schemes can apply to different types of sensor network applications and there is no single scheme that fits all applications over sensor networks.

Since sensor nodes are normally deployed in a terrain of close proximity and they are observing same signal source(s), the signal sample correlation across nodes is high. Furthermore, since sensors observe an event or events in a continuous manner, the observed discrete signal samples as a time series also exhibit high temporal correlation. Therefore, it should be possible to achieve better compression ratios for sensor data by exploiting this spatio-temporal correlation present in sensor network applications. Applying wavelet transformation to sensor data and compressing them has been used for storage reduction in a sensor node [15]. In this work, we propose a new data compression scheme, called ESPIHT, for sensor network applications, that exploits both spatial and temporal correlations to reduce the bits to send. In-network processing is deemed to be efficient because of the reduction on data bits to communicate, however, sending raw sensing data locally is required for many important signal processing applications in wireless sensor networks.

Before we go into the ESPIHT scheme, we give a high level description on discrete wavelet transform (DWT) and SPIHT compression scheme of wavelet coefficients and we refer the readers to [17, 23, 16] for more details on these topics. DWT is a technique to decompose signals into different high-pass or low-pass subbands (expressed by a number of wavelet coefficients) which represent the original signal at different temporal and frequency resolutions. These wavelet coefficients preserve the energy of the original signal; however, energy distribution after DWT is more compact in certain subbands. For the case of low-pass signal, the low-pass subbands contain more energy than the high-pass subbands. To approximate the original signal, we can code large coefficients with high precision while leaving out small coefficients so that good compression can be achieved. SPIHT is one of wavelet compression algorithms. SPIHT is based on a so-called set partitioning principle on the coefficients. In each iteration, it partitions the coefficients into two categories, significant or insignificant based on a threshold. It outputs one bit plane from all outstanding significant coefficients at each iteration. By this way, SPIHT can allocate the available bits to most important bits of most important coefficients so that the reconstructed signal has high SNR based on a bit budget.

Our proposed ESPIHT scheme performs: (a) a wavelet transformation on a frame (i.e. a fixed number of samples) to obtain subbands; (b) compression on the coefficients with quantization using Trellis Coded Quantization (TCQ) Technique and source coding (DSC). The wavelet transformation and wavelet packet decomposition [16] on a frame of data result in wavelet coefficients. After the transformation, the signal energy is concentrated into a few subbands which facilitates compression. Note that this subband transformation does not change the spatial correlation presented in the data. As in SPIHT algorithm [17], ESPIHT separates coefficients into two categories: significant and insignificant. To code significant coefficients, we apply distributed source coding that exploits the spatial correlation. It differs from SPIHT in three major ways, (1) it uses constrained depth-first search on binary tree traversal for finding significant coefficients; (2) it uses a dynamic programming approach on marking the visited nodes and (3) TCQ and DSC have been combined into an efficient coding of significant coefficients. ESPIHT outperforms SPIHT in efficiency and SNR gain and requires less memory compared to SPIHT. It uses a queue to store isolated wavelet coefficients so that it can handle non-smooth and stationary signal data better than SPIHT in terms of SNR gain.

10.1.2 DEPLOYMENT ISSUES

In the deployment of distributed source coding scheme to exploit the spatial correlation, there are two general related problems, namely, synchronization of sensor clocks and calibration of sensor readings. We emphasize here that calibration is still needed even in a totally synchronized network. The need for calibration is due to the fact that the wavefront normal direction is not always orthogonal to the sensor baseline as shown in Fig. 10.2.

Synchronization on the other hand could help calibration or correlation tracking. If sensor readings are synchronized upto a sampling interval level, the search space of a calibration algorithm or a tracking algorithm is reduced to the maximal lag between two sensors plus one (lag for difference in samples). In other words, synchronization helps to reduce the processing overhead of calibration or tracking while the overhead is a concern for low power sensor node.

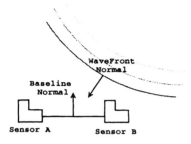

Figure 10.2. Normal Directions of Wavefront and Sensors Baseline

This is certainly worthwhile especially when a lightweight time synchronization algorithm is available [18].

Figure 10.3 shows the cross-plot of two nearby sensor readings. It shows the spatial correlation vs different lags. The best result is obtained when the lag is 3 for the frames shown. However, the calibration lags are changing slowly from frame to frame. This number has to be estimated periodically in order for the scheme to have a high performance. ESPIHT requires both sensor calibration and correlation tracking. However, for some compression scheme aimed on spatial correlation, calibration can be integrated into tracking via a prediction model.

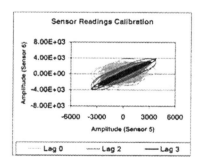

Figure 10.3. Spatial Correlation at 139 Second with Three Different Lags

In what follows, we first present the algorithm ESPIHT which includes distributed source coding and correlation tracking. After the algorithm, we report results of our experimental studies which includes how the signal distortion affects the application level performance. Next we compare the performance of ESPIHT with other known source coding schemes for wireless sensor networks. We then draw some conclusions for this chapter.

10.2 SPATIO-TEMPORAL COMPRESSION SCHEME: ESPIHT

The main goal in ESPIHT is to fully exploit the spatial and temporal correlation present in sensor data to achieve good compression ratio while maintaining acceptable signal-to-noise (SNR) ratio. The scheme is designed to operate on sensor data at a node on a frame by frame basis. A frame consists of n samples of data collected per second by a sensor node. ESPIHT performs two major steps: (a) wavelet transformation on a frame data to obtain subband coefficients; (b) compression of the coefficients with a combined quantization and source coding. Wavelet transformation on a frame of data compacts signal energy into a few subbands. This signal energy concentration facilitates data compression, since subbands with less energy can be approximated without sacrificing much in SNR. As shown in [19], trellis and lattice based code can be used to exploit spatial correlation. ESPIHT employs a source coding that uses trellis coded quantization and trellis code to exploit both spatial and temporal correlation.

In ESPIHT, the coefficients are represented by a binary tree as we have 1-D data and this representation is similar to quadtree used for 2-D image data. This binary tree as shown in Fig. 10.4, where subtrees in gray with insignificant contributions may be absorbed in a single parent node. Nodes at higher levels of this binary tree normally indicate higher power coefficients. In the encoding process, an iterative procedure is used to pick significant coefficients in the subbands. ESPIHT uses the same set partitioning principles as those used by SPIHT and further partitions the coefficients into three types, namely, ISO, LIP, and LSP. ISO coefficients have at least one significant descendant. LIP coefficients are not significant and neither are their descendants. LSP coefficients are significant in the current iteration. For each coefficient, there are two marks, i.e. mark bit and d-mark bit, denoting the significance of the coefficients and the binary tree rooted at the coefficient, respectively. ESPIHT also uses a stack denoted by LIS as a placeholder for subtrees which need to be checked for significance in each iteration.

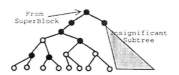

Figure 10.4. "Binary Tree" used in ESPIHT (Gray Part for Insignificance)

ESPIHT has the following major differences compared to SPIHT: (1) explicit partitioning of the coefficients into ISO, LIP, and LSP coefficients; (2) introduction of two marks for pre-traversal; (3) traversal changed from breadth-first like search to depth-first like search; (4) removal of the bit extraction operations in the refinement pass; (5) dynamic programming approach for the d-mark evaluation; (6) applying a scalar quantizer on the coefficients before enqueuing the LSP coefficients; and (7) seamless integration with distributed source coding (to be explained later). This scalar quantizer is coupled with TCQ, it generates partition indices which are the input to TCQ. Since the fluctuation of the signal over time is quite large, (6) also helps to maintain a steady SNR performance over frames.

With this partition, the bitstream consists of three chunks: ISO bits, LIP bits and LSP bits. ISO bits consist of mark bits and sign bits of ISO coefficients. LIP bits consists of mark bit, d-mark bits and sign bits. LSP bits consist of unsigned coefficient magnitude bits. ISO bits and LSP bits present no dependency, and therefore, as in Gaussian source, any coding on these bits give little compression payoff. LIP bits present dependency due to the similarity of binary trees (binary tree in 1-D case is the counterpart of quadtree in 2-D case). We use adaptive Huffman coding on these bits. It uses only one round of traversal on the path information bits of LIP bits, where path information bits are used to denote the traversal path from LIP as root from previous iteration to the LIP in this iteration.

In ESPIHT, DSC is combined with Trellis Coded Quantization (TCQ) for coding LSP coefficients. TCQ has superior performance on Gaussian source over other scalar quantizers. However, regular TCQ requires traversals of a trellis diagram which could be costly. Instead, we use small constraint length trellis diagram with table look-up rather than using conventional add-compare-select (ACS) steps. Further, only a single ACS step is required once at the initialization step to set up the tables. In order to keep the tables smaller, normalization procedure is employed during the trellis traversal [20]. Unlike in DISCUS [12] where trellis code is employed at the decoder, ESPIHT uses it at the encoder side. The coding cost is almost negligible since no ACS arithmetic is performed after initialization.

To decode ESPIHT encoded data at the receiver node, a reference frame is required. Since ESPIHT enqueues LSP coefficients based on the frame other than the reference frame, to refer the same LSP queue items of the reference frame could be devastating if one item in the reference is mismatched with that in the encoding frame. To overcome this, we use the corresponding absolute position to find the reference quantized sample while decoding a sample since decoder keeps track of position of the next decoding coefficient. In this way, decoding is actually simpler, and only one table look-up operation and convolutional coding are required for DSC decoding provided that the reference frame already is quantized the same way as that for the encoder.

The pseudocode of the ESPIHT scheme is shown below. We use the following notations in the pseudocode,

$$\mathcal{O}(E) = \{E' \mid E' \to E\}$$
$$\mathcal{D}(E) = \{E' \mid E' \Rightarrow E\}$$
$$\mathcal{U}(E) = \mathcal{D}(E) \cup \{E\}$$

where $E' \to E$ means E' is a child of E and $E' \Rightarrow E$ means E' is a descendant of E. $S_n(E)$ is the significance test function and y_m for $m = 2^R \cdots 1$ is the m-th scalar quantization level.

Pseudocode of ESPIHT
$\emptyset$ *denotes empty set. "$\leftarrow$" means to evaluate on the right-value and assign to left value, "$\Leftarrow$" means not to evaluate on the right-value and instead directly fetch the computed value and assign it to left-value. Also in the mark assignment statements, we can fetch either both of them or one of them based on the requirements. $\mathcal{M}^d(E)$ means the mark of the descendants of node E, i.e. at least one of the descendants is significant in the current iteration.*
(1) **Initialization:** $n = R$, LSP $= \emptyset$
  LIS $= \emptyset$, ISO $= \emptyset$
  push all points in the superblock to LIP
(2) **Sorting Pass:**
(2-a) dequeue E from ISO
$\mathcal{M} \leftarrow S_n(E)$
if E is not root then
  output $\mathcal{M}$
if $\mathcal{M} = 1$ then
  quantize c_E and enqueue $Q(c_E)$ to LSP
else
  enqueue E to ISO
repeat (2-a) until ISO is empty or old queue tail is processed

(2-b) pop E from LIP
$(\mathcal{M}, \mathcal{M}^d) \leftarrow S_n(\mathcal{U}(E))$
start path compression
if E not root then [1]
  output $\mathcal{M}$
  if $\mathcal{M} = 1$ then
    quantize c_E and enqueue $Q(c_E)$ to LSP
output $\mathcal{M}^d$

---

[1] The output of the encoding bits are first written to a buffer and the buffer is emptied when it is full, the same method is used for decoding when reading bits from bitstream, this reduces the I/O access.

if $\mathcal{M}^d = 1$ then
  push E to LIS
  processLIS
  if $\mathcal{M} = 0$ and E not root then
    enqueue E to ISO
else
  if E is not root then
    enqueue E to LIP
repeat (2-b) until LIP is empty

(2-c) end path compression
  encode path strings

(3) **Refinement Pass:** select TCQ bitrate r ($r < R$)
  quantizes k index residues using TCQ   /* *k: LSP length* */
  output k TCQ trellis bits
  Run DSC on k $(r - 1)$-bit symbols
  output k DSC trellis bits
(4) **Quantization-Step Update:** decrement n by 1 and repeat Step 2

In the ESPIHT pseudocode with additional routines shown in the Appendix, two types of data structures are used: queue and stack. They are efficient than using linked list structure especially when resources are limited. One salient advantage is the reduction of memory requirement. Compared to SPIHT, the memory used by ESPIHT is reduced by 80% in the same worst case scenarios for both schemes.

In the ESPIHT pseudocode, we only present the encoding part. For decoding, the only differences are (1) all bit output routines are replaced by reading bits from the bitstream, (2) DSC encoding is replaced by DSC decoding in the Refinement Pass, (3) scalar quantization is replaced by corresponding de-quantization and (4) path compression is replaced by path decompression which is part of adaptive Huffman coding (we use the FGK algorithm in the code implementation).

In ESPIHT, the compression scheme is lossy, this is due to the truncation error on forward and inverse wavelet transformation, TCQ quantization, buit-in scalar quantization, and distributed source coding on significant coefficients. However, it can be simply modified to support lossless compression. The approach can be taken by (1) using a wavelet lifting scheme so that there is no loss in transformation and (2) running ESPIHT on the coefficients (integers) without quantization on coefficients of LSP, and the iteration goes from $n = \log(M)$, where M is the largest coefficient, to 2. Iteration for coefficients of 1's can be inferred from the list of LSP at the decoder so the values themselves are not needed to encode into bitstream. After all non zero coefficients are extracted, the rest are zeros. In the DSC part, only the first n iterations, where the coefficients in the corresponding LSP have at least n bits, can use DSC coding. The bit length n is determined so that no error happens in DSC decoding for coefficients larger than or equal to 2^{n-1} (additional steps could be taken using error checksum before the encoding). When DSC is applied, after the partitioning, instead of running through a trellis using VA, the partition subset index is directly encoded into the bitstream to eliminate the distortion introduced by bit flips in VA outputs. Based on our simulation experiments, the compression ratio is at least 3x.

However, we can achieve a compression factor of more than 3x, while SNR gain is above 60dB for lossy case. For applications requiring high fidelity, but can tolerate some loss on the reconstructed signal, ESPIHT scheme would be useful as well. Further, ESPIHT is flexible to adapt to tradeoffs in SNR gain requirements and compression ratio.

10.2.1 DISTRIBUTED SOURCE CODING IN ESPIHT

We have omitted the detailed steps of DSC in the pseudocode for clarity. It involves a coset partitioning and a forward and backward traversals of a trellis with constraint length of 3. DSC design in ESPIHT differs from previous methods in DSC e.g. [12] which are based in time domain signal samples, for details, refer to [12, 21]. At first glance, DSC in ESPIHT seems to be based on the wavelet coefficients, however, it actually is based on these residual values with the most significant bits and sign bits removed. The residuals are not zero mean, since all the residuals are positive and sign bits are output before the encoding of LSP coefficients. The scalar quantizer used in each iteration step almost have no effect on the DSC design, and DSC parameters are invariant on different iterations. This property simplifies the design dramatically. In wavelet based compression scheme, SNR decreases gravely if all bits are allocated to a few large coefficients (they are going to be significant in first few iterations). In each iteration of ESPIHT, in encoding of LSPs' coefficients, fixed number of bits are allocated to them, in our experiment, we use only two bits per residual. In other application, however, this parameter can be adjusted based on the signal statistics and application SNR requirement and this can be easily done via a training dataset. The pseudocodes of DSC are shown below.

DSC_Encode(n, LSP)
n is the number of LSP coefficients in current iteration, LSP is an array of LSP coefficients to encode. Trellis termination is done explicitly by use the length parameter n which is also available to decoder.
for step k of trellis traversal do

normalizing u_i^k, for $i = 1, 2, 3, 4$
$i \leftarrow$ TCQ partition index
$u_i^k \leftarrow$ normalized cumulative metrics, where $i = 1, 2, 3, 4$ [2]
$u_i^k = T_t(u_i^k, i)$
$A_k = u_1^k + (u_2^k << 1) + (u_3^k << 2) + (u_4^k << 3)$
$P \leftarrow \min\{u_1^n, u_2^n, u_3^n, u_4^n\}$
$TBA(P) \rightarrow$ survive path   // trellis bit is output inside TBA.

DSC_Decode(n)
Use of convolutional code to recover the TCQ coset indices, n is known at DSC decoding stage and it is computed at the Sorting Pass.
for each step of trellis traversal
 $b \leftarrow$ current trellis bit
 $p \leftarrow$ current node index on trellis
 $d \leftarrow T_c(p, b)$
 $x \leftarrow$ absolute reference index   // computed by sorting pass
 $r \leftarrow$ reference TCQ partition index
 $i \leftarrow T_d(d, r)$ // decoded TCQ index

The trellis diagram used is a 4-state Ungerböck trellis diagram [19]. In the DSC part of ESPIHT, since we use only trellis of small constraint length, we are able to improve its computation efficiency significantly. The techniques are precomputation and building tables for run-time look-up. It eliminates almost all arithmetical operations in the trellis traversals. The forward traversal of trellis uses only one look-up operation per coefficient, e.g. T_t trellis table, and Trace-Back Algorithm (TBA) outputs the trellis bits on tracing back with one comparison operation per trellis step. For normalization, it is not needed for each step, it is done every other 20 steps, it requires six comparison operations in worst case and four subtract operations. The purpose of normalization is to help to keep the trellis table small. In the decoding side, two table look-ups are needed, e.g. T_c for convolutional code and T_d for TCQ partition index recovery (for detail on the implementation using Viterbi algorithm, refer to [21]).

10.2.2   SENSOR READING CALIBRATION AND CORRELATION TRACKING IN ESPIHT

Correlation tracking is critical for the performance of a distributed source coding scheme if the sensor readings are not calibrated. In general, correlation tracking also depends on the sensor clock synchronization. If the sensors'

---

[2] In the implementation code, only four parameters for cumulative metrics are used, namely, u_1, u_2, u_3, u_4, and here a superscript on them is for notational convenience.

clocks are not synchronized, it is more costly and difficult to track the correlations. As shown in the algorithm, the distributed source coding is applied to the residuals of the wavelet coefficients; therefore, it is generally different from tracking in time domain. In the multiresolution domain, the wavelet coefficients have little temporal correlation, but the spatial correlation as shown in the time domain is still present in multiresolution domain. To measure the performance of a correlation tracking algorithm, these factors are considered: (1) processing overhead of the tracking algorithm; (2) communication cost; and (3) tracking accuracy.

The correlation tracking in ESPIHT is performed in two steps: (1) time domain least square estimation for sensor calibration; (2) adaptive partitioning for residual correlation tracking. We shall explain step (2) later. For step (1) to be effective, the readings from different sensors have to be synchronized with the skewness upto one sampling interval. The number of samples needed to calibrate two sensor readings changes slowly; therefore, the estimation in step (1) is conducted periodically. Figure 10.5 shows the skewness of two sensors in a course of 200 seconds with sampling rate at 1 KHz. In order to perform the least square estimation, we require a number of samples compressed only temporally (with respect to past and future samples) to be sent to the receiver and this number is much smaller than the frame length. From Fig. 10.5, to run the estimation, we need about 10 samples in every 5 seconds; this overhead is negligible when amortized to many frames. Note that this parameter is different for different pair of sensors or for the same pair of sensors with different sampling rate. The algorithm is simply to compute the mean squared error with different lags from both sensors and pick the smallest and the corresponding lag is the number of samples to use for calibration.

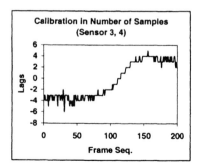

Figure 10.5. Change of Calibration Lags

Before we go into tracking part of the step (2) above. Some comments on the residuals may help. What makes the tracking in ESPIHT different from the time-domain tracking is that we are working on the residuals of wavelet coefficients instead of the samples directly. As will be shown in the experimental evaluation section, the wavelet coefficients between two sensors are also correlated; roughly speaking, the spatial correlation in sensor networks is invariant to a wavelet transform. When the wavelet coefficients are correlated, so are the residuals. This is because the residuals are only a constant offset the corresponding coefficients in each round of ESPIHT (one round here means one iteration of ESPIHT on a given threshold).

From the encoding point of view, the tracking processing boils down to one parameter, i.e. TCQ partition interval in one round. Note that the partition intervals may be different in different rounds. The tracking task is to select a proper interval for each round (the number of round is fixed in ESPIHT and we use uniform partitioning). We require one frame out of a fixed number of frames to be sent encoded with SPIHT, i.e. excluding the DSC part of ESPIHT. Sensors use this frame to estimate the proper interval for each round and this parameter gets updated on next SPIHT encoded frame. We here do not distinguish the sender from the receiver since the receiver nodes are taking turns in a sensor array; furthermore, each node maintains the current partitioning intervals to use for each threshold level. One noticeable feature is that the partitioning interval does not change neither among sensor pairs inside one sensor array nor at different sampling rates, and it largely depends on the distance between the array and the signal source, and the propagation loss of the signal. This is different from what we observed on the calibration case. With this feature, the partitioning intervals for different rounds can be updated periodically at the array level and they are used by all communicating pairs inside the array.

The processing overhead on calibration and correlation tracking is only the least square estimation on a short length of samples. The overall communication cost on calibration and tracking is less than 1% for an array of 7 sensors. The accuracy on calibration is perfect while that on tracking depends on update frequency.

10.3 EXPERIMENTAL EVALUATION

In this section, we first analyze the datasets to show that the correlations are present and the calibration and correlation tracking is needed for a source scheme to function properly in a wireless sensor network. We then shall study what is the impact of SNR loss from application standpoint with an example application of ATR. Finally, we present the comparison results on efficiency and performance of proposed ESPIHT with other known schemes.

We base our experimental study on field datasets gathered by Army Research Laboratory (ARL) using acoustic sensors with objects moving in the field [22]. From the ground truth, the sensor is about 10 feet apart and the distance from the target to sensor array from 50 meters to 1000 meters. The field data consists of samples from seven acoustic sensors at a sampling rate of 1 KHz. We call the "1024 samples" collected by a sensor in one second a *frame*. These datasets are collected with seven sensors (called an sensor array) distributed in a close terrain. The diameter of a typical array is about 20 feets.

Figure 10.6 shows signal amplitude fluctuation over time and Figure 10.7 shows the spatial correlation of the wavelet coefficients from two different sensors (i.e. sensor 3 and sensor 4) at the 100th second. In Fig. 10.7, the curve in dark color is the difference curve in wavelet coefficients on the two frames which can be considered as the spatial noise, and the grey curve is the coefficient plot of sensor 4. The coefficient curve of sensor 3 at 100th second does not show in the plot for clarity. The SNR formula used in this study is computed as follows:

$$10 \log \frac{\sum_{i=1}^{1024} s_i^2}{\sum_{i=1}^{1024} (s_i - s_i')^2},$$

where s_i is the source signal amplitude and s_i' is the reconstructed signal sample amplitude at time i.

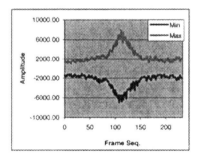

Figure 10.6. Signal Amplitude Fluctuation

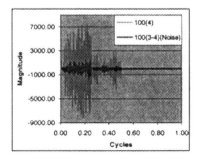

Figure 10.7. Spatial Correlation of Wavelet Coefficients (4 levels)

10.3.1 SIGNAL DISTORTION FROM APPLICATION STANDPOINT

In this subsection, we shall study how the signal distortion affects the signal processing results. Specifically, we have measured the signal distortion and its effects on object tracking. This measurement could help us on selecting proper parameters for these algorithms to yield high performance. In our experiments, we use the benchmark program developed at U.S. Army Research Laboratory and Information Science Institute of University of Southern California. We used three different scenarios: (I) one target moves along an almost straight trajectory; (II) one target moves along a trajectory with a maneuver; (III) two targets move and cross each other. Figures 10.8 - 10.10 show the Direction of Arrival (DoA) using two reconstructed signals with different SNR gains for the three scenarios, respectively. For Scenario (I) and (II), the signal distortion has almost no effect on the DoA estimation as long as the SNR gain is above 18dB. For scenario (III), the two curves has some discrepancies for 35dB and 21.7dB (with bitrate at 2.4bps). (Note that the discrepancies do not necessarily mean a false alarm on the tracking result, since it is only one DoA from one sensor and the result has to be fused with neighbor sensor nodes to determine the actual target movement. Note that the two DOA's in each curve are lumped together, it is beyond the scope of this work to recover the individual DoA from them.)

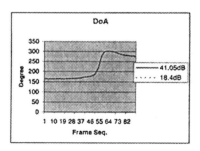

Figure 10.8. SNR Loss on DoA Estimation Effects: 1 target with an almost straight trajectory (Scenario I)

10.3.2 PERFORMANCE OF ESPIHT

In this subsection, we shall report the measured coding efficiency based on implementation on iPAQ with SA-1100 processor and comparison study with other known schemes for wireless sensor networks.

For a sensor to use a coding scheme, the processing overhead is a concern because of the low power nature of these devices. We give experimentally measured times on execution based on both fixed point arithmetic and float point

Compression Techniques for Wireless Sensor Networks 223

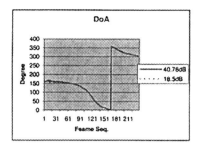

Figure 10.9. SNR Loss on DoA Estimation Effects: 1 target with maneuvering (Scenario II)

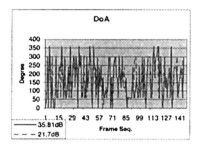

Figure 10.10. SNR Loss on DoA Estimation Effects: 2 target crossing (Scenario III)

arithmetic. We choose Coiflet alike basis [23] which is biorthogonal (regular coiflet bases are orthogonal) for fixed point implementation and orthogonal basis, DAUB-12, for float point implementation. Other basis may also can be used, however, based on our experiment, gain is not significantly different on bases of the same complexity. The filter parameters in the selected biorthogonal basis are all rational numbers so we are able to implement the wavelet transform using only fixed point arithmetic with some manageable information loss. Although lifted wavelets can transfer integer to integer, float point arithmetic is required during the transform. For the case of Coiflet alike basis, the wavelet analysis and synthesis part in our implementation is based on the following algorithm [23],

$$c_{m,n} = \sum_k g_{2n-k} a_{m-1,k} \qquad (10.1)$$

$$a_{m,n} = \sum_k h_{2n-k} a_{m-1,k} \qquad (10.2)$$

where $g_i \neq 0$ for $i = -k_g, \cdots, m_g$ and 0 otherwise, and they are the corresponding high-pass Finite Impulse Response (FIR) filter coefficients, $m_g + k_g$ is the tape length of the filter. $h_i \neq 0$ for $i = -k_h, \cdots, m_h$ and 0 otherwise, and they are the corresponding low-pass FIR filter coefficients, $m_h + k_h$ is the

n	-3	-2	-1	0	1	2	3	4
h	0	$-\frac{1}{20}$	$\frac{1}{4}$	$\frac{3}{5}$	$\frac{1}{4}$	$-\frac{1}{20}$	0	0
$\tilde{h}$	$-\frac{3}{280}$	$-\frac{3}{56}$	$\frac{73}{280}$	$\frac{17}{28}$	$\frac{73}{280}$	$-\frac{3}{56}$	$-\frac{3}{280}$	0
g	0	$-\frac{3}{280}$	$-\frac{3}{56}$	$\frac{73}{280}$	$-\frac{17}{28}$	$\frac{73}{280}$	$-\frac{3}{56}$	$-\frac{3}{280}$
$\tilde{g}$	0	0	$\frac{1}{20}$	$\frac{1}{4}$	$-\frac{3}{5}$	$\frac{1}{4}$	$\frac{1}{20}$	0

Table 10.1. Filter Coefficients Used for the Biorthogonal Basis Implementation.

tape length of the filter. $c_{m,n}$'s are the coarser resolution coefficients of high-pass subband and $a_{m,n}$'s are the coarser resolution coefficients of low-pass subband at level m. The synthesis algorithm is as follows:

$$a_{m-1,l} = \sum_n [\tilde{h}_{2n-l} a_{m,n} + \tilde{g}_{2n-l} c_{m,n}],$$

where the filter coefficients $\tilde{h}_i$ and $\tilde{g}_i$ are different from h_i and g_i in Eq. (10.1 – 10.2), but they are associated by the following equations,

$$\tilde{g}_i = (-1)^i h_{-i+1}$$

$$g_i = (-1)^i \tilde{h}_{-i+1}.$$

For Coiflet-3 basis, the FIR filter coefficients are shown in Table 10.1.

The platform we used in this implementation is iPAQ with LINUX armv4l (kernel version 2.4.0). The codec was developed in LINUX based on Pentium and cross-complied to run on the platform. Table 10.2 shows the ESPIHT encoding, ESPIHT decoding, Discrete Wavelet Transform (DWT) and Inverse Discrete Wavelet Transform (IDWT) times with DAUB-12 basis and Table 10.3 shows the same with Coiflet-3 basis. The first column is the SNR performance in dB and the corresponding bitrate in bits per sample. Note that the time in the tables includes I/O and these are the times required during regular task execution after field initialization. From the data in Table 10.2, for a node without hardware support of floating arithmetic, the processing overhead is tremendous. Figure 10.11 shows the overall encoding times for all available frames using Coiflet-3 based transformation. In the figure, the decoding time is omitted. However, in the comparison study that follows: we give the overhead including the both. One thing to note here is that in ESPIHT, the time for decoding is always shorter than that for encoding.

In the remainder of this section, we shall report results on comparing ES-PIHT with two other source coding schemes, i.e. DISCUS, and SPIHT. Based on our experiment, by the plots of the cross-correlation and autocorrelation of the reconstructed frames, we found DISCUS and SPIHT can effectively exploit

Compression Techniques for Wireless Sensor Networks 225

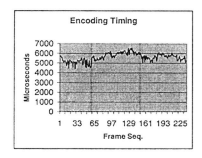

Figure 10.11. Encoding Timing over Frames

Table 10.2. Timing with DAUB-12 basis

Time(μs)	DWT	IDWT	Enc.	Dec.
26.88dB, 3.2bps	248334	215774	2235	1267
21.43dB, 2.8bps	248436	213864	2110	1120
18.51dB, 2.3bps	248435	212587	2014	1020

Table 10.3. Timing with Coiflet-3 basis

Time(μs)	DWT	IDWT	Enc.	Dec.
24.62dB, 3.1bps	2990	2598	2224	1271
20.43dB, 2.7bps	2988	2626	2112	1126
17.35dB, 2.1bps	2956	2583	2010	1021

the spatial correlation and temporal correlation of the sensor data, respectively. We shall compare ESPIHT with temporal only and spatial only schemes and demonstrate when the spatio-temporal scheme is needed and gives better performance. Roughly speaking, DISCUS outperforms the ESPIHT and SPIHT when the sampling rate is low and ESPIHT and SPIHT outperforms DISCUS when the sampling rate is high.

We have implemented the other two schemes on iPAQ LINUX platform. For DISCUS implementation, we used the version of 4-state trellis-based coset formation with Lloyd-max scalar quantization of 8 levels. For ESPIHT and SPIHT, the wavelet transform is 6 levels. The number of iterations is determined at run-time so that they can give a specified bitrate output. We first take the frames from the dataset which has a sampling rate of 1024 Hz. The left plot of Fig. 10.12 shows the SNR gains for the three schemes at one bitrate. In our comparison, we only compare bitrate at 2.4 bits-per-sample (bps) which gives a practical SNR from signal processing application point of view. Since when bitrate is below this point, the received data can cause a high false alarm rate on the DoA estimation for the certain type of object moving. In the SNR performance comparison, for SPIHT, the bitrate is controlled precisely at 2.4 bps while the bitrates for the other two schemes are controlled close to 2.4 bps at each frame. From the figure, ESPIHT outperforms in average both DISCUS and SPIHT by 1dB, 7dB margin, respectively. Next, we downsample the frame to 256 samples per frame to create a dataset with sampling frequency at 256 Hz. The right plot of Fig. 10.12 shows the SNR gains of the three schemes. In this comparison, we have set the bitrate at 3.0 bits-per-sample in order to obtain a sustainable SNR performance for application and the same method as before is used to control the bitrates for the three schemes. From the plot, we can see that DISCUS is comparable to the other two schemes in terms of SNR performance, since its processing overhead is low (see below), it is wise to use DISCUS for low sampling rate.

Figure 10.13 shows the computation overheads associated with each scheme, the times obtained are based on the runs in which all three schemes generate bitrates close to 2.4 bps. The overhead in time in this comparison includes encoding time, decoding time, I/O time and forward and inverse transformation times if apply. The initialization time is excluded in the overhead calculation since it is only activated once. The basis used in the transformation of ESPIHT and SPIHT is Coiflet-3. From the figure, ESPIHT has similar overhead with SPIHT in spite that ESPIHT has an embedded distributed source coding. The overhead of DISCUS is the least among the three schemes largely due to the the lack of transformation. Although ESPIHT addes some extra overhead, we believe it is worthwhile to use ESPIHT because of its large margin in SNR gain.

From the experiment, we can see DISCUS, SPIHT and ESPIHT fits into different situations, namely, network density, sampling rate. When sampling rate is high, SPIHT is favorable, and DISCUS is favorable when sensors are close and sampling rate is low, ESPIHT is favorable when sensors are close deployed and the sampling rate is relatively high. Other issues associated with DISCUS and ESPIHT is the correlation tracking and the need to calibrate sensor readings. These are some extra overhead should be taken into account to select a scheme to use.

10.4 CONCLUSIONS

In this chapter, we have reviewed a number of source coding schemes for wireless sensor networks, and we have proposed a novel data compression scheme, called ESPIHT, to exploit both spatial and temporal correlation of sensor signal data. The experiments on sensor field data from an ATR application show that ESPIHT achieves a factor of 8 compression with SNR of 20dB or higher. It is efficient and flexible to meet different SNR requirements. The

Compression Techniques for Wireless Sensor Networks 227

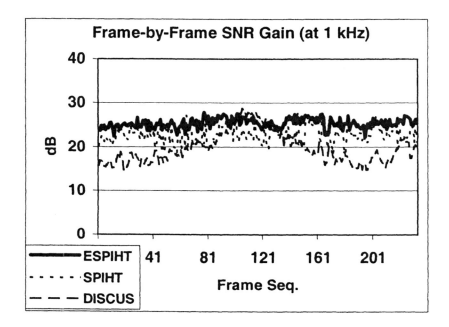

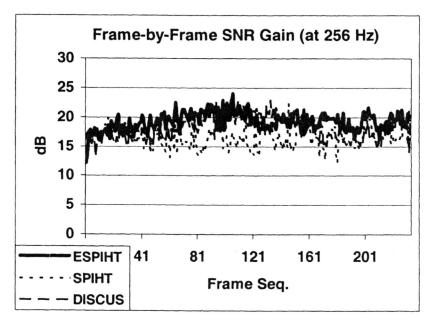

Figure 10.12. SNR Performance Comparison.

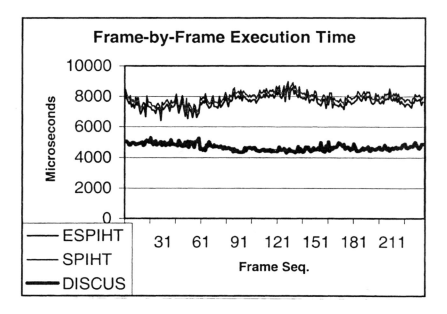

Figure 10.13. Coding Time Comparison.

SNR performance is superior to other known schemes with negligible added processing overhead. The scheme can also be extended for lossless compression. It has been demonstrated to give superior performance to ATR applications, we believe it fits well into more broad wireless sensor network applications.

ACKNOWLEDGMENTS

This research is partially supported by the DARPA under contract F33615-00-C-1633 in the Power Aware Computing and Communications Program (PAC/C) and by the DARPA and Air Force Research Laboratory, Air Force Material Command, USAF, under agreement number F30602-01-2-0549. The views and conclusions contained herein are those of the authors and should not be interpreted as necessarily representing the official policies or endorsements, either expressed or implied, of the Defense Advanced Research Projects Agency (DARPA), the Air Force Research Laboratory, or the U.S. Government.

REFERENCES

[1] D. Estrin, R. Govindan, J. Heidemann, and S. Kumar, "Next century challenges: Scalable coordination in sensor networks," *Proc. of the Fifth Annual International Conference on Mobile Computing and Networks (MOBICOM'99), Seattle, WA, USA*, pp. 263–270, Aug.

1999.

[2] J. Kulik, W. Rabiner, and H. Balakrishnan, "Adaptive protocols for information dissemination in wireless sensor networks," *Proc. of the Fifth Annual International Conference on Mobile Computing and Networks (MOBICOM'99), Seattle, WA, USA*, pp. 263–270, Aug. 1999.

[3] Sensoria, "Sensoria sensor node [online]," *Available: http://www.sensoria.com*, 2002.

[4] Rockwell Scientific Center, "Hidra sensor node [online]," *Available: http://wins.rockwellscientific.com*, 2002.

[5] Inc. Crossbow Technology, "Cots dust motes [online]," *Available: http://www.xbow.com*, 2001.

[6] University of California at Berkeley and etc, "Habitat monitoring on great duck island, maine [online]," *Available: http://www.greatduckisland.net*, 2002.

[7] A. Mainwaring, J. Polastre, R. Szewczyk, D. Culler, and J. Anderson, "Wireless sensor networks for habitat monitoring," *First ACM International Workshop on Wireless Sensor Networks and Applications (WSNA '02)*, Sept. 2002.

[8] J. C. Chen, R. E. Hudson, and K. Yao, "Source localization and beamforming," *IEEE Signal Processing Magazine*, vol. 19, pp. 30–39, Mar. 2002.

[9] E. Shih, S. Cho, N. Ickes, R. Min, A. Sinha, A. Wang, and A. Chandrakasan, "Physical layer driven algorithm and protocol design for energy-efficient wireless sensor networks," *Proc. of the Fifth Annual International Conference on Mobile Computing and Networks (MOBICOM'99), Rome, Italy*, pp. 272–286, July 1999.

[10] D. Slepian and J. K. Wolf, "Noiseless coding of correlated information sources," *IEEE Transactions on Information Theory*, vol. IT-19, pp. 471–480, July 1973.

[11] A. D. Wyner and J. Ziv, "The rate-distortion function for source coding with side information at the decoder," *IEEE Transactions on Information Theory*, vol. IT-22, pp. 1–10, Jan. 1976.

[12] S. S. Pradhan and K. Ramchandran, "Distributed source coding using syndromes (DISCUS): Design and construction," *IEEE Transactions on Information Theory*, vol. 49, pp. 626–643, Mar. 2003.

[13] T. J. Flynn and R. M. Gray, "Encoding of correlated observations," *IEEE Transactions on Information Theory*, vol. IT-33, pp. 773–787, Nov. 1987.

[14] S. S. Pradhan, J. Kusuma, and K. Ramchandran, "Distributed compression in a dense sensor network," *IEEE Signal Processing Magazine*, vol. 19, pp. 51–60, Mar. 2003.

[15] D. Ganesan, D. Estrin, and J. Heidemann, "DIMENSIONS: Why do we need a new data handling architecture for sensor networks?," *First Workshop on Hot Topics in Networks (HotNets-1)*, Oct. 2002.

[16] M. V. Wicherhauser, "INRIA lectures on wavelet packet algorithms [online]," *Available: http://www.math.wustl.edu/~ victor/papers/lwpa.pdf*, Nov. 1991.

[17] A. Said and W. A. Pearlman, "A new, fast, and efficient image codec using set partitioning hierarchical trees," *IEEE Transactions on Circuits and Systems for Video Technology*, vol. 6, pp. 243–250, June 1996.

[18] J. V. Greunen and J. Rabaey, "Lightweight time synchronization for sensor networks," *Second ACM International Workshop on Wireless Sensor Networks and Applications (WSNA '03)*, Oct. 2003.

[19] M. W. Marcellin and T. R. Fischer, "Trellis coded quantization of memoryless and Gauss-Markov sources," *IEEE Transactions on Communications*, vol. 38, pp. 82–92, Jan. 1990.

[20] C. Rader, "Memory management in a viterbi decoder," *IEEE Transactions on Communications*, vol. com-29, pp. 1399–1401, Sept. 1981.

[21] C. Tang, C. S. Raghavendra, and V. K. Prasanna, "An energy efficient adaptive distributed source coding scheme in wireless sensor networks," *Proc. of IEEE International Conference on Communications (ICC'03), Anchorage, AK, USA*, pp. 732– 737, May 2003.

[22] Army Research Laboratory and DARPA, "ATR field datasets [online]," Available: http://pasta.east.isi.edu, 2000-2002.

[23] I. Daubechies, *Ten Lectures on Wavelets*, SIAM, 1992.

APPENDIX: PSEUDOCODE OF ESPIHT ROUTINES

Pseudocode of ProcessLIS
(a) pop E from LIS
for each $E' \in \mathcal{O}(E)$ do
   $(\mathcal{M}, \mathcal{M}^d) \Leftarrow S_n(\mathcal{U}(E'))$
   output $\mathcal{M}$
   if $\mathcal{O}(E') \neq \emptyset$ then
     output $\mathcal{M}^d$
   if $\mathcal{M} = 1$ then
     quantize $c_{E'}$ and enqueue $Q(c_{E'})$ to LSP
   else if $\mathcal{O}(E') = \emptyset$ then
     enqueue E' to ISO and continue for-loop
   if $\mathcal{M} = 0$ then
     enqueue E' to ISO /* E' is an isolate node */
   if $\mathcal{M}^d = 1$ then
     push E' to LIS
   else /* tree is pruned here */
     if $\mathcal{M} = 0$ then
       push E' to LIP
     else
       for each $\tilde{E} \in \mathcal{O}(E')$ do
         if $\mathcal{O}(\tilde{E}) \neq \emptyset$ then
           push $\tilde{E}$ to LIP
         else
           enqueue $\tilde{E}$ to ISO
repeat (a) until LIS is empty

Pseudocode of $S_n(E)$
E is an isolate node or a LIP node (including roots of binary tree).
if E is a node then
   $m \leftarrow 2^n$
   return $\mid c_E \mid > \frac{y_m + y_{m-1}}{2}$
else
   return $\hat{S}_n(E)$

Pseudocode of $\hat{S}_n(E)$
E^d is marked at node E if any node in $\mathcal{D}(E)$ is significant.
$m \leftarrow 2^n$
for next unvisited child E' of E do
    $\hat{S}_n(E')$
if $| c_E | > \frac{y_m + y_{m-1}}{2}$ then
    mark E
if any child of E is marked or d-marked then
    mark E^d
return marks of (E, E^d)

Note in above, level m in $S_n(E)$ is actually computed only once in the main loop and passed as a parameter to the mark evaluation routine, it is the same for the $\hat{S}_n(E)$.

Chapter 11

FUNDAMENTAL LIMITS OF NETWORKED SENSING

The Flow Optimization Framework

Bhaskar Krishnamachari
Department of Electrical Engineering-Systems
University of Southern California
Los Angeles, CA, USA 90089
bkrishna@usc.edu

Fernando Ordóñez
Department of Industrial Systems Engineering
University of Southern California
Los Angeles, CA, USA 90089
fordon@usc.edu

Abstract We describe a useful theoretical approach — the flow optimization framework — that can be used to identify the fundamental performance limits on information routing in energy-limited wireless sensor networks. We discuss the relevant recent literature, and present both linear constant-rate and non-linear adaptive rate models that optimize the tradeoff between the total information extracted (Bits) and the total energy used (Joules) for a given sensor network scenario. We also illustrate the utility of this approach through examples, and indicate possible extensions.

Keywords: Wireless Sensor Networks, Optimization, Network Flows, Fundamental Limits

11.1 INTRODUCTION

Because of the unique characteristics of wireless sensor networks (severe energy constraints, unattended operation, many-to-one flows, data-centric operation), information routing in these systems presents novel design challenges.

Several protocols have been proposed for querying, routing and data gathering in these sensor networks, that have been primarily validated via simulations and limited experimentation. These include cluster-based and chain-based data gathering techniques [1, 2], attribute-based routing [3, 4], indexed storage and retrieval [5], database-style querying [6, 7], and active query techniques [8, 9]. Given the severe resource constraints, application-specificity, and need for robust performance in sensor networks, it is clearly crucial to complement these ongoing routing protocol development efforts by the concurrent development of a strong theoretical understanding.

There are significant challenges inherent in developing such a theory — traditional tools such as queuing theory [10] can be used to analyze throughput and delay issues, but these are of secondary importance in energy-limited sensor networks. Network information theoretic approaches involving exact analysis have traditionally resulted in limited progress [11], but more recently there has been an attempt to provide useful results by focusing on asymptotic behavior. For example, Gupta and Kumar [12] and Xie and Kumar [13] have analyzed the asymptotic capacity of multihop wireless networks; work that has been extended to consider mobility [14] as well as directional antennae [15]. The capacity of wireless sensor networks has been addressed by taking into account spatial correlations in data from nearby nodes [16, 17]. Other theoretical efforts in the area of sensor networks have been focused on understanding the complexity and optimality of data-aggregation [18, 19], and first-order mathematical modeling of specific querying and routing protocols such as ACQUIRE [9] and Directed Diffusion [20].

In this chapter, we describe a useful theoretical approach — the flow optimization framework — that can be used to identify the fundamental performance limits on information routing for a specified network. In particular, we focus on a data-gathering application, consisting of n nodes with given locations and finite remaining energies, and explore how information should be routed in the network to maximize the total information extractable from the network. We first survey the recent literature on variants of flow optimization models that have been proposed for wireless networks by several authors in recent years. We then present two models for the flow optimization approach – the first is an optimization model with non-linear convex constraints permitting rate adaptation through transmit power control; the second is a simpler linear optimization model in which the rate is kept fixed on all links. We then illustrate the usefulness of such models through some numerical examples, showing how the information extracted varies with available energy, how the two models (linear and non-linear) compare, and how reception costs impact optimal routing behavior.

The key feature of flow optimization-based modeling of sensor networks is that it is a computational framework. Thus, it yields performance bounds not in the form of closed-form expressions, but rather numerically, for specified scenarios. As we show, this framework is well suited to explore the impact of different design variables such as node location, energy distribution, rate-adaptation etc. The framework is also useful as a benchmark for comparing the performance of implementable protocols on a test-suite of scenarios. The basic optimization models we present are for single-sink data gathering scenarios, and incorporate energy constraints and costs for transmission, reception and sensing, channel capacity constraints, as well as information constraints including fairness. These models assume a (TDMA/FDMA like) scheduled medium access scheme with no interference. However, as we shall discuss, these models can be extended in principle to incorporate soft interference, data aggregation, richer energy models, and to some extent even mobility.

11.2 FLOW OPTIMIZATION IN WIRELESS NETWORKS

Network flow optimization, which forms the foundation of the approach we describe in this chapter, is an established area of Operations Research and is described in detail in the book by Ahuja, Magnanti and Orlin [21]. In the simplest max-flow problem, a graph $G = (V, E)$ is given with a specified source node s and a sink node t. The edges of the graph have capacities C_{ij} and the objective is to determine the flows on the edges f_{ij} with capacity and flow conservation constraints that maximize the total flow from s to t. A generalization of this is the multi-commodity flow problem where the goal is to maximize the sending of several different commodities (each possibly having different sources and sinks) over a network with restricted capacity. In recent years, flow optimization has been applied by several researchers to the analysis of multi-hop wireless networks.

Toumpis and Goldsmith analyze capacity regions for general-purpose multi-hop wireless networks [22, 23]. Using a linear-programming optimization based formulation (that is equivalent to a network flow problem), the authors study the characteristics of the maximum information throughput that can be obtained in a wireless network with arbitrary topology. Non-linear constraints are considered in the optimization models for wireless networks discussed in [24, 25]. In these works the authors consider jointly optimizing the routing as well as rate-adaptive power control and bandwidth allocation. They also treat the constraints imposed by interference in their models. It should be noted however that all of these models do not consider energy constraints that are important in sensor networks.

Chang *et al.* use the flow optimization formulation to maximize the lifetime of an ad hoc network in [26]. They propose a class of flow augmentation and flow redirection algorithms that balance the energy consumption rates across nodes based on the remaining battery power of these nodes. Optimization models have also been used to study maximum lifetime conditions for sensor networks. Bhardwaj and Chandrakasan [27] develop upper bounds on the network lifetime based on optimum role assignments to sensors (e.g. whether they should act as routers or aggregators). Kalpakis *et al.* examine the MLDA (Maximum Lifetime Data Aggregation) problem and the MLDR (Maximum Lifetime Data Routing) problem in [28], again formulating it using network flows. They use the solution obtained by solving the LP to construct an optimal data gathering schedule. Ordóñez and Krishnamachari have developed nonlinear models (permitting rate-adaptation) for maximizing information extraction in wireless sensor networks subject to energy and fairness constraints [29, 30].

Recent research has also looked into obtaining implementable algorithms (based on flow optimization) that provide near-optimal performance. Garg and Konemann propose and present an excellent discussion of fast approximation techniques for solving the multi-commodity flow problem [31]. Chang, et al apply the Garg-Konemann algorithm to the problem of maximizing the network lifetime of an ad hoc network in [32]. Sadagopan and Krishnamachari [33] extend the Garg-Konemann to provide approximation algorithms for a maximum information extraction problem in sensor networks, and also present faster, implementable, heuristics that are also shown to be near-optimal.

We will now describe both non-linear and linear flow optimization models for wireless sensor networks, and present some illustrations of the utility of this framework.

11.3 OPTIMIZATION MODELS FOR WIRELESS SENSOR NETWORKS

We now present flow optimization models that investigate the trade-off between maximum information extraction and minimum energy requirements for a given network topology. We begin by looking at non-linear models with rate-adaptation, in which the flow rates on each link can be adapted by varying transmit powers. We then examine simpler linear models in which the rates are kept constant. In both models, we assume that there are n source nodes, and a sink numbered $n + 1$, located with pairwise distances d_{ij} in a given area. In the basic models it is assumed that there is an overall energy budget of E_{tot} (Joules) to distribute among the sensor nodes (this could be easily modified to a per-node energy budget of E_{tot}/n if needed). The transmit power on link (i, j) is P_{ij} (Joules/sec), while β and C are the sensing and reception energy costs (Joules/bit). It is assumed that the relation between the flow rate f_{ij} and transmission power P_{ij} on a link is given by Shannon's capacity:

$$f_{ij} \leq \log\left(1 + \frac{P_{ij}d_{ij}^{-2}}{\eta}\right) \quad (11.1)$$

This expression assumes that the decay factor of the medium is d_{ij}^{-2}, the communication channel has a noise power of η, and that all transmissions are scheduled (e.g. via TDMA/FDMA) such that they are non-interfering.

11.3.1 NON-LINEAR ADAPTIVE RATE MODELS

In this model the link rates f_{ij} and transmission powers P_{ij} are design variables, and T is total time duration of (in seconds) communication on each link. Thus $\sum_{j=1}^{n} f_{j,n+1}T$ represents the total number of bits extracted by the sink from the network. The objective of our first model is therefore to find the coordinated operation of all nodes by setting transmission powers and flow rates in order to maximize the amount of information that reaches the sink:

$$\max \quad \sum_{j=1}^{n} f_{jn+1}T$$

$$\text{s.t.} \quad \sum_{j=1}^{n+1} f_{ij} - \sum_{j=1}^{n} f_{ji} \geq 0$$

$$\sum_{j=1}^{n+1} f_{ij} - \sum_{j=1}^{n} f_{ji} \leq \alpha_i \sum_{j=1}^{n} f_{jn+1} \quad (11.2)$$

$$\sum_{i=1}^{n} (\beta f_{in+1} + P_{in+1})T + \sum_{i=1}^{n}\sum_{j=1}^{n}(Cf_{ij} + P_{ij})T \leq E_{\text{tot}}$$

$$f_{ij} \leq \log\left(1 + \frac{P_{ij}d_{ij}^{-2}}{\eta}\right)$$

$$f_{ij} \geq 0, P_{ij} \geq 0$$

Since we do not consider aggregation in this simple model, the first constraint ensures that outgoing flow from a node is never less than the incoming flow. The second constraint imposes the fairness requirement that source i may not contribute more than a fraction α_i of the total flow to the sink. The third constraint imposes a network-wide energy constraint on the weighted costs of transmissions, receptions and sensing. Note that we do not explicitly constrain the communication between any pairs of nodes here, since we assume that we can make use of an ideal non-interfering schedule of transmissions; however it would be trivial to incorporate an additional constraint that limits the maximum transmit power of all nodes to minimize interference.

We simplify this problem using the arc-incidence matrix N, which for a network with $n + 1$ nodes and m arcs, is a $n + 1$ by m matrix with coefficients equal to 0, 1 or -1. The matrix is defined by

$$N_{i(k,l)} = \begin{cases} 1 & \text{if } i = k \\ -1 & \text{if } i = l \\ 0 & \text{otherwise} . \end{cases}$$

With this notation we can show that the problem is equivalent to:

$$\max \sum_{j=1}^{n} f_{jn+1} T$$

$$\text{s.t.} \quad 0 \leq Nf \leq \alpha \sum_{j=1}^{n} f_{jn+1}$$

$$\sum_{i=1}^{n} \sum_{j=1}^{n+1} \kappa_j f_{ij} T + \eta d_{ij}^2 \left(e^{f_{ij}} - 1 \right) T = E_{\text{tot}}$$

$$f \geq 0,$$

(11.3)

where we define $\kappa_j = C$ if $j = 1 : n$ and $\kappa_{n+1} = \beta$. A related dual problem to the problem above, which minimizes the energy to obtain a certain amount of information b_{out} (in bits), is

$$\min \sum_{i=1}^{n} \sum_{j=1}^{n+1} \left(\kappa_j f_{ij} + \eta d_{ij}^2 (e^{f_{ij}} - 1) \right)$$

$$\text{s.t.} \quad 0 \leq Nf \leq \alpha \sum_{j=1}^{n} f_{jn+1}$$

$$\sum_{j=1}^{n} f_{jn+1} T = b_{\text{out}}$$

$$f \geq 0 .$$

(11.4)

We can show that the optimal solutions for Problems (11.3) and (11.4) are in fact related, a relationship that illustrated with the following example. In Figure 11.1 we plot both the maximal information extracted as a function of the energy bound and minimum energy needed as a function of the information bound. The experiments that originated these results considered the same WSN with all nodes placed in a straight line, the sink node at one end, 10 sensor nodes uniformly distributed from a distance 1 to 10 of the sink, and the following values for other problem parameters: $\beta = 0.00001$, $C = 0.00005$, $\eta = 0.0001$, $T = 1$, and $\alpha_i = 0.20$ for all i. The minimum information bound was varied from $b_{\text{out}} = 1$, to $b_{\text{out}} = 20$ when solving Problem (11.4), and the maximum energy bound was varied from $E_{\text{tot}} = 0.01$ to $E_{\text{tot}} = 0.2$ when solving Problem (11.3).

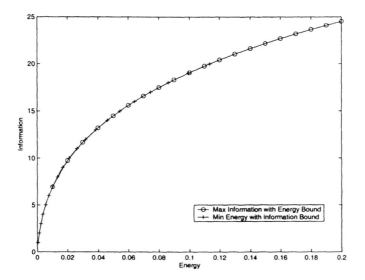

Figure 11.1. Optimal energy versus information

11.3.2 LINEAR CONSTANT RATE MODEL

An alternative, computationally simpler, linear flow optimization model is obtained if we do not permit rate adaptation and assume that there is a fixed transmission rate $f_{ij} = R$ (bits/sec) for each link in the network. The transmission powers are therefore also fixed and given by $P_{ij} = \eta d_{ij}(e^R - 1)$ (in J/sec). In this model our decision variables are how many bits to send from i to j, b_{ij}. Given that the transmission rate is fixed the time taken for transmission on a given link is therefore variable and depends on the number of bits sent (recall that this was a constant T for all links in the previous model).

The corresponding problem to Problem (11.3) in which the goal is to maximize the bits extracted for a given total energy budget is

$$\max \quad \sum_{j=1}^{n} b_{jn+1}$$
$$\text{s.t.} \quad 0 \leq Nb \leq \alpha \sum_{j=1}^{n} b_{jn+1}$$
$$\sum_{i=1}^{n} \sum_{j=1}^{n+1} \kappa_j b_{ij} + \frac{\eta d_{ij}^2}{R}\left(e^R - 1\right) b_{ij} = E_{\text{tot}}$$
$$b \geq 0 \,.$$

(11.5)

The corresponding problem to Problem (11.4) in which we minimize the energy usage subject to a given information requirement b_{out} is

$$\min \sum_{i=1}^{n} \sum_{j=1}^{n+1} \kappa_j b_{ij} + \frac{\eta d_{ij}^2}{R} \left(e^R - 1\right) b_{ij}$$
$$\text{s.t.} \quad 0 \leq Nb \leq \alpha b_{\text{out}}$$
$$\sum_{j=1}^{n} b_{jn+1} = b_{\text{out}} \qquad (11.6)$$
$$b \geq 0 .$$

We will now undertake a comparison of the adaptive and constant rate models. The latter model, involving a linear program is computationally more tractable, but as one may expect, we find that the loss of a degree of freedom (rate adaptation) results in inefficiency.

11.4 A COMPARISON OF THE NON-LINEAR AND LINEAR MODELS

We consider Problems (11.3) and (11.5). In other words we will compare how much information b_{out} (bits) can be extracted from a given sensor network according to each model, when we are given a limited budget of overall energy E_{tot} (Joules). We will compare the total information that can be extracted with the same budget of energy for each model. For the comparisons we will tune the model parameter T for the non-linear adaptive rate model, and the parameter R for the linear constant rate model. These parameters in effect tune the throughput (bits/second) with which information is extracted from the network in each model.

11.4.1 SIMPLE EXAMPLE

As an illustration, we first consider a simple problem with two source nodes in line with the sink; one of the nodes provides all the information (i.e. $\alpha_1 = 0$ and $\alpha_2 = 1$). We assume that all the information is originating from the node furthest from the sink. In this example we compare the performance of the non-linear and linear models by studying which outputs more data for a given amount of energy E_{tot}. We assume that the sink is at $(0,0)$ and the nodes at $(1,0)$ and $(2,0)$, and that we have a sensing cost β (in J/bit), reception cost C (in J/bit) and a noise power η. It is straightforward to show that for the non-linear model the optimal flow rate f for this example satisfies

$$CfT + \beta fT + 2\eta \left(e^f - 1\right) T = E_{\text{tot}}$$
$$\Rightarrow (C + \beta)f + 2\eta \left(e^f - 1\right) = \frac{E_{\text{tot}}}{T} \qquad (11.7)$$

and the optimal number of bits b for the linear model in turn satisfies the relation

$$Cb + \beta b + 2\eta \left(e^R - 1\right) \frac{b}{R} = E_{\text{tot}}$$

$$b = \frac{E_{\text{tot}} R}{2\eta \left(e^R - 1\right) + R(C + \beta)} \quad (11.8)$$

Note that the optimal solution in both cases depend on a tunable parameter T for the non-linear model and R for the linear model. We study now how the optimal solutions vary with these parameters.

We considered additional problem parameters as $E_{\text{tot}} = 1$, $\eta = 0.01$, $\beta = 0.001$, and $C = 0.001$. We compute the total amount of bits that can be extracted from the linear and non-linear models (b and $f * T$ respectively) for different values of their respective tunable parameters (R and T respectively). In Figure 11.2 we plot the total bits that are extracted versus the flow rate in doing so for both models:

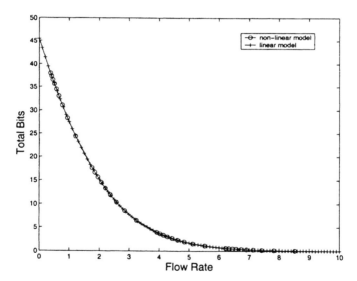

Figure 11.2. Total bits sent to sink versus flow rate for linear and non-linear models, simple example

Analytically it is easy to show that $\lim_{R \to 0} b(R) = \frac{E_{\text{tot}}}{2\eta + C + \beta}$, and $\lim_{R \to \infty} b(R) = 0$.

The function $b(R)$ is maximized for the rate that satisfies $e^R(1 - R) = 1$, which occurs when R is zero, i.e. as small as possible.

As there is no explicit analytical expression of the optimal solution for the optimal flow rate f, we can only obtain approximate limits for the total flow extracted at T tends to 0 and ∞. As $T \to \infty$ the expression $(C + \beta)f + 2\eta(e^f - 1) \to 0$, therefore $(C+\beta)f + 2\eta(e^f - 1) = (C+\beta)f + 2\eta \sum_{k=1}^{\infty} \frac{f^k}{k!} \approx (C + \beta + 2\eta)f$, and thus $\lim_{T \to \infty} Tf(T) \approx \frac{E_{tot}}{2\eta+C+\beta}$. Likewise as $T \to 0$, then $(C + \beta)f + 2\eta(e^f - 1) \to \infty$ and thus $(C + \beta)f + 2\eta(e^f - 1) \approx 2\eta e^f$, which implies that $Tf(T) \approx T\log(\frac{E_{tot}}{2\eta T}) \to 0$. Thus for the non-linear model the total information extracted is maximized for a large T. For both the linear and non-linear models, these observations are consistent with figure 11.2 which shows that the maximum total information is extracted when the overall throughput is kept as low as possible (i.e. to the left of the curve).

11.4.2 GENERAL CASE

We consider a square grid scenario, with a 3×3 uniform square grid of sensor nodes in a $[0, 10]^2$ square sending information to a sink located outside the square, located at $(-3, 5)$, other problem parameters were set as $\alpha_i = 0.15$ for all nodes, $\beta = 0.00001$, $\eta = 0.0001$, $E_{\text{tot}} = 10$, and cost $C = 0.001$.

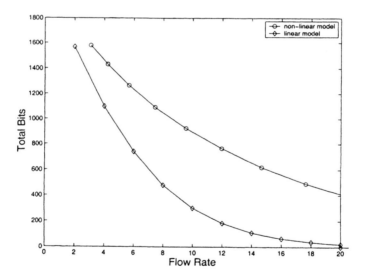

Figure 11.3. Total bits sent to sink versus flow rate for linear and non-linear models, square example

In figure 11.3 we compare the total bits extracted for each the linear and non-linear models respectively as a function of the total rate to the sink. Note that the non-linear model outputs much more information for the same level of energy, at all rate levels. This shows that the computational tractability of the constant rate linear models comes at the expense of some inefficiency. Rate adaptation can provide significant additional information for the same budget (it is nearly an order of magnitude higher in this scenario).

11.5 MULTI-HOP BEHAVIOR

Here we investigate how varying the reception cost C affects the hopping behavior of the optimal solution. In this section we are simply considering the non-linear model. Clearly a very high reception cost will make preferable to send the information directly to the sink, while a very inexpensive reception will allow for hopping in the optimal solution. An approach is to use an optimization model to see for which values of C the optimal routing hops and for which it does not. We first consider a very simple example that is also amenable to an analytical solution and then see how the insights can be generalized.

11.5.1 SIMPLE EXAMPLE

To provide an initial solution to this question we consider a very simplified problem consisting of only two sensor nodes, one of which provides all the information (that is $\alpha_1 = 0$ and $\alpha_2 = 1$). The question is to try to predict when node 2 will prefer to send the information directly to the sink and when it will prefer to route it through node 1. In order to avoid a trivial solution we place node 1 closer to the sink than node two, in fact for simplicity we place it exactly mid-way between node 2 and the sink. Assume also that we have $b_{\text{out}} = 1$, that we have a sensing cost of β, and noise parameter of η. Clearly the decision of whether to route information or not will be affected for different reception costs C. We note that for small values of C node 2 will find more attractive to route its information through node 1. In this case node two has the alternative to route part of the information and send the rest directly to the sink. For high values of C the network will decide that it's too expensive to route information through node 1 and node 2 will send everything directly. Here we investigate for what values of C will the network decide to hop and for which to send the information directly.

With the use of optimization models we solve for the optimal routing behavior given different values of the reception cost C. In Figure 11.4 we plot the total value of flow that is sent directly to the sink from node 2 for different reception cost values. This computational example additionally has the following parameters values $\beta = 0.00001$ and $\eta = 0.1$. We note that for reception costs higher than a critical value (plotted as a dashed vertical line) node 2 sends all the information directly to the sink. We also note that it is never optimal to hop all the information, as for any reception cost there is some fraction of the information being sent directly. We finally note that there is a dramatic change in the type of the routing solution as C varies from 10^{-2} to 0.3.

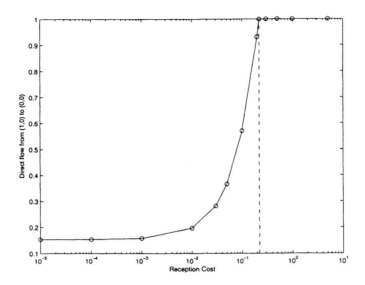

Figure 11.4. Flow to the sink without hopping as a function of the reception cost C

Due to the simplicity of this example, we can analyze the results further. We are simply comparing the solution in which we route all the information directly at a cost

$$h_s = \beta + \eta(e - 1)$$

with the case in which we send f_1 from node 2 to node 1 and then to the sink, and $f_2 = 1 - f_1$ directly form node 2 to the sink, at a cost

$$\begin{aligned} h_c(f_1) &= \beta + \frac{1}{4}\eta(e^{f_1} - 1) + Cf_1 + \frac{1}{4}\eta(e^{f_1} - 1) + \eta(e^{1-f_1} - 1) \\ &= \beta + Cf_1 + \frac{1}{2}\eta(e^{f_1} - 1) + \eta(e^{1-f_1} - 1). \end{aligned}$$

The amount of information that will be routed will be the minimizer of function $h_c(f_1)$ on the domain $[0, 1]$.

We need to determine for what values of C will $h_c(f_1) < h_s$ for some $f_1 \in (0, 1]$, which means that it is more convenient to route f_1 than to send everything directly. Equivalently we will determine for what value of C, the function $H_C(f_1) = h_c(f_1) - h_s \geq 0$ for all $f_1 \in [0, 1]$, these are the values of C that will make it more convenient to send directly rather than route the information. It is easy to show that $H_C(f_1)$ is a convex function and $H_C(0) = 0$, therefore to guarantee that $H_C(f_1) \geq 0$ for all $f_1 \geq 0$ it is sufficient to show that $H'_C(f_1) \geq 0$. This last condition reduces to $C \geq \eta(e - \frac{1}{2})$. This critical value for the reception cost is $C = 0.221828$ for the problem parameters of this example. We plot this value a vertical line in Figure 11.4. Note that the the optimal solution routes all the information from sources directly to the sink precisely at that critical value.

11.5.2   GENERAL CASE

To illustrate the general case we considered a 3×3 uniform square grid of sensor nodes sending information to a sink located outside the square. The example considers 9 nodes uniformly distributed in a $[0, 10]^2$ square with the sink located at $(-3, 5)$, other problem parameters were set as $\alpha_i = 0.15$ for all nodes, $\beta = 0.00001$, $\eta = 0.0001$, and $b_{\text{out}} = 10$. The reception cost C was varied between 10^{-5} and 1.

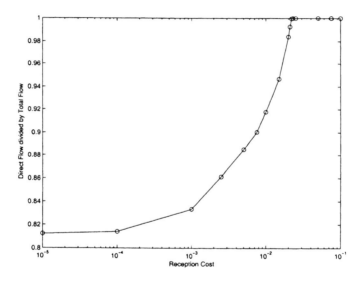

Figure 11.5. Fraction of total flow on the network that goes directly to the sink as a function of the reception cost C

To analyze the result for a general WSN we also construct a function $H_C(f)$ that quantifies the difference between a hopping solution and a non-hopping solution. Let $\hat{f}$ be the optimal routing of information that sends all information directly. That is it solves Problem (11.4) with the additional constraint that $\sum_{i,j=1}^{n} f_{ij} = 0$. Then the minimal energy needed to send directly to the sink a given amount of information b_{out} is

$$h_s = \beta b_{\text{out}} + \sum_{i=1}^{n} \eta d_{in+1}^2 (e^{\hat{f}_{in+1}} - 1).$$

We define $H_C(f)$ for any feasible flow f to be the difference between the energy consumed to send b_{out} to the sink by routing through f and h_s, that is

$$H_C(f) = \sum_{i=1}^{n}\sum_{j=1}^{n+1} \kappa_j f_{ij} + \eta d_{ij}^2\left(e^{f_{ij}} - 1\right) - h_s$$

$$= \sum_{i,j=1}^{n} C f_{ij} + \sum_{i=1}^{n}\sum_{j=1}^{n+1} \eta d_{ij}^2 \left(e^{f_{ij}} - e^{\hat{f}_{ij}}\right).$$

The condition that the reception cost must satisfy, in order to indicate that it forces direct routing as the optimal solution, is that for every feasible solution f we have $H_C(f) \geq 0$. It is easy to see that if $\tilde{f}$ is a routing solution that sends all the information directly to the sink, then $H_C(\tilde{f}) \geq 0$, and $H_C(\hat{f}) = 0$. Therefore C induces a no-hopping optimal routing solution if $\nabla H_C(\hat{f})^t(f - \hat{f}) \geq 0$ for all feasible solutions f. Taking derivatives we have that

$$\nabla H_C(\hat{f})^t(f - \hat{f}) = \sum_{i,j=1}^{n}(C + \eta d_{ij}^2)f_{ij} + \sum_{i=1}^{n}\eta d_{in+1}^2 e^{\hat{f}_{in+1}}\left(f_{in+1} - \hat{f}_{in+1}\right)$$

$$= C\sum_{i,j=1}^{n} f_{ij} + \sum_{i=1}^{n}\sum_{j=1}^{n+1} \eta d_{ij}^2 e^{\hat{f}_{ij}}\left(f_{ij} - \hat{f}_{ij}\right).$$

If f is a non-hopping feasible flow then we can show that $\nabla H_C(\hat{f})^t(f - \hat{f}) \geq 0$, from the KKT conditions of $\hat{f}$. If f is a solution that does some hopping, that is $\sum_{i,j=1}^{n} f_{ij} > 0$, then the condition that C has to satisfy to force a non-hopping solution is

$$C \geq \frac{\sum_{i=1}^{n}\sum_{j=1}^{n+1} \eta d_{ij}^2 e^{\hat{f}_{ij}}\left(\hat{f}_{ij} - f_{ij}\right)}{\sum_{i,j=1}^{n} f_{ij}}.$$

This condition implies that the threshold value for the reception cost C^*, above which it is preferable to send all the information directly, is the solution to the following maximization problem over an open domain:

$$\begin{aligned} C^* = \max_{f,\kappa} \quad & \tfrac{1}{\kappa}\sum_{i=1}^{n}\sum_{j=1}^{n+1} \eta d_{ij}^2 e^{\hat{f}_{ij}}\left(\hat{f}_{ij} - f_{ij}\right) \\ \text{s.t.} \quad & 0 \leq Nf \leq \alpha \sum_{j=1}^{n} f_{jn+1} \\ & \sum_{i,j=1}^{n} f_{ij} = \kappa \\ & f \geq 0 \\ & \kappa > 0 \end{aligned}$$

This example illustrates the possible use of the flow optimization models to examine qualitative issues such the use of multi-path routing versus direct transmission to a base station.

11.6 CONCLUSIONS

In this chapter, we have described the flow optimization framework for analyzing the fundamental limits of sensor network performance. We reviewed some of the relevant recent literature on the subject, and illustrated the framework by presenting several models including both non-linear adaptive rate models as well as linear constant rate models. We illustrated the utility of this framework by studying the energy-information tradeoffs in sensor network, investigating the gains that are possible through rate adaptation, and exploring the conditions under which multi-hop routes are present in the optimal solution. Another important use of the flow optimization framework is that the numerical fundamental performance limits that it provides can be used as a benchmark for measuring the performance of practical implementations.

There are a number of directions in which the flow optimization models we have presented can be extended. The models above assume conservation of flows at each node. If nodes can aggregate data, this constraint must be changed. Data aggregation constraints can be best modelled by incorporating multi-commodity flows. The models above assume that the network has scheduled communications on all links. For a CDMA-like environment, interference poses a non-convex constraint. There are some techniques that can be used to handle such constraints approximately [25]. Models can also consider the possibility of mobile nodes, in which locations and therefore inter-node distances can be varied as a design parameter at the expense of some energy for motion. However, this may also introduce a non-convex constraint. Another area of open research is in exploiting the structure of these optimization problems to develop constructive, implementable algorithms that obtain near-optimal performance in practice.

ACKNOWLEDGEMENT

The work described in this article has been made possible through a 2003 USC Zumberge Interdisciplinary Research Grant.

REFERENCES

[1] W. Heinzelman, A. Chandrakasan, and H. Balakrishnan, "Energy-Efficient Communication Protocol for Wireless Microsensor Networks," *Proc. Hawaii Conference on System Sciences*, Jan. 2000.

[2] S. Lindsey, C. S. Raghavendra, "PEGASIS: Power Efficient GAthering in Sensor Information Systems," *Proc. ICC*, 2001.

[3] C. Intanagonwiwat, R. Govindan, and D. Estrin, "Directed Diffusion: A Scalable and Robust Communication Paradigm for Sensor Networks," *Proc. ACM Mobicom*, Boston MA, August 2000.

[4] F. Ye, H. Luo, J. Cheng, S. Lu, and L. Zhang, "A two-tier data dissemination model for large-scale wireless sensor networks," *Proc. Eighth ACM/IEEE International Conference on Mobile Computing and Networking (MobiCom '02)*, Atlanta, Georgia, USA, September 2002.

[5] S. Ratnasamy, et al. "GHT: A Geographic Hash Table for Data-Centric Storage," *Proc. First ACM International Workshop on Wireless Sensor Networks and Applications (WSNA '02)*, September 2002.

[6] P. Bonnet, J. E. Gehrke, and P. Seshadri, "Querying the Physical World," *IEEE Personal Communications*, vol. 7, no. 5, October 2000.

[7] R. Govindan, J. Hellerstein, W. Hong, S. Madden, M. Franklin, S. Shenker, "The Sensor Network as a Database,", *Technical Report 02-771*, Computer Science Department, University of Southern California, September 2002.

[8] M. Chu, H. Haussecker, F. Zhao, "Scalable information-driven sensor querying and routing for ad hoc heterogeneous sensor networks." *International Journal of High Performance Computing Applications*, 2002.

[9] N. Sadagopan, B. Krishnamachari, and A. Helmy, "The ACQUIRE Mechanism for Efficient Querying in Sensor Networks," *First IEEE International Workshop on Sensor Network Protocols and Applications (SNPA'03)*, May 2003.

[10] D. Bertsekas and R. Gallager *Data Networks*, Prentice Hall, 1987.

[11] A. Ephremides and B. Hajek, "Information theory and communication networks: An unconsummated union," *IEEE Transactions on Information Theory (Commemorative Issue)*, 44:2384–2415, 1998.

[12] P. Gupta and P.R. Kumar, "The Capacity of Wireless Networks," *IEEE Transactions on Information Theory*, vol. IT-46, no. 2, pp. 388-404, March 2000.

[13] L.-L. Xie and P. R. Kumar, "A Network Information Theory for Wireless Communication: Scaling Laws and Optimal Operation." Submitted to IEEE Transactions on Information Theory, April 2002. Available online at <http://decision.csl.uiuc.edu/~prkumar/publications.html>.

[14] M. Grossglauser, D. Tse, "Mobility Increases the Capacity of Ad-hoc Wireless Networks," *INFOCOM*, pp. 1360-69, 2001.

[15] S. Yi, Y. Pei, S. Kalyanaraman, "On the Capacity Improvement of Ad Hoc Wireless Networks Using Directional Antennas," *Mobihoc '03*, Annapolis, Maryland, 2003.

[16] A. Scaglione, S. D. Servetto, "On the Interdependence of Routing and Data Compression in Multi-Hop Sensor Networks," *8th ACM International Conference on Mobile Computing and Networking (MobiCom)*, Atlanta, GA, September 2002.

[17] D. Marco, E. Duarte-Melo, M. Liu and D. L. Neuhoff, "On the many-to-one transport capacity of a dense wireless sensor network and the compressibility of its data", in *Proc. International Workshop on Information Processing in Sensor Networks (IPSN)*, April 2003.

[18] B. Krishanamachari, D. Estrin and S. Wicker, "The Impact of Data Aggregation in Wireless Sensor Networks," *International Workshop of Distributed Event Based Systems (DEBS)*, Vienna, Austria, July 2002.

[19] Ashish Goel and Deborah Estrin, " Simultaneous Optimization for Concave Costs: Single Sink Aggregation or Single Source Buy-at-Bulk," *Proc. ACM-SIAM Symposium on Discrete Algorithms*, 2003.

[20] B. Krishnamachari and J. Heidemann, "Application-specific Modelling of Information Routing in Sensor Networks," *unpublished manuscript*, 2003. Available online at <http://ceng.usc.edu/~bkrishna>.

[21] R. K. Ahuja, T. Magnanti, and J. B. Orlin. *Network Flows: Theory, Algorithms, and Applications.* Prentice Hall, New Jersey, 1993.

[22] S. Toumpis and A.J. Goldsmith, "Capacity regions for wireless ad hoc networks," *International Symposium on Communication Theory and Applications*, 2001

[23] S. Toumpis and A.J. Goldsmith, "Capacity Regions for Wireless Ad Hoc Networks," To appear in the *IEEE Transactions on Wireless Communications*, 2003.

[24] L. Xiao, M. Johansson, and S. Boyd, "Simultaneous routing and resource allocation via dual decomposition," *Proceedings of 4th Asian Control Conference*, September 25-27, 2002.

[25] M. Johansson, L. Xiao and S. Boyd, "Simultaneous routing and resource allocation in CDMA wireless data networks," *IEEE International Conference on Communications*, Anchorage, Alaska, May 2003.

[26] J. Chang and L. Tassiulas. "Energy Conserving Routing in Wireless Ad Hoc Networks," *IEEE Infocom 2000*.

[27] M. Bhardwaj and A.P. Chandrakasan, "Bounding the Lifetime of Sensor Networks Via Optimal Role Assignments", *Proceedings of INFOCOM 2002*, pp. 1587-1596, New York, June 2002.

[28] K. Kalpakis, K. Dasgupta, and P. Namjoshi, "Maximum Lifetime Data Gathering and Aggregation in Wireless Sensor Networks," In the *Proceedings of the 2002 IEEE International Conference on Networking (ICN'02)*, Atlanta, Georgia, August 26-29, 2002.

[29] B. Krishnamachari and F. Ordóñez, "Analysis of Energy-Efficient, Fair Routing in Wireless Sensor Networks through Non-linear Optimization," *Workshop on Wireless Ad hoc, Sensor, and Wearable Networks, in IEEE Vehicular Technology Conference - Fall*, Orlando, Florida, October 2003.

[30] B. Krishnamachari and F. Ordóñez, "Optimal Information Extraction in Energy-Limited Wireless Sensor Networks," *unpublished manuscript*, June 2003. Available online at <http://ceng.usc.edu/~bkrishna>.

[31] N. Garg, J. Konemann, "Faster and Simpler Algorithms for Multicommodity Flow and Other Fractional Packing Problems," *FOCS* 1998.

[32] J. H. Chang and L. Tassiulas, "Fast Approximate Algorithm for Maximum Lifetime Routing in Wireless Ad-hoc Networks," *Networking 2000, vol. 1815 of Lecture Notes in Computer Science*, pp. 702-713, Paris, France, May 2000.

[33] N. Sadagopan and B. Krishnamachari, "Maximizing Data Extraction in Energy-Limited Sensor Networks," *unpublished manuscript*, June 2003. Available online at <http://ceng.usc.edu/~bkrishna>.

IV

SECURITY

Chapter 12

SECURITY FOR WIRELESS SENSOR NETWORKS

Sasikanth Avancha, Jeffrey Undercoffer, Anupam Joshi and John Pinkston

University of Maryland, Baltimore County
Department of Computer Science and Electrical Engineering
1000 Hilltop Circle, Baltimore, MD 21250
{ savanc1, junder2, joshi, pinkston } @cs.umbc.edu

Abstract   This chapter identifies the vulnerabilities associated with the operational paradigms currently employed by Wireless Sensor Networks. A survey of current WSN security research is presented. The security issues of Mobile Ad-Hoc Networks and infrastructure supported wireless networks are briefly compared and contrasted to the security concerns of Wireless Sensor Networks. A framework for implementing security in WSNs, which identifies the security measures necessary to mitigate the identified vulnerabilities is defined.

Keywords: Sensor Networks, Security, Wireless, WSN.

12.1   INTRODUCTION

Information technology has dramatically changed the manner in which societies protect its citizens. For example, recent media reports [14] detail the Department of Energy's (DOE) "SensorNet" project and National Oceanic and Atmospheric Administration's (NOAA) "DCNet". SensorNet consists of sensors that measure wind direction in order to forecast how urban "wind fields" might disperse fallout from a weapon of mass destruction. DCNet consists of sensors that measure gamma-radiation. Although disjoint, the sensors of each network are co-located. In their current configuration, SensorNet and DCNet transmit their data to a network of fixed and mobile relay collection stations where the data is processed. The DOE and NOAA also have similar initiatives in Manhattan. Although these prototypical "Wireless Sensor Networks" (WSNs) are limited to measurements of wind plumes and gamma radiation, it is only a matter of time until they are extended to include sensors that are able to detect vibrations, chemicals, biological agents, explosives, footsteps, and voices.

Technology, however, is a double-edged sword. Just as nations and societies employ technology to protect themselves, their adversaries employ technology to counter and mitigate the security afforded by these new and innovative protective measures. Consequently, steps need to be taken to ensure the security of these protective technologies. In the case of a WSN, the underlying data network, the physical sensors, and the protocols used by the WSN all need to be secured.

To exemplify this double-edged sword, consider the omnipresent threat to distributed control systems (DCS) [24] and supervisory control and data acquisition systems (SCADA) [6]. These control systems are often wireless and share some similarities with WSNs. The simplest of these systems collect data associated with some metric and, based upon its value, cause some event to occur. Events include the closing or opening of railway switches, the cycling of circuit breakers, and the opening and closing of valves. Complex DCSs and SCADAs accept data from multiple sensors and using controllers, govern a wide array of devices and events in response to the measured data.

On April 23, 2000, Vitek Boden was arrested in Queensland, Australia and was eventually found guilty of computer hacking, theft and causing significant environmental damage [33]. This criminal case is of particular interest because, to date, it is the only known case where someone has employed a digital control system to deliberately and maliciously cause damage. Using a transmitter and receiver tuned to the same frequencies as the SCADA controlling Queensland's municipal water system and a laptop computer he effectively became the master controller for the municipality's water system. He then proceeded to wreak havoc upon the city's water supply by causing the release of raw sewage into local waterways and green spaces.

DCSs and SCADAs were not designed with public access in mind, consequently they lack even rudimentary security controls. Moreover, if DCSs and SCADAs were constrained by security controls, they may fail to work properly because their timing and functionality are predicated upon unfettered communications between system components.

Similarly, the security requirements of a WSN impose costly constraints and overhead due to a sensor node's limited power supply and computational resources. In a manner not unlike the Boden case, a WSN bereft of security controls will most likely suffer the same fate as did Queensland's municipal water system.

The goal of this chapter is to present a framework for implementing security in WSNs. Existing WSN research, which is minimal, is identified and is presented. The threats to a WSN are identified, the operational paradigms utilized by WSNs and their associated vulnerabilities are presented. The security models applicable to other types of wireless networks, (e.g.: 802.11b or Bluetooth-based wireless Internet and MANETs), are compared and contrasted to the security models that are specified for WSNs. The WSN security state-of-the-art is detailed, identifying where the state-of-the-art falls short of the ideal. Ideas are presented for moving the state-of-the-art closer to the ideal, recommendations are made, and conclusions are drawn.

12.2 THREATS TO A WSN

There are many vulnerabilities and threats to a WSN. They include outages due to equipment breakdown and power failures, non-deliberate damage from environmental factors, physical tampering, and information gathering. We have identified the following threats to a WSN:

Passive Information Gathering: If communications between sensors, or between sensors and intermediate nodes or collection points are in the clear, then an intruder with an appropriately powerful receiver and well designed antenna can passively pick off the data stream.

Subversion of a Node: If a sensor node is captured it may be tampered with, electronically interrogated and perhaps compromised. Once compromised, the sensor node may disclose its cryptographic keying material and access to higher levels of communication and sensor functionality may be available to the attacker. Secure sensor nodes, therefore, must be designed to be tamper proof and should react to tampering in a *fail complete manner* where cryptographic keys and program memory are erased. Moreover, the secure sensor needs to be designed so that its emanations do not cause sensitive information to leak from the sensor.

False Node: An intruder might "add" a node to a system and feed false data or block the passage of true data. Typically, a false node is a computationally robust device that impersonates a sensor node.

While such problems with malicious hosts have been studied in distributed systems, as well as ad-hoc networking, the solutions proposed there (group key agreements, quorums and per hop authentication) are in general too computationally demanding to work for sensors.

Node Malfunction: A node in a WSN may malfunction and generate inaccurate or false data. Moreover, if the node serves as an intermediary, forwarding data on behalf of other nodes, it may drop or garble packets in transit. Detecting and culling these nodes from the WSN becomes an issue.

Node Outage: If a node serves as an intermediary or collection and aggregation point, what happens if the node stops functioning? The protocols employed by the WSN need to be robust enough mitigate the effects of outages by providing alternate routes.

Message Corruption: Attacks against the integrity of a message occur when an intruder inserts themselves between the source and destination and modify the contents of a message.

Denial of Service: A denial of service attack on a WSN may take several forms. Such an attack may consist of a jamming the radio link or could it could exhaust resources or misroute data. Karlof and Wagner [21] identify several DoS attacks including: "Black Hole", "Resource Exhaustion", "Sinkholes", "Induced Routing Loops", "Wormholes", and "Flooding" that are directed against the routing protocol employed by the WSN..

Traffic Analysis: Although communications might be encrypted, an analysis of cause and effect, communications patterns and sensor activity might reveal enough information to enable an adversary to defeat or subvert the mission of WSN. Addressing and routing information transmitted in the clear often contributes to traffic analysis. We further address traffic analysis in the following subsection.

Traffic analysis is an issue that has occasionally attracted the attention of authors writing about security in networks [11, 12, 31]. Traffic analysis is the term used for the process of inferring information about the communications of an encrypted target network. Although unable to read the encrypted message contents, the analyst examines the externals - which station is sending, to whom messages are sent, and the patterns of activity. Sometimes, the identities of the correspondents are contained in unencrypted message headers, while at other times they can be inferred by direction finding techniques on radio signals. The order of battle can be deduced using traffic analysis - i.e., which station is the command headquarters, which ones are at intermediate levels, and which are at the lowest echelons. Patterns of activity can also be used to deduce the target's alert status - e.g., routine, heightened, attack imminent.

Classically, traffic analysis is countered by communication systems employing traffic flow security. In this mode, the system transmits an encrypted stream continuously, encrypting idle messages when there is no valid traffic to be sent. In this way, the unauthorized listener can not tell when the parties are actually communicating and when they are not, and is thus unable to make traffic analysis deductions. With radio links in years past, traffic flow security was often employed. Doing so did not add significant expense - it occupied a radio frequency continuously, but radio spectrum was not at a premium, nor was electric power for the transmitter a concern.

With WSNs having limited-energy nodes, the practicality of traffic flow security becomes quite problematic. Now, the cost of sending dummy traffic results directly in a reduction in the lifetime of the node. The energy required for communications is typically the dominant factor in battery lifetime of a node, so the lifetime is reduced by the ratio of dummy traffic to real traffic, which must be substantially greater than 1, to provide any useful traffic flow security protection. This will be intolerable in virtually all cases; consequently traffic flow security for WSNs is in need of further research. The effects of traffic analysis may be partially mitigated by encrypting the message header that contains addressing information, however further research is needed.

12.3 WSN OPERATIONAL PARADIGMS AND THEIR CORRESPONDING VULNERABILITIES

We categorize WSNs according to its operational paradigm. Some models of operation are simple; the sensor takes some measurement and blindly transmits the data. Other operational models are complex and include algorithms for data aggregation and data processing. In order to discuss security measures for a WSN sensibly, one must know the threats that must be defended, and equally important, those that need not be provided for. It makes no sense to even attempt to design protection against an all-powerful adversary, or even against an adversary who gets the last move in a spy-vs-spy, move-countermove game. One must select a model of the adversary's capabilities and work to that.

A security architecture will depend on our assumptions about the integrity of the hardware and software of the base stations and outstation nodes, and our assumptions about the capability of an adversary to eavesdrop, intrude on the network, or obtain physical access to a node.

What do we need to protect the WSN? Is it the content of the data being communicated (confidentiality)? Do we need to ensure the correctness of this data (data integrity)? Does the system need to be robust and survivable when confronted by a sophisticated adversary? What do we assume about the adversary's capability to physically capture and subvert one of the deployed nodes, and to extract sensitive information such as cryptographic keys from it? The answers to these questions will guide the designer to incorporate the right capabilities.

Typically, to ensure the desired level of system operation, one must have confidence in the correctness and completeness of the data that is collected and forwarded to the point(s) where decisions are made and responses initiated. One may or may not care if anyone else knows that the network is there, or can understand the traffic.

The following briefly describes the operational paradigms that a WSN may use. In each case, we assume the presence of a base station, or controller.

Simple Collection and Transmittal. The sensors take periodic measurements and transmit the associated data directly to the collection point. Transmission occurs either immediately following data collection or is scheduled at some periodic interval. In this paradigm each node is only concerned with its transmission to the base station, which is assumed to be within range. Thus, any notion of routing or co-operation among nodes is absent from this paradigm.

This operational paradigm is vulnerable to attacks directed against the Link Layer. Denial-of-service attacks include *jamming* the radio frequency and *collision induction*. It is also vulnerable to *spoofing* attacks in which a counterfeit data source broadcasts spurious information. If the data is considered to be sensitive and it is not encrypted, then a loss of confidentiality may occur if someone passively monitors the transmissions emanating from the WSN. This paradigm and all the following are also susceptible to physical attacks - capture of a node and subsequent subversion. Such threats are countered by tamper resistant technologies, which may transmit an alert and/or self destruct when tampering is detected. Discussion of these techniques is beyond the scope of this chapter. Replay attacks in which an adversary transmits old and/or false data to nodes in the WSN can also be mounted on the six paradigms discussed here.

Forwarding. Sensors collect and transmit data to one or more neighboring sensors that lie on a path to the controller. In turn, the intermediate sensors forward the data to the collection point or to additional neighbors. Regardless of the length of the path, the data eventually reaches the collection point. Unlike the first paradigm, co-operation among nodes in "routing" the data to the base station is part of this paradigm. That is, a node that receives data intended for the base station attempts to transmit the same toward the latter, instead of throwing the data away.

In addition to being vulnerable to the attacks identified under the Simple Collection and Transmittal paradigm, this method is also vulnerable to *Black Hole*, *Data Corruption* and *Resource Exhaustion* attacks. In a Black Hole attack, the sensor node that is responsible for forwarding the data drops packets instead of forwarding them. A Data Corruption attack occurs when the intermediate node modifies transient data prior to forwarding it. These attacks require that the node be subverted or that a foreign, malicious node be successfully inserted into the network. A Resource Exhaustion attack occurs when an attacker maliciously transmits an inordinate amount of data to be forwarded, consequently causing the intermediate node(s) to exhaust their power supply.

Receive and Process Commands. In this paradigm, sensors receive commands from a controller, either directly or via forwarding, and configure or re-configure themselves based on the commands. This ability to process commands is in addition to that of transmitting unsolicited data to the controller and helps in controlling the amount of data handled by the WSN. In this model, the communication paradigm changes from being exclusively many-to-one to now include one-to-many communication which means that whereas in the former, the data transmitted was intended *only* for the base station, in the latter, the data (i.e., command) is applicable to one or more sensor nodes. Commands may be broadcast to the entire WSN or may be unicast to a single sensor. If unicast messaging is employed, then some form of addressing of each individual node needs to be employed. However, no guarantees on the unicast message actually reaching the intended recipient can be given, because none of the nodes in the WSN may be aware of either route(s) to the recipient or the topology of the WSN.

In addition to being vulnerable to all of the aforementioned attacks, the Receive and Process Commands paradigm is also vulnerable to attacks where an adversary impersonates the controller and issues spurious commands.

Self-Organization. Upon deployment, the WSN self organizes, and a central controller(s) learns the network topology. Knowledge of the topology may remain at the controller (e.g. base station) or it may be shared, in whole or in part, with the nodes of the WSN. This paradigm may include the use of more powerful sensors that serve as cluster heads for small coalitions within the WSN.

This paradigm requires a strong notion of routing, therefore, in addition to being vulnerable to all of the previously listed attacks, this paradigm is vulnerable to attacks against the routing protocol. These attacks include *Induced Routing Loops, Sinkholes, Wormholes* and *HELLO Flooding*. They are presented and discussed in detail in [21].

Data Aggregation. Nodes in the WSN aggregate data from downstream nodes, incorporating their own data with the incoming data. The composite data is then forwarded to a collection point.

This paradigm is particularly vulnerable to replay attacks because the *authentication* of its downstream peers becomes an issue. In the previous paradigms, the authentication of the sensor node was left to the controller, which is not an issue because controllers are robust and considerably more powerful than the sensor nodes. In this paradigm, each sensor node that utilizes data from another sensor node now can not just forward the data as received, and therefore must ensure that the data is provided by an authorized member of the WSN.

Optimization: Flexibility and Adaptation. Predicated upon their own measurements and upon the values of incoming data, this paradigm requires that the sensors in the WSN make decisions. For example, a decision may be whether to perform a calculation, or if the cost is less, acquire the needed value from a peer, provided that the peer has the value and that knowledge is known in advance by the requester.

This operational paradigm shares the same security concerns and issues as does the Data Aggregation paradigm.

12.4 TOWARD A GENERIC SECURITY MODEL FOR WSN

The previous section illustrated that as functional and communication complexity increases, so do the vulnerabilities. The following section examines the manner in which these vulnerabilities are mitigated by the application of security measures. We show that as the communication model becomes more complex, the security measures become non-trivial and require increasingly sophisticated solutions. We conclude this section by proposing a generic security model that is applicable across all WSN operational paradigms.

The overlap between security issues and their associated solutions in infrastructure supported wireless networks and mobile ad-hoc networks (MANETs) is minimal and usually quite different from those in WSNs. Before discussing WSN security models in detail, we briefly describe the security models applied to infrastructure-supported wireless networks and MANETs noting where they are applicable to WSNs.

Infrastructure supported wireless networks, e.g., wireless LANs, use a centralized, point-to-point communication model and are user-centric – users with wireless devices access the wired infrastructure via a *wireless base station*. The base station is the sole interface between users and the wired infrastructure and is responsible for security at the link layer, although additional security measures at the network and application layer may be applied. The primary security measure implemented in the base station is link-level encryption. User authentication and access control may be applied at the base station or it may be applied at points further upstream from the base station. Regardless, these functions are invoked only when a user attempts to access the wired infrastructure via the base station. Due to the user-centric nature of these networks, authentication and access control are of greater concern than confidentiality. Additionally, the point-to-point communication model ensures that end-to-end security is enabled. The centralized nature of these networks makes them vulnerable to attacks such as spoofing, passive data gathering and denial-of service, most of which are directed at the base station. When encryption is implemented, the goal of the attacker becomes the compromise of the encryption key. The travails of the Wired Equivalent Privacy (WEP) protocol in 802.11b for wireless LANs [13] are well-documented and serve to highlight the security problems of these networks.

In contrast to wireless networks, MANETs employ a distributed, multi-hop, node-centric communication model. In a MANET, users control their wireless devices, however the device itself has some degree of autonomy. The autonomy is best illustrated by a device's choosing of the most appropriate set of neighboring devices to contact based on user requirements. Thus, communication in MANETs is node-centric, rather than user-centric. We may immediately observe that device authentication and data confidentiality are much more important than access control, which may not be relevant in certain situations. Unlike infrastructure-supported wireless networks, the security problem confronted by a MANET is to mitigate the actions of *malicious* users who may attempt to disrupt communication in the network. Thus, security protocols in MANETs employ mechanisms such as certificates to authenticate users and encrypt data using symmetric or asymmetric algorithms. Due to the fact that MANETs allow multi-hop communications, most attacks are directed against the protocols that route data between intermediate nodes on the path from source to destination. Thus, network-level security is the focus of attention in security research related to MANETs. Protocols and methods designed to address this issue include SEAD [16], Ariadne [17], security enhancements in AODV [5], secure position aided routing [8], secure ad hoc routing [29] and an on-demand secure routing protocol resilient to Byzantine failures [3]. End-to-end security is a non-trivial problem in MANETs because security protocols must rely on intermediate nodes which, depending upon their individual capabilities, may not contain all of the mechanisms required by the protocols.

Let us now explore the security requirements, features, problems and possible solutions for each of the six WSN operational paradigms described in Section 12.1.

12.4.1 SIMPLE COLLECTION AND TRANSMITTAL

In this operational paradigm, sensor nodes sense and transmit the associated data to the collection point (e.g.: base station; controller). This paradigm, therefore, uses a centralized, point-to-point, single-hop communication model similar to that in infrastructure-supported wireless networks. The primary security requirements of this paradigm are data access, authentication and data confidentiality. Confidentiality can be ensured by the use of data encryption. Symmetric key encryption methods, such as DES [28], are suitable for this paradigm. Authentication is implicitly ensured by the use of pre-deployed keys that are shared between, and unique to, the collection point and each individual sensor node [2]. Each node uses its key to encrypt data before transmission; the collection point decrypts the data using the shared key corresponding to that node.

Spread spectrum communications may be used to offset efforts to jam the frequency band. Error correcting codes offer some relief to collision inductions. End-to-end security is trivially enabled by encrypting the data frame; because each sensor node can communicate directly with the collection point.

12.4.2 FORWARDING

A straightforward solution to the "radio range" problem presented in the Simple Collection and Transmittal paradigm is to allow sensor nodes far away from the collection point to transmit data to neighboring sensor nodes, which in turn *forward* the data toward the collection point. The forwarding process may span multiple sensor nodes on the path between the source node and the collection point. Thus, this paradigm uses a centralized, multi-hop communication model. The overarching requirement to minimize energy consumption complicates possible security mechanisms.

We observe that a single shared key between the collection point and a sensor node is no longer sufficient to ensure reliable transmission of a node's data to the collection point. Why? Consider, for example, the case of a sensor node positioned two hops away from the collection point. It shares a key with the collection point and encrypts its data using this key. The sender requires one of its neighbors to forward the encrypted data to the collection point. However, due to random deployment, it is quite unlikely that the originating node is aware of the identity of any of its neighbors. Therefore, it must resort to limited broadcasting. Now, the encrypted data arrives at some intermediate node one hop away from the collection point. The question is, how does this node know that the message contains valid data and is not part of an attack mounted by some malicious node? On a similar note, how can the originating sensor node be confident that the intermediate sensor node is not malicious and will actually forward the encrypted data to the collection point instead of dropping it?

To resolve these issues, Avancha et al. [2] introduce a system that utilizes pre-built headers encrypted under the intermediate node's key. At the origin, the entire frame intended for the controller is encrypted under the senders key and inserted into another frame that is prepended with the pre-built header and broadcast. When the intermediate node receives the broadcast, it strips off the prepended header and re-broadcasts the frame. It is then received by the controller and decrypted. We note that this solution possesses limited scalability in terms of number of hops; as the number of hops increases, the number of pre-built headers to prepend also increases leading to increased message size.

Avancha et al. [2] also present a mechanism to mitigate the effects of a *Black Hole* and *Data Corruption* attacks, which could possibly be effected by a compromised intermediate node. The mechanism is in the form of an algorithm that employs counters and timers. It tracks the absence of expected data from each sensor, quantifies the amount of corrupted data received at the controller from each sensor, and compares those values to acceptable statistical norms. If the controller determines a sensor node to be aberrant, it is culled from the WSN.

12.4.3 RECEIVE AND PROCESS COMMANDS

The paradigms discussed in sections 12.4.1 and 12.4.2 above described a "many-to-one" communication model, designed exclusively for *unsolicited* data transmissions. (We define unsolicited data transmissions as those that emanate from sensor nodes without any external stimulus, such as a command, and are directed toward the controller.) As is well-known, data transmission is the most expensive of all operations in a WSN. Thus, unsolicited data transmissions, especially if they are unnecessary (e.g.: 100 sensor nodes reporting directly to a controller that the temperature in the region is 90 F), may reduce the lifetime of the WSN significantly. If transmissions could be regulated appropriately, then the lifetime of the WSN could be increased without affecting the quality of information reaching the controller. This calls for the use of a "one-to-many" communication model, where the controller transmits commands to the sensor nodes.

Consider, for example, a group of sensor nodes that are deployed in order to monitor temperature. Upon deployment, all nodes begin operating in the *idle mode*, which is a low-power mode. The controller broadcasts a *wakeup* command to a set of sensor nodes, which react to this command by transitioning to the *active* state. Subsequently, the controller broadcasts a *getdata* command, which solicits data from the sensor nodes. Finally, the controller instructs the sensor nodes to *idle* and the cycle repeats periodically. Thus, the combination of the *one-to-many* and *many-to-one* communication models is more energy-efficient than simply using the latter for unsolicited data transmission.

As expected, the cost of improved energy efficiency in this paradigm requires a more complicated security model. In the previous paradigm, the main security issue was to build *trust* among sensor nodes so that they could cooperate in data transmission to the collection point. This issue is now extended to the controller. How does a sensor node authenticate the command it received as being broadcast by the controller? This issue is further complicated by the use of multi-hop communications. A sensor node that is two hops away from the controller must depend upon some set of "intermediate neighbors". How does such a sensor node verify the integrity of the message it receives from its neighbors? How can it satisfy itself that the message was not tampered with by an intermediate node along the path from the controller?

The first issue can be addressed by the use of *broadcast authentication* as discussed in the SPINS project [30] or by the use of shared secrets between the controller and the individual sensor nodes. The second issue can be resolved by distributing encrypted identities of sensor nodes within radio range of the controller among those sensor nodes that are beyond radio range of the controller by using pre-built headers, presented in the Section 12.4.2.

12.4.4 SELF-ORGANIZATION

The operating paradigm presented in the Section 12.4.3 describes secure bi-directional communications between the controller and the sensor nodes. This paradigm in this section extends the bi-directional communication model by introducing the concept of *self-organization*. The self-organization paradigm requires that the WSN achieve organizational structure without human intervention. It consists of three primary tasks: *node discovery, route establishment* and *topology maintenance*. The accomplishment of these three tasks leads to the formation of a true WSN. While the previous paradigms used a centralized communication model, this and the following paradigms seek to employ a combination of centralized and distributed communication in order to allow the WSN to perform as efficiently and securely as possible.

In the node discovery process, the controller or a sensor node broadcasts a discovery message, e.g., a *HELLO* message. In response to this message, nodes unicast a message indicating their presence in the proximity of the broadcaster, e.g., a *HELLO-REPLY* message. This message sequence is sufficient for establishing a secure single-hop WSN. For a multi-hop WSN, this sequence must be augmented by the encrypted node identities or some other mechanism. An important point to note is that node discovery itself must be performed in a secure, authenticated manner to mitigate the effects of *traffic analysis* and replay attacks as much as possible.

Following node discovery, routes between the sensor nodes and the controller must be established. In order to ensure continuous connectivity, multiple routes between a pair of nodes may be established. An important question that arises in this context is: Should security in the WSN be end-to-end or be restricted to pair wise security between nodes? As discussed Karlof and Wagner in [21], most existing routing protocols for WSNs are vulnerable to a host of attacks including flooding , wormhole [18], sinkhole and Sybil [9] attacks. Consequently, the routing protocol is extremely important and needs to be secure.

In order to protect against attacks to the routing protocol, all routing information that is distributed throughout the WSN needs to be encrypted, be protected by an anti-replay mechanism and its source needs to be authenticated. Sensors are so resource constrained that they cannot maintain key tables containing all of their neighbors keys, nor can they use asymmetric encryption. Currently, a key shared across the WSN is the most viable solution to protecting routing information (i.e.: source and destination addresses). Moreover, as is the case whenever a sensor contains cryptographic keying material, the physical sensor needs to be configured so that tampering will erase the keys and render the sensor inoperable. The sensor also needs to be fabricated so that keys are not leaked via electronic emanations.

Topology maintenance in a WSN is unlike that in any other wireless network. In WSNs, nodes are stationary. Therefore, the topology, once established, usually does not change. However, as nodes perform their assigned tasks they deplete and eventually exhaust their energy store, causing them to die. The WSN may be "refreshed" by the periodic addition of new nodes to the WSN. The addition of new nodes to the existing network also implies additional security concerns. Both the node discovery and route establishment algorithms need to be re-run, and whether the algorithms are centralized or distributed is cause for additional concerns. If the WSN is centralized and shared secrets are used for encryption, then the keys of the new nodes must be deployed on the controller using either a secure key distribution algorithms or programmed with the same key schedule as in SPINS. If the WSN is distributed, then key management procedures must be invoked to ensure that the new nodes possess the relevant encryption keys. This is accomplished by loading the controller and the new sensor node with the shared key.

12.4.5 DATA AGGREGATION

So far, our discussion of WSNs and security has assumed that all sensor nodes transmit their data directly to the base station in either a solicited or unsolicited manner. Under this assumption, the sensors are not dependent upon the integrity or authenticity of the data. This results in the hundreds or perhaps thousands of independent *data streams*. An important problem in WSNs is to control these data streams so that unnecessary data transmissions can be eliminated and the collection point can be prevented from becoming a bottleneck.

A substantial body of research on wireless sensor networks is devoted to researching the problem of controlling data streams [19, 22, 27, 26, 34]. The prevailing solution to this problem is to *aggregate* or *fuse* data within the WSN and transmitting an aggregate of the data to the controller. The idea therefore, is to allow a sensor node to transmit its data to its neighbors, or some subset thereof. In turn, some algorithm controls which node will combine the data received from its neighbors and forward it toward the controller. This data aggregation process results in a substantial energy savings in the WSN. Typical aggregation operations include MAX, MIN, AVG, SUM and many other well-known database management techniques.

The following exemplifies the averaging (AVG) methodology. Consider the case of 100 sensor nodes deployed to measure temperature and transmit the collected data to the controller. Without data aggregation, 100 temperature readings must be transmitted, possibly multiple times, to the collection point. With data aggregation, if 1 in 10 nodes performs the AVG operation on data received from its immediate neighbors, the total number of transmissions to the collection point is reduced from 100 to only 10. The problem of controlling data streams is now reduced to choosing the most appropriate *aggregation points*, i.e., the subset of nodes to perform aggregation. A number of solutions to this problem are discussed in literature, including [22, 20, 25].

However, the implicit assumption is that the sensor nodes trust each other so that any pair of nodes can exchange data and that a node can incorporate the incoming data with its own. The problem with this assumption is that it does not take into account the very real possibility that an adversary may have deployed malicious data sources in the WSN for nefarious purposes. For example, a malicious node may endeavor to have itself elected as an aggregation point and then throw away all of the data that it receives from its neighbors, or even worse, transmit corrupt or fictitious data to upstream neighbors. Again, this operational paradigm requires an even more complex set of security controls. At first glance, the problem may seem trivial, given that secure routes would have been established during the self-organization process described in Section 12.4.4. However, the secure routes were established without considering data aggregation and the choice of aggregation points. Thus, if end-to-end security were established between the controller and each sensor node in the WSN, then data aggregation is not possible. On the other hand, point-to-point security allows complete flexibility in the choice of aggregation points, but is not a scalable solution because each node would have to keep cryptographic information for all other nodes in the WSN. Thus, a simple solution to the *secure data aggregation* problem is the use of a key that is common to the WSN. The intrinsic weakness in this approach is the use of a common key, if it is somehow compromised then so is the entire WSN. In [15] Hu and Evans discuss a time delayed protocol to securely aggregate data in wireless networks.

12.4.6 OPTIMIZATION: FLEXIBILITY AND ADAPTABILITY

The WSN paradigms thus far considered focus on the data gathering and reporting functions of a WSN. The nodes in the WSN are not concerned with the *semantics* of the data they have obtained through the sensing task. The sole concern is that it must be transmitted elsewhere, possibly for further analysis. To this end, all the nodes execute a fixed set of protocols, one protocol per the link, network and security layers, irrespective of the environmental and security conditions affecting the WSN. Inflexibility in the choice of protocols and the inability to adapt to changing conditions could render a WSN inoperable or cause it to function sub-optimally. Here, optimality encompasses both energy usage and security. Consider, for example, a WSN that uses a centralized, point-to-point communication model as described in Section 12.4.1. Given the cost of transmitting at full power, the utility of this type of WSN is limited to an application that requires surveillance and monitoring in a relatively small area, such as a single room in a building. By implication, the same WSN cannot be used for monitoring purposes over large areas, such as a bridge or stadium. On the other hand, an agile WSN, one that is capable of functioning in either a centralized mode or a distributed mode can be deployed without the constraints imposed by the area of coverage. Furthermore, if the WSN is provided with multiple combinations of link, network, security and aggregation protocols, then it can dynamically choose a particular combination based upon existing environmental and security conditions. Finally, permitting the WSN to use the data that it collects from the environment, to make on-the-fly decisions regarding protocol execution will enable it to self-optimize in terms of energy usage and security profile. For example, if the WSN senses harmful chemicals in the environment along with a sudden increase in radio noise levels (indicating a Denial of Service attempt by some adversary), it can take appropriate countermeasures by re-routing around the jammed areas. Thus, the WSN exhibits both flexibility (in terms of protocol execution) and adaptability (in terms of protecting itself), which are most desirable attributes of a WSN.

12.4.7 A GENERIC WSN SECURITY MODEL

We now describe the principal components of an ideal, generic security model for wireless sensor networks. Some components of this model, such as communication security and key management, have been and continue to be topics of active research; others, such as data aggregation and self-healing, have yet to receive a considerable amount of attention.

Communication model: Hybrid communication employing both centralized and distributed models; the centralized model is used when one or more powerful nodes exist, around which less-powerful sensor nodes can *cluster* and the distributed model is employed when no powerful nodes exist. These models can be used together at the same time to form a hierarchical WSN – the centralized model to first form clusters and the distributed model for *inter-cluster* communication.

Communication security: As was the case with the communication model, the mechanisms to secure communication between nodes are also deployed in a hybrid manner. In the case where more powerful nodes exist and clusters can be formed, end-to-end communication security between the designated *clusterhead* and each individual sensor node in the cluster should be used. Subsequently, inter-cluster communication security, i.e., communication between clusterheads, should be pair wise. In the absence of more powerful nodes, as the WSN is formed in a distributed manner, it is appropriate to employ pair wise security, but only for a fixed number of pairs. This is because pair wise security is not scalable as the number of nodes in the WSN increases. Thus, pure pair wise security is not feasible in the ideal security model.

Key management: Due to the fact that most sensor nodes in a WSN have limited amount of energy, public-key cryptographic mechanisms, which are expensive in terms of energy consumption, are not suitable to WSNs. Private-key cryptography, on the hand, is quite applicable to WSNs due to its low energy requirements. However, in a hybrid WSN that consists of nodes of varying capabilities and resources, it is feasible to employ both public-key and private key mechanisms for security. Thus, intra-cluster communications are secured by private-key cryptography and inter-cluster communications via public-key cryptography. An additional problem in WSNs, although not unique to them, is key distribution. The principal mechanisms to solve this problem are pre-deployed keys (i.e., offline key distribution), group keying and arbitrated keying [7]. In the ideal model, all three mechanisms are interchangeably used predicated upon the exact composition of the WSN. If the WSN consists of clusters, then either arbitrated or group keying is appropriate. A flat topology in the WSN calls for pre-deployed keys so that either pair wise or end-to-end security may be employed by the nodes.

Data aggregation: In the ideal security model, data aggregation can be performed as often as required and in a manner that conforms to the security requirements. Additionally, based on chosen communication model, different aggregation algorithms can be executed at different points during the lifetime of the WSN. For example, at start up, the WSN may have a single controller and many simple sensor nodes. In this case, data aggregation may be performed by the controller itself. This calls for a data aggregation algorithm that is suitable to the centralized communication model. As additional nodes are added, the WSN may switch to the distributed communication model; data aggregation should be performed either by multiple designated or elected nodes based on the security mechanism, i.e., end-to-end or pair wise, that is in effect at that point in time.

Self-healing: Security models for WSNs face problems not only in the form of external attacks on the network, but also in the form of breakdowns due to node failure, especially due to energy exhaustion. The ideal model is able to withstand the various types of attacks detailed in Section 12.3, employing both passive and active mechanisms. Passive mechanisms include data encryption and node authentication, while active mechanisms include key revocation and removing offending nodes from the WSN. In the ideal model, security countermeasures exist at every layer: spread-spectrum techniques at the link-layer, encryption at the network and application layers, authentication at the application layer and aberrant behavior detection at the network and application layers. It may not be feasible or required to activate all these mechanisms at the same time, rather they are invoked in an application-specific and environment-specific manner. Node failure causes route breakdown and may cause failure of end-to-end or pair wise communication security, possibly leading to network partition. The ideal security model consists of mechanisms to monitor and track the health of all nodes in the network, thereby enabling quick (re)-establishment of secure routes around nodes that provide indications of imminent failure. Based on the communication model, this may require invoking procedures to distribute keys to neighboring nodes as required.

12.5 WSN SECURITY: STATE-OF-THE-ART

Security aspects of wireless sensor networks have received little attention compared to other aspects. Key management in sensor networks has been dealt with to a certain extent, but research and development of security architectures has been less extensive. In this section, we present a brief overview of various key management protocols and security architectures for WSNs, including our contributions.

12.5.1 KEY MANAGEMENT

Basagni et al. [4] present a key management scheme for pebblenets, defined as large ad hoc networks consisting of nodes of limited size called pebbles. The key management scheme uses symmetric cryptography. In this scheme, each pebble is equipped with a group identity key which enables it to participate in key management. Data traffic is secured using a global key shared by all nodes, called the Traffic Encryption Key (TEK). TEKs are periodically refreshed. TEK generation and distribution requires selection of a key manager to perform these tasks. The goal of this work, therefore, is to select a key manager. The protocol designed to achieve this goal consists of two phases. First, pebbles organize into a cluster with a clusterhead. The clusterheads subsequently organize into a backbone. Finally, a fraction of the clusterheads of the backbone is selected among which a pebble is chosen as the new key manager.

Carman et al. [7] have conducted a detailed study of various keying protocols applicable to distributed sensor networks. They classify these protocols under pre-deployed keying, arbitrated protocols, self-enforcing autonomous keying protocols and hybrid approaches. The authors also present detailed comparisons between various keying protocols in terms of energy consumption.

Eschenauer and Gligor [10] present a key-management scheme for sensor network security and operation. The scheme includes selective distribution and revocation of keys to sensor nodes as well as node re-keying without substantial computation and communication capabilities. It relies on probabilistic key sharing among the nodes of a random graph and uses simple protocols for shared-key discovery and path-key establishment, and for key revocation, re-keying and incremental addition of nodes. The security and network connectivity characteristics supported by the key-management scheme are discussed.

12.5.2 SECURITY ARCHITECTURES

The Security Protocols for Sensor Networks (SPINS) project [30] consists of two main threads of work: an encryption protocol for SmartDust motes called Secure Network Encryption Protocol (SNEP) and a broadcast authentication protocol called micro-Timed, Efficient, Streaming, Loss-tolerant Authentication (μTESLA). In SPINS, each sensor node shares a unique master key with the base station. Other keys required by the SNEP and the μTESLA protocols are derived from this master key. SNEP is based on Cipher Block Chaining implemented in the Counter mode (CBC-CTR), with the assumption that the initial value of the counter in the sender and receiver is the same. Thus, the sender increments the counter after sending an encrypted message and the receiver after receiving and decrypting it. To achieve authenticated broadcasts, μTESLA uses a time-released key chain. The basic idea resolves around the unidirectionality of one-way functions. There are two requirements for correct functioning of this protocol: (i) the owner of the key release schedule has to have enough storage for all the keys in the key chain (ii) every node in the network has to at least be loosely time synchronized, i.e. with minor drifts. The time-released key chain ensures that messages can be authenticated only after receiving the appropriate key in the correct time slot.

Karlof and Wagner [21] consider routing security in sensor networks. This work proposes security goals for sensor networks, presents classes of attacks and analyzes the security of well-known sensor network routing protocols and energy-conserving topology maintenance algorithms. The authors conclude that all the protocols and algorithms are insecure and suggest potential countermeasures. The attacks discussed in this work include bogus routing information, selective forwarding, sinkholes, Sybil, wormholes and HELLO flooding.

Communication security in wireless sensor networks is addressed in [32]. The approach in this work is to classify the types of data that typically exist in sensor networks and to identify possible communication security threats according to that classification. The authors propose a scheme in which each type of data is secured by a corresponding security mechanism. This multi-tiered security architecture where each mechanism has different resource requirements, is expected to enable efficient resource management.

Law et al. [23] discuss security aspects of the EYES project, which is concerned with self-organizing, collaborative, energy-efficient sensor networks. This work contains three contributions. The first is a survey that discusses the dominant issues of energy-security trade-off in network protocol and key management design. This survey is used to chart future research directions for the security framework in EYES. Second, the authors propose an assessment framework based on a system profile that enables application classification. Third, some well-known cryptographic algorithms for typical sensor nodes are benchmarked. This work also investigates resource requirements of symmetric key algorithms RC5 and TEA.

Our contributions to WSN security include two security architectures that employ centralized and distributed security models respectively. The first architecture [2], useful for applications such as perimeter protection, assumes the existence of a single controller in the WSN. Keys are pre-deployed on all sensor nodes; each sensor node shares a unique key with the controller. We use DES with 64-bit keys for encryption. The controller is responsible for secure node discovery, route establishment and topology maintenance. A unique feature of this architecture is its invulnerability to traffic analysis due to end-to-end encryption. Additionally, the architecture consists of a network repair protocol that detects and eliminates from the network, aberrant nodes – those that have either been compromised by some adversary or have exhausted their energy. The second architecture [1], employs a clustering approach to form a secure WSN. This architecture assumes the existence of a few powerful nodes around which other sensor nodes cluster. The clustering protocol is a modification of the centralized protocol used in the first architecture. The powerful nodes, called clusterheads, are responsible for secure self-organization within the cluster. After the formation of clusters, the clusterheads form a chain in order to be able to exchange and aggregate data generated in individual clusters. The controller may be either static or mobile and receives the final aggregated data from one of the clusterheads. Intra-cluster security is end-to-end, inter-cluster security is pair wise in this architecture.

12.6 STATE-OF-THE-ART TO IDEAL

In this section, we discuss the research effort required to bridge the gap between the state-of-the-art in WSN security and the ideal, in terms of the principal components discussed in Section 12.4.7.

Let us consider the current security architectures for WSNs and the requirements of communication models with integrated communication security. It is evident that the principal focus of WSN security has been on centralized approaches; there is a need to develop distributed approaches and combine them with the centralized approaches to design robust hybrid models for communication security. Our work [1] attempts to make progress in this direction. The use of hybrid models for communication security will enable easier integration of data aggregation and key management algorithms. This will also ensure flexibility and adaptability of the WSN; self-healing mechanisms and the ability to react to changing conditions can also be easily integrated.

Both centralized and distributed key management techniques for WSNs have been discussed in literature. Research efforts directed toward this problem have shown that key distribution in WSNs can be energy-efficient and secure under certain conditions. However, we observe that distributed key management techniques, as discussed in literature, are completely independent of any security architecture. For example, the work on secure pebblenets [4] is mainly concerned with choosing a key manager in every round, but does not address the issues involved in using the key for encryption, authentication or other security functions. On the other hand, the SPINS project [30] and our efforts [2, 1] assume pre-deployed keying in the entire security architecture. The ideal security model will consist of a combination of robust, energy-efficient, secure key distribution mechanisms with well-defined, comprehensive security architectures.

A similar situation is observed when security architectures and data aggregation algorithms are considered. Current security architectures do not really consider the issue of integrating data aggregation algorithms; rather they assume that designated nodes, such as the controller or clusterheads, will perform the required aggregation function(s). On the other hand, data aggregation algorithms assume complete and unhindered co-operation among all sensor nodes in the WSN as far as performing the aggregation function(s) is concerned. This assumption is non-trivial; a security protocol that supports such a co-operation model will not be scalable because *pair wise* communication security is required across the WSN. Thus, integration of well-known, energy-efficient data aggregation algorithms [22, 20, 25] with robust security architectures is essential in designing the ideal security model.

Flexibility, adaptability and self-healing mechanisms are essential to the functioning of a WSN and optimal resource use during its lifetime. None of the existing security architectures use the data associated with sensed environmental conditions to help detect the beginnings of attacks or of aberrant behavior by nodes. This reduces the ability of the WSN to protect itself from attacks mounted within and outside the network. In fact, detecting and preventing attacks from within, i.e., attacks mounted by compromised nodes, is a harder problem than preventing external attacks such as jamming. Thus, the move toward the ideal security model calls for the design and development of compact, lightweight mechanisms to capture and *reason* over data describing environmental and security conditions.

12.7 CONCLUSIONS

Improvements in wireless networking and micro-electro-mechanical systems (MEMS) are contributing to the formation of a new computing domain – distributed sensor networks. These ad-hoc networks of small, fully programmable sensors will be used in a variety of applications: on the battlefield, as medical devices, in equipment maintenance and in perimeter security systems.

Unless security is considered during the design of the physical sensor, its protocols and operational models, sensor networks will remain vulnerable to attacks at several different levels.

This chapter identified the vulnerabilities of the operational paradigms currently used by Wireless Sensor Networks and presented security mechanisms to mitigate and lessen those vulnerabilities. An all encompassing generic security model, operating across all operational paradigms, was proffered and suggestions for moving the current state of the art WSN security to that model was suggested.

REFERENCES

[1] S. Avancha, J. Undercoffer, A. Joshi, and J. Pinkston. A Clustering Approach to Secure Sensor Networks. Technical report, University of Maryland Baltimore County, 2003.

[2] S. Avancha, J. Undercoffer, A. Joshi, and J. Pinkston. Secure Sensor Networks for Perimeter Protection. *accepted for publication in Special Issue Computer Networks on Wireless Sensor Networks*, 2003.

[3] B. Awerbuch, D. Holmer, C. Nita-Rotaru, and H. Rubens. An On-Demand Secure Routing Protocol Resilent to Byzantine Failures. In *ACM Workshop on Wireless Security (WiSe)*, September 2002.

[4] S. Basagni, K. Herrin, D. Bruschi, and E. Rosti. Secure Pebblenets. In *Proc. of MobiHOC '01*, pages 156–163, October 2001.

[5] S. Bhargava and D. P. Agrawal. Security Enhancements in AODV proocol for Wireless Ad hoc Networks. In *Vehicular Technology Conference*, May 2001.

[6] Stuart Boyer. *SACDA: Supervisory Control and Data Acquisition*. ISA, Triangle Park, NC, January 1999.

[7] D. W. Carman, P. S. Kruus, and B. J. Matt. Constraints and Approaches for Distributed Sensor Network Security (Final). Technical Report 00-010, NAI Labs, 2000.

[8] S. Carter and A. Yasinsac. Secure Position Aided Ad hoc Routing Protocol. In *Proceedings of the IASTED International Conference on Communications and Computer Networks (CCN02)*, November 2002.

[9] J. R. Douceur. The Sybil Attack. In *Proc. IPTPS '02*, March 2002.

[10] L. Eschenauer and V. Gligor. A Key Management Scheme for Distributed Sensor Networks. In *Proc. of ACM CCS2002*, November 2002.

[11] X. Fu, B. Graham, R. Bettati, and W. Zhao. On Countermeasures to Traffic Analysis Attack. In *Fourth IEEE SMC Information Assurance Workshop*, 2003.

[12] Y. Guan, X. Fu, D. Xuan, P. U. Shenoy, R. Bettati, and W. Zhao. Netcamo: Camouflaging network traffic for QOS-guaranteed critical applications. *IEEE Trans. on Systems, Man, and Cybernetics Part a: Systems and Humans, Special Issue on Information Assurance*, pages 253–265, July 2001.

[13] R. Housley and W. Arbaugh. Security problems in 802.11-based networks. *Communications of the ACM*, 46(5):31–34, 2003.

[14] Spencer S. Hsu. Sensors may track terror's fallout. The Washington Post. Page A01, June 2, 2003.

[15] L. Hu and D. Evans. Secure aggregation for wireless networks. In *Workshop on Security and Assurance in Ad hoc Networks*, pages 384 – 391, January 2003.

[16] Y. Hu, D. B. Johnson, and A. Perrig. SEAD: Secure Efficient Distance Vector Routing for Mobile Wireless Ad Hoc Networks. In *Proceedings of the 4th IEEE Workshop on Mobile Computing Systems and Applications (WMCSA 2002)*, pages 3–13, June 2002.

[17] Y. Hu, A. Perrig, and D. B. Johnson. Ariadne: A Secure On-Demand Routing Protocol for Ad hoc Networks. In *MobiCom 2002*, September 2002.

[18] Y. Hu, A. Perrig, and D. B. Johnson. Wormhole Detection in Wireless ad hoc Networks. Technical Report TR01-384, Department of Computer Science, Rice University, June 2002.

[19] C. Intanagonwiwat, R. Govindan, and D. Estrin. Directed Diffusion: A Scalable and Robust Communication Paradigm for Sensor Networks. In *Proc. of MobiCom '00*, August 2000.

[20] K. Kalpakis, K. Dasgupta, and P. Namjoshi. Efficient Algorithms for Maximum Lifetime Data Gathering and Aggregation in Wireless Sensor Networks. Technical Report TR-CS-02-13, University of Maryland Baltimore County, 2002.

[21] C. Karlof and D. Wagner. Secure Routing in Sensor Networks: Attacks and Countermeasures. In *Proc. of First IEEE International Workshop on Sensor Network Protocols and Applications*, May 2003.

[22] B. Krishnamachari, D. Estrin, and S. B. Wicker. Modeling Data-Centric Routing in Wireless Sensor Networks. Technical Report CENG 02-14, Dept. of Computer Engineering, USC, 2002.

[23] Y. W. Law, S. Dulman, S. Etalle, and P. Havinga. Assessing Security-Critical Energy-Efficient Sensor Networks. In *Proc. of 18th IFIP International Information Security Conference*, May 2003.

[24] Robert Lewis. *Modelling Distributed Control Systems Using IEC 61499*. IEE, UK, February 2001.

[25] S. Lindsey, C. S. Raghavendra, and K. M. Sivalingam. Data Gathering Algorithms in Sensor Networks Using Energy Metrics. *IEEE Transactions on Parallel and Distributed Systems*, 13(9):924–935, September 2002.

[26] S. R. Madden and M. J. Franklin. Fjording the Stream: An Architecture for Queries over Streaming Sensor Data. In *Proc. of 18th International Conference on Data Engineering*, February 2002.

[27] S. R. Madden, M. J. Franklin, J. M. Hellerstein, and W. Hong. TAG: A Tiny AGgregation Service for Ad-Hoc Sensor Networks. In *Proc. of Fifth Symposium on Operating Systems Design and Implementation (USENIX - OSDI '02)*, December 2002.

[28] National Institute of Standards and Technology. *FIPS 46-2; Data Encryption Standard*, December 1993.

[29] P. Papadimitratos and Z. J. Haas. Secure Routing for Mobile Ad hoc Networks. In *Communication Networks and Distributed Systems Modeling and Simulation Conference (CNDS 2002)*, January 2002.

[30] A. Perrig, R. Szewczyk, V. Wen, D. Culler, and J. D. Tygar. SPINS: Security Protocols for Sensor Networks. *Wireless Networks Journal (WINET)*, 8(5):521–534, September 2002.

[31] J. Raymond. Traffic Analysis: Protocols, attacks, design issues and open problems. In H. Federrath, editor, *Designing Privacy Enhancing Technologies: Proceedings of International Workshop on Design Issues in Anonymity and Unobservability*, volume 2009 of *LNCS*, pages 10–29, 2001.

[32] S. Slijepcevic, M. Potkonjak, V. Tsiatsis, S. Zimbeck, and M. B. Srivastava. On Communication Security in Wireless Ad-Hoc Sensor Networks. In *Proc. of WETICE*, pages 139–144, 2002.

[33] Tony Wilson. Cybercrime's New Foe. The Gold Coast Bulletin (Australia). Page 14, October 25, 2002.

[34] Y. Yao and J. E. Gehrke. The Cougar Approach to In-Network Query Processing in Sensor Networks. *SIGMOD Record*, 31(3):9–18, September 2002.

Chapter 13

KEY DISTRIBUTION TECHNIQUES FOR SENSOR NETWORKS

Haowen Chan, Adrian Perrig, and Dawn Song
Carnegie Mellon University, Pittsburgh, PA 15213
{ haowen,perrig,dawnsong} @cmu.edu

Abstract This chapter reviews several key distribution and key establishment techniques for sensor networks. We briefly describe several well known key establishment schemes, and provide a more detailed discussion of our work on random key distribution in particular.

Keywords: Sensor network, key distribution, random key predistribution, key establishment, authentication.

13.1 INTRODUCTION

The general *key distribution* problem refers to the task of distributing secret keys between communicating parties in order to facilitate security properties such as communication secrecy and authentication.

In sensor networks, key distribution is usually combined with initial communication establishment to bootstrap a secure communication infrastructure from a collection of deployed sensor nodes. These nodes may have been pre-initialized with some secret information but would have had no prior direct contact with each other. This combined problem of key distribution and secure communications establishment is sometimes called the *bootstrapping problem*. A bootstrapping protocol must not only enable a newly deployed sensor network to initiate a secure infrastructure, but it must also allow nodes deployed at a later time to join the network securely. This is a highly challenging problem due to the many limitations of sensor network hardware and software.

In this chapter, several well-known methods of key distribution will be discussed and evaluated. Besides these, an in-depth study on one particular method (random key predistribution) will be presented.

13.2 SENSOR NETWORK LIMITATIONS

The following characteristics of sensor networks complicate the design of secure protocols for sensor networks, and make the bootstrapping problem highly challenging.

- *Vulnerability of nodes to physical capture.* Sensor nodes may be deployed in public or hostile locations (such as public buildings or forward battle areas) in many applications. Furthermore, the large number of nodes that are deployed implies that each sensor node must be low-cost, which makes it difficult for manufacturers to make them tamper-resistant. This exposes sensor nodes to physical attacks by an adversary. In the worst case, an adversary may be able to undetectably take control of a sensor node and compromise the cryptographic keys.

- *Lack of a-priori knowledge of post-deployment configuration.* If a sensor network is deployed via random scattering (e.g. from an airplane), the sensor network protocols cannot know beforehand which nodes will be within communication range of each other after deployment. Even if the nodes are deployed by hand, the large number of nodes involved makes it costly to pre-determine the location of every individual node. Hence, a security protocol should not assume prior knowledge of which nodes will be neighbors in a network.

- *Limited bandwidth and transmission power.* Typical sensor network platforms have very low bandwidth. For example, the UC Berkeley Mica platform's transmitter has a bandwidth of 10 Kbps, and a packet size of about 30 bytes. Transmission reliability is often low, making the communication of large blocks of data particularly expensive.

13.3 THE PROBLEM OF BOOTSTRAPPING SECURITY IN SENSOR NETWORKS

A bootstrapping scheme for sensor networks needs to satisfy the following requirements:

- Deployed nodes must be able to establish secure node-to-node communication.

- Additional legitimate nodes deployed at a later time can form secure connections with already-deployed nodes.

- Unauthorized nodes should not be able to gain entry into the network, either through packet injection or masquerading as a legitimate node.

- The scheme must work without prior knowledge of which nodes will come into communication range of each other after deployment.

- The computational and storage requirement of the scheme must be low, and the scheme should be robust to DoS attacks from out-of-network sources.

13.3.1 EVALUATION METRICS

Sensor networks have many characteristics that make them more vulnerable to attack than conventional computing equipment. Simply assessing a bootstrapping scheme based on its ability to provide secrecy is insufficient. Listed below are several criteria that represent desirable characteristics in a bootstrapping scheme for sensor networks.

- *Resilience against node capture.* It is assumed the adversary can mount a physical attack on a sensor node after it is deployed and read secret information from its memory. A scheme's resilience toward node capture is calculated by estimating the fraction of total network communications that are compromised by a capture of x nodes *not including* the communications in which the compromised nodes are directly involved.

- *Resistance against node replication.* Whether the adversary can insert additional hostile nodes into the network after obtaining some secret information (e.g. through node capture or infiltration). This is a serious attack since the compromise of even a single node might allow an adversary to populate the network with clones of the captured node to such an extent that legitimate nodes could be outnumbered and the adversary can thus gain full control of the network.

- *Revocation.* Whether a detected misbehaving node can be dynamically removed from the system.

- *Scalability.* As the number of nodes in the network grows, the security characteristics mentioned above may be weakened.

Each bootstrapping protocol usually involves several steps. An *initialization* procedure is performed to initialize sensor nodes before they are deployed. After the sensor nodes are deployed, a *key setup* procedure is performed by the nodes to set up shared secret keys between some of the neighboring nodes to establish a secure link.

13.4 USING A SINGLE NETWORK-WIDE KEY

The simplest method of key distribution is to pre-load a single network-wide key onto all nodes before deployment. After deployment, nodes establish communications with any neighboring nodes that also possess the shared network key. This can be achieved simply by encrypting all communications in the shared network-wide key and appending *message authentication codes* (MACs) to ensure integrity.

The properties of the single network-wide key approach are as follows:

- *Minimal memory storage required.* Only a single cryptographic key is needed to be stored in memory.

- *No additional protocol steps are necessary.* The protocol works without needing to perform key discovery or key exchange. This represents savings in code size, node hardware complexity and communication energy costs.

- *Resistant against DoS, packet injection.* The MACs guard against arbitrary packet injection by an adversary that does not know the network-wide key k. Replay attacks can be prevented by including the source, destination, and a timestamp in the message. A worst case denial of service attack would be to replay large numbers of packets in order to force nodes to perform many MAC verifications. However, symmetric cryptographic MAC verifications are extremely fast and this kind of DoS attack would not be very effective at preventing normal operation.

The main drawback of the network-wide key approach is that if just a single node is compromised, the entire network loses all its security properties, since the network-wide key is now known to the adversary. This makes it impractical except in two possible scenarios.

- *The nodes are tamper-resistant.* In this case, tamper resistance is incorporated into the sensor nodes such that it becomes impractical for an adversary to attempt extraction of the network-wide key. In general, this is impractical for critical sensor systems (such as security or safety applications) since adversaries attacking these systems will have a large incentive to defeat the tamper-resistance. However, low cost tamper-resistance may actually be feasible for non-critical sensor applications such as domestic temperature monitoring.

- *No new nodes are ever added to the system after deployment.* In this case, the sensor nodes use the network wide key to encrypt unique *link keys* which are exchanged with each of their neighbors. For example, node A might generate a unique key k_{AB} and send it to B under encryption by the network-wide key. Once the link keys are in place, all communications are encrypted using the appropriate link keys and the network-wide key is erased from the memory of the nodes. Any node which is subsequently compromised thus reveals no secret information about the rest of the network. In general such an approach is also somewhat impractical since it is usually desirable to have the ability to add new nodes to the network to replace failed or exhausted nodes. A possible way to address this would be to perform a large-scale audit of all sensor nodes prior to every phase of adding new nodes. The audit would have to ensure that every node's hardware and software has not been altered maliciously. Once this has been ascertained, a new network-wide key could be distributed to the nodes to enable addition of the new nodes. However, such a comprehensive audit would probably be too costly to be practical in most applications.

13.5 USING ASYMMETRIC CRYPTOGRAPHY

The favored method of key distribution in most modern computer systems is via asymmetric cryptography, also known as public key methods. If sensor node hardware is able to support asymmetric cryptographic operations, then this is a potentially viable method of key distribution.

A brief outline of a possible public-key method for sensor networks is as follows. Prior to deployment, a master public/private keypair, (K_M, K_M^{-1}) is first generated. Then, for every node A, its public/private keypair (K_A, K_A^{-1}) is generated. This keypair is stored in node A's memory along with the master public key K_M and the master key's signature on A's public key. Once all the nodes are initialized in this fashion, they are ready for deployment.

Once the nodes have been deployed, they perform key exchange. Nodes exchange their respective public keys and master key signatures. Each node's public key is verified as legitimate by verifying the master key's signature using the master public key which is known to every node in the network. Once the public key of a node has been received, a symmetric link key can be generated and sent to it, encrypted by its public key. Upon reception of the session key, key establishment is complete and the two nodes can communicate using the symmetric link key.

The properties of this approach are as follows:

- *Perfectly resilient against node capture.* Capture of any number of nodes does not expose any additional communications in the network, since these nodes will have no knowledge of any secret link keys besides the ones that they are actively using.

- *Possible to revoke known compromised keypairs.* Revocation can be performed by broadcasting the entire revoked keypair, signed by the master key. Nodes receiving the broadcast can authenticate it as coming from the central authority and ignore any future communications purporting to originate from the revoked keypair. If a large number of keypairs are to be revoked, the master keypair itself could be updated by broadcasting an updated new master key signed by the old master key, then unicasting the new master key's signature on each of the legitimate public keys (the unicast is encrypted in the respective nodes' public keys). The revoked keypairs will not be signed by the new master key and will thus be rejected as invalid by all nodes in the network.

- *Fully scalable.* Signature schemes function just as effectively regardless of the number of nodes in the network.

However, using asymmetric cryptography has its disadvantages, as follows:

- *Dependence on asymmetric key cryptographic hardware or software.* All known asymmetric cryptographic schemes involve computationally intensive mathematical functions, for example, modular exponentiation of large numbers. Basic sensor node CPU hardware, however, is extremely limited, often lacking even an integer multiply instruction. In order to implement asymmetric cryptography on sensor nodes, it is necessary to either implement dedicated cryptographic hardware on a sensor node, thus increasing its hardware cost, or encode the mathematical functions in software, thus reducing the amount of code space and memory for sensing functions. Given that asymmetric cryptography is only used for key setup upon node deployment, this represents a tiny fraction of the sensor node's lifetime. Significantly increasing the cost of a node is thus difficult to justify.

- *Vulnerability to denial-of-service.* Asymmetric cryptographic operations involve significant amounts of computation and it can take up to minutes of intense processing for a sensor node to complete verification of a single signature. Thus, the nodes are vulnerable to a battery exhaustion denial of service attack where they are continuously flooded with illegal signatures. The sensor nodes will attempt to verify each signature, consuming valuable battery power in the process, while not being able to establish connections with the legitimate nodes. Since this denial-of-service can only occur during the key establishment phase, the sensor network is only vulnerable when it is newly deployed or when additional nodes are being added to the network. Increased monitoring of the site for adversarial transmissions during those times could alleviate the DoS problem. However, such DoS attacks make it infeasible for asymmetric cryptography to be used in sensor networks deployed in large or unmonitored areas such as battlefields.

- *No resistance against node replication.* Once a node is captured, its keypair can be used to set up links with every single one of the nodes in the network, effectively making the node "omnipresent". This could potentially put it in a position to subvert the routing infrastructure of alter sensor network operation. Such an attack is not difficult to prevent with some added countermeasures. For example, each time a node forms a connection with another node, they could both report the event to a base station encrypted using their master public key. Any node that has an exceptionally high degree could then be immediately revoked.

13.6 USING PAIRWISE KEYS

In this approach, every node in the sensor network shares a unique symmetric key with every other node in the network. Hence, in a network of n nodes, there are a total of $^nC_2$ unique keys. Every node stores $n - 1$ keys, one for each of the other nodes in the network.

After deployment, nodes must perform key discovery to verify the identity of the node that they are communicating with. This can be accomplished with a challenge/response protocol.

The properties of this approach are as follows:

- *Perfect resilience to node capture.* Similar to the asymmetric cryptography scheme, any node that is captured reveals no information about the communications being performed in any other part of the network. Its pairwise keys could be used to perform a node replication attack throughout the network, but this could be countered using the same method as described for asymmetric cryptography in the previous section.

- *Compromised keys can be revoked.* If a node is detected to be compromised, its entire set of $n - 1$ pairwise keys is simply broadcast to the network. No authentication is necessary. Any node that hears a key in its set of pairwise keys broadcast in the open immediately stops using it. This effectively cuts off the revoked node from the network.

- *Only uses symmetric cryptography.* The pairwise keys scheme achieves many of the benefits of using asymmetric cryptography without needing dedicated hardware or software to before the more complex asymmetric cryptographic primitives. This not only makes the nodes cheaper but also makes the network less vulnerable to energy-sapping denial of service attacks.

The main problem with the pairwise keys scheme is poor scalability. The number of keys that must be stored in each node is proportional to the total number of nodes in the network. With an 80 bit key, a hundred node network will require 1kB of storage on each node for keys alone. This means it will probably be prohibitively expensive to scale this scheme up to thousands of nodes.

13.7 BOOTSTRAPPING SECURITY OFF THE TRUSTED BASE STATION

This method of key distribution uses a trusted, secure base station as an arbiter to provide link keys to sensor nodes. The sensor nodes authenticate themselves to the base station, after which the base station generates a link key and sends it securely to both parties. An example of such a protocol is contained in SPINS [20].

A sketch of the events of the protocol could be as follows. Prior to deployment, a unique symmetric key is generated for each node in the network. This node key is stored in the node's memory and will serve as the authenticator for the node as well as facilitate encrypted communications between the node and the base station. The base station has access to all the node keys either directly (they are stored in its memory) or indirectly (the base station relays all communications to a secured workstation off site).

This method, unlike the other methods mentioned previously, assumes some level of reliable transport is available between the node and the base station before any key establishment has taken place. Since this transport occurs before any security primitives are in place, it will necessarily have to be assumed as insecure, however, as long as it is reliable in a way such that a small number of malicious nodes are unable to prevent the transmission of messages to and from the base station then the protocol presented here is viable. *Flooding* (where nodes naively re-broadcast any heard messages until the entire network is reached) is a simple example of such a reliable method of transport.

Now assume that after deployment, the node A wants to establish a shared secret session key SK_{AB} with node B. Since A and B do not share any secrets, they need to use a trusted third party S, which is the base station in our case. Both A and B share a master secret key with the base station, $\mathcal{X}_{AS}$ and $\mathcal{X}_{BS}$, respectively.

The properties of this method of key establishment are as follows.

- *Small memory requirement.* For every node, a secret symmetric key shared with the base station is needed, as well as one unique link key for each one of its neighbors. This is a very small set of keys. Further the code involved is extremely simple and requires only communication and symmetric cryptographic primitives.

- *Perfect resilience to node capture.* Any node that is captured divulges no secret information about the rest of the network.

- *Revocation of nodes is simple.* The base station has a record of all nodes that have established a link key with any given node. If a node is to be revoked, the base station securely transmits the revocation message to all the nodes that may be in communication with the revoked node. This revocation message is encrypted with the secret key that is shared only between the recipient node and the base station, and hence secrecy and authentication are ensured. To prevent any other nodes from establishing links with the revoked node, the base station simply needs to reject requests that involve the revoked node as a principal.

- *Node replication is easily controlled.* Since all key establishment activity takes place through the central base station, auditing becomes trivial. For example, the base station can immediately tell the current degree (number of established secure links) of any node. If this number is too high, then it refuses to generate any new link keys for that node. Thus node replication is effectively restricted.

However, key establishment through a base station has its disadvantages, as follows.

- *Not scalable – significant communication overhead.* If any two nodes wish to establish a secure communications, they must first communicate directly with the base station. In a large network, the base station may be many hops away, thus incurring a significant cost in communication. Worse still, since all large number of communications has the same source or destination (i.e. the base station), this may lead to contention and congestion at the nodes closest to the base station. The problem is further exacerbated by the fact that the transport must be able to function reliably without having any security primitives to rely on. This generally means that multiple transmission paths have to be used to prevent a malicious node from blocking a transmission. The communication overhead is thus further increased.

- *The base station becomes a target for compromise.* Since the base station has access to all the secret node keys in the sensor network, compromise of the base station's key store will expose the secrecy of all links that are established after the time of the compromise. This may not be a problem if the communications base station merely acts as a gateway to a workstation at a remote, secured site, since the adversaries would have to successfully attack the secure workstation in order to gain the node keys. However, such a set up is not feasible for all applications, particularly sensor networks deployed far afield, for example, for seismic or wildlife monitoring. In such a situation the node keys have to be stored on the base station itself, which exposes them to attack.

13.8 RANDOM KEY PREDISTRIBUTION - NOTATION

The rest of the chapter will focus on one particular method of key distribution known as random key predistribution. For clarity, we list the symbols used in the following sections in the table below:

c	desired confidence level (probability) that the sensor network is connected after completing the connection protocol.
d	the expected degree of a node – *i.e.* the expected number of secure links a node can establish during key-setup.
m	number of keys in a node's key ring
n	network size, in nodes
n'	the expected number of neighbor nodes within communication radius of a given node
p	probability that two neighbor nodes can set up a secure link during the key-setup phase.
q	for the q-composite scheme, required amount of key overlap
S	key pool (set of keys randomly chosen from the total key space)
$\|S\|$	size of the key pool.
t	threshold number of votes after which a node will be revoked.

13.9 THE BASIC RANDOM KEY PREDISTRIBUTION SCHEME

Eschenauer and Gligor first proposed a random key-predistribution scheme [11]. In the remainder of this chapter, we refer to their approach as the *basic scheme*. Let m denote the number of distinct cryptographic keys that can be stored on a sensor node. The basic scheme works as follows. Before sensor nodes are deployed, an *initialization phase* is performed. In the initialization phase, the basic scheme picks a random pool (set) of keys S out of the total possible key space. For each node, m keys are randomly selected from the key pool S and stored into the node's memory. This set of m keys is called the node's *key ring*. The number of keys in the key pool, $|S|$, is chosen such that two random subsets of size m in S will share at least one key with some probability p.

After the sensor nodes are deployed, a *key-setup phase* is performed. The nodes first perform key-discovery to find out with which of their neighbors they share a key. Such key discovery can be performed by assigning a short identifier to each key prior to deployment, and having each node broadcast its set of identifiers. Nodes which discover that they contain a shared key in their key rings can then verify that their neighbor actually holds the key through a challenge-response protocol. The shared key then becomes the key for that link.

After key-setup is complete, a connected graph of secure links is formed. Nodes can then set up *path keys* with nodes in their vicinity whom they did not happen to share keys with in their key rings. If the graph is connected, a path can be found from a source node to its neighbor. The source node can then generate a path key and send it securely via the path to the target node.

One needs to pick the right parameters such that the graph generated during the key-setup phase is connected. Consider a random graph $G(n, p_l)$, a graph of n nodes for which the probability that a link exists between any two nodes is p_l. Erdös and Rényi showed that for monotone properties of a graph $G(n, p_l)$, there exists a value of p_l over which the property exhibits a "phase transition", i.e. it abruptly transitions from "likely false" to "likely true" [21]. Hence, it is possible to calculate some expected degree d for the vertices in the graph such that the graph is connected with some high probability c, where $c = 0.999$, for example. Eschenauer and Gligor calculate the necessary expected node degree d in terms of the size of the network n as:

$$d = \left(\frac{n-1}{n}\right)(\ln(n) - \ln(-\ln(c))) \qquad (13.1)$$

From the formula, $d = O(\log n)$. In our examples we expect d to be in the range of 20 to 50.

For a given density of sensor network deployment, let n' be the expected number of neighbors within communication range of a node. Since the expected node degree must be at least d as calculated, the required probability p of successfully performing key-setup with some neighbor is:

$$p = \frac{d}{n'} \qquad (13.2)$$

Since the models of connectivity are probabilistic, there is always the chance that the graph may not be fully connected. This chance is increased if the deployment pattern is irregular or the deployment area has unpredictable physical obstacles to communication. It is difficult to anticipate such scenarios prior to knowing the specifics of the deployment area. To address this, if the network detects that it is disconnected, sensor nodes should perform *range extension*. This may involve increasing their transmission power, or sending a request to their neighbors to forward their communications for a certain number of hops. Range extension may be gradually increased until a connected graph is formed after key-setup. A useful way for a node to detect if a network is connected is by checking if it can perform multi-hop communication with all base stations. If not, range extension should be performed.

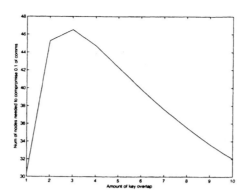

Figure 13.1. The expected number of nodes an adversary needs to capture before it is able to eavesdrop on any link with probability 0.1, for various amounts of key overlap q. Key ring size $m = 200$ keys, probability of connection $p = 0.5$.

13.10 Q-COMPOSITE RANDOM KEY PREDISTRIBUTION SCHEME

In the basic scheme, any two neighboring nodes need to find a single common key from their key rings to establish a secure link in the key-setup phase. We propose a modification to the basic scheme where q common keys ($q > 1$) are needed, instead of just one. By increasing the amount of key overlap required for key-setup, we increase the resilience of the network against node capture.

Figure 13.10 reflects the motivation for the q-composite keys scheme. As the amount of required key overlap increases, it becomes exponentially harder for an attacker with a given key set to break a link. However, to preserve the given probability p of two nodes sharing sufficient keys to establish a secure link, it is necessary to reduce the size of the key pool $|S|$. This allows the attacker to gain a larger sample of S by breaking fewer nodes. The interplay of these two opposing factors results in an optimal amount of key overlap to pose the greatest obstacle to an attacker for some desired probability of eavesdropping on a link.

13.10.1 DESCRIPTION OF THE Q-COMPOSITE KEYS SCHEME

Initialization and key setup. The operation of the q-composite keys scheme is similar to that of the basic scheme, differing only in the size of the key pool S and the fact that multiple keys are used to establish communications instead of just one.

Key Distribution Techniques for Sensor Networks 289

In the initialization phase, we pick a set S of random keys out of the total key space, where $|S|$ is computed as described later in Section 13.10.1.0. For each node, we select m random keys from S (where m is the number of keys each node can carry in its key ring) and store them into the node's key ring.

In the key-setup phase, each node must discover all common keys it possesses with each of its neighbors. This can be accomplished with a simple local broadcast of all key identifiers that a node possesses. While broadcast-based key discovery is straightforward to implement, it has the disadvantage that a casual eavesdropper can identify the key sets of all the nodes in a network and thus pick an optimal set of nodes to compromise in order to discover a large subset of the key pool S. A more secure, but slower, method of key discovery could utilize client puzzles such as a Merkle puzzle [18]. Each node could issue m client puzzles (one for each of the m keys) to each neighboring node. Any node that responds with the correct answer to the client puzzle is thus identified as knowing the associated key.

After key discovery, each node can identify every neighbor node with which it shares at least q keys. Let the number of actual keys shared be q', where $q' \geq q$. A new communication link key K is generated as the hash of *all* shared keys, e.g. $K = hash(k_1 || k_2 || \ldots || k_{q'})$. The keys are hashed in some canonical order, for example, based on the order they occur in the original key pool S. Key-setup is not performed between nodes that share fewer than q keys.

Computation of key pool size. We assume that we are required to take the sensor network's physical characteristics as a given parameter. Specifically, we are provided with a probability of full network connectivity c and the expected number of neighbors of each node n'. Via Equation 13.1, we first calculate d, the expected degree of any given node. This can be input to Equation 13.2 to calculate p, the desired probability that any two nodes can perform key-setup.

We now need to calculate the critical parameter $|S|$, the size of the key pool. If the key pool size is too large, then the probability of any two nodes sharing at least q keys would be less than p, and the network may not be connected after bootstrapping is complete. If the key pool size is too small, then we are unnecessarily sacrificing security. We would like to choose a key pool size such that the probability of any two nodes sharing at least q keys is $\geq p$. Let m be the number of keys that any node can hold in its key ring. We would like to find the largest S such that any two random samples of size m from S have at least q elements in common, with a probability of at least p.

We compute $|S|$ as follows. Let $p(i)$ be the probability that any two nodes have exactly i keys in common. Any given node has $\binom{|S|}{m}$ different ways of picking its m keys from the key pool of size $|S|$. Hence, the total number of ways for both nodes to pick m keys each is $\binom{|S|}{m}^2$. Suppose the two nodes have

i keys in common. There are $\binom{|S|}{i}$ ways to pick the i common keys. After the i common keys have been picked, there remain $2(m-i)$ distinct keys in the two key rings that have to be picked from the remaining pool of $|S|-i$ keys. The number of ways to do this is $\binom{|S|-i}{2(m-i)}$. The $2(m-i)$ distinct keys must then be partitioned between the two nodes equally. The number of such equal partitions is $\binom{2(m-i)}{m-i}$. Hence the total number of ways to choose two key rings with i keys in common is the product of the aforementioned terms, i.e. $\binom{|S|}{i}\binom{|S|-i}{2(m-i)}\binom{2(m-i)}{m-i}$. Hence, we have

$$p(i) = \frac{\binom{|S|}{i}\binom{|S|-i}{2(m-i)}\binom{2(m-i)}{m-i}}{\binom{|S|}{m}^2} \quad (13.3)$$

Let $p_{connect}$ be the probability of any two nodes sharing sufficient keys to form a secure connection. $p_{connect} = 1-$ (probability that the two nodes share insufficient keys to form a connection), hence

$$p_{connect} = 1 - (p(0) + p(1) + \cdots + p(q-1)) \quad (13.4)$$

For a given key ring size m, minimum key overlap q, and minimum connection probability p, we choose the largest $|S|$ such that $p_{connect} \geq p$.

13.10.2 EVALUATION OF THE Q-COMPOSITE RANDOM KEY DISTRIBUTION SCHEME

We evaluate the q-composite random key distribution scheme in terms of resilience against node capture and the maximum network size supported. We note that this scheme has no resistance against node replication since node degree is not constrained and there is no limit on the number of times each key can be used. The scheme can support node revocation via a trusted base station similar to the approach in [11].

Resilience against node capture in q-composite keys schemes. In this section we evaluate how the q-composite scheme improves a sensor network's resilience in the face of a node capture attack by calculating the fraction of links in the network that an attacker is able to eavesdrop on *indirectly* as a result of recovering keys from captured nodes. That is, we attempt to answer the question: For any two nodes A and B in the network, where neither A nor B have been captured by the attacker, what is the probability that the attacker can decrypt their communications using the subset of the key pool that was recovered from the nodes that were compromised.

We show that the q-composite key scheme strengthens the network's resilience against node capture when the number of nodes captured is low. Let the number of captured nodes be x. Since each node contains m keys, the probability that a given key has not been compromised is $(1-\frac{m}{|S|})^x$. The expected

Key Distribution Techniques for Sensor Networks

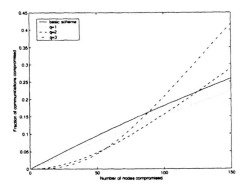

Figure 13.2. Probability that a specific random communication link between two random nodes A, B can be decrypted by the adversary when the adversary has captured some set of x nodes that does not include A or B. Key ring size $m = 200$, probability of key-setup $p = 0.33$.

fraction of total keys compromised is thus $1 - (1 - \frac{m}{|S|})^x$. For any communication link between two nodes, if its link key was the hash of i shared keys, then the probability of that link being compromised is $(1 - (1 - \frac{m}{|S|})^x)^i$. The probability of setting up a secure link is $p = p(q) + p(q+1) + \ldots + p(m)$. Hence, we have that the probability that any secure link setup in the key-setup phase between two uncompromised nodes is compromised when x nodes have been captured is

$$\sum_{i=q}^{m} \left(1 - \left(1 - \frac{m}{|S|}\right)^x\right)^i \frac{p(i)}{p}$$

This equation also represents the fraction of additional communications (*i.e.* external communications in the network independent of the captured nodes) that an adversary can compromise based on the information retrieved from x number of captured nodes. Figure 13.2 shows how it varies with the number of nodes captured by the attacker.

We note that the scale of the x-axis shows absolute numbers of nodes compromised (*i.e.* independent of the actual total size of the network) while the y-axis is the fraction of the total network communications compromised. Hence, the schemes are not infinitely scalable - a compromise of x number of nodes will always reveal a fixed fraction of the total communications in the network regardless of network size. A method to estimate the largest supportable network size of the various schemes is discussed in Section 13.10.2.0.

The q-composite keys scheme offers greater resilience against node capture when the number of nodes captured is small. For example, in Figure 13.2a, for $q = 2$, the amount of additional communications compromised when 50 nodes have been compromised is 4.74%, as opposed to 9.52% for the basic scheme. However, when large numbers of nodes have been compromised, the

q-composite keys schemes tend to reveal larger fractions of the network to the adversary. By increasing q, we make it harder for an adversary to obtain small amounts of initial information from the network via a small number of initial node captures. This comes at the cost of making the network more vulnerable once a large number of nodes have been breached. This may be a desirable trade-off because small scale attacks are cheaper to mount and much harder to detect than large scale attacks. It is easy to mask an attack on a single node as a communications breakdown due to occlusion or interference; it is much harder to disguise an attack on many nodes as a natural occurrence.

The q-composite scheme removes the incentive for small scale attacks since the amount of additional information revealed in the rest of the network is greatly reduced. It forces the attacker to attempt large scale attacks which are expensive and more easily detectable.

Maximum supportable network sizes for the q-composite keys scheme.
In this section we assess the scalability of the random key schemes we have presented thus far.

Since a fixed number of compromised nodes causes a *fraction* of the remaining network to become insecure, these random-key distribution schemes cannot be used for arbitrarily large networks. For example, based on Figure 13.2a, in the basic scheme, the capture of 50 nodes compromises approximately 9.5% of communications in the network. For a network of 10,000 nodes this translates to an approximate payoff of 10% of communications compromised for a cost to the attacker of capturing just 0.5% of total nodes, representing a relatively modest investment for a high payoff.

We can estimate a network's maximum supported size by framing the following requirement:

> *Limited global payoff requirement*: Suppose the adversary has captured some nodes, but is only able to break some fraction $f \leq f_m$ of all communications. We require that each subsequent node that is compromised to the enemy allows them to break as many links in the rest of the network, on expectation, as the average connectivity degree of a single node.

In other words, given that the network is still mostly secure ($f \leq f_m$), we would like that, on average, after capturing some node, the adversary does not learn more about the rest of the network than they learn about the communications of the node itself. Via this requirement, smaller scale attacks on a network must be mainly economically justified by the value of the individual nodes compromised rather than the amount of information that the captured keys can reveal in the rest of the network, thus limiting the incentive of an adversary to begin an attack. The maximum compromise threshold f_m intuitively represents the level of compromise past where the adversary gains an unacceptably high confidence of guessing the sensor readings of the entire net-

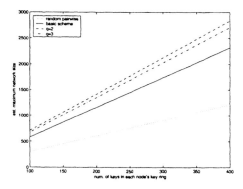

Figure 13.3. Maximum network sizes $(p = 0.33, f_m = 0.1)$

work, and thus the network must be considered exposed and no longer secret. f_m will vary depending on the application and the correlation of different sensor readings.

Using the definition of limited global payoff, we can estimate the maximum allowable sizes for the networks such that our requirement holds true. For any number x of nodes compromised, we know that some fraction $f(x)$ of the remaining secure links created after key-setup have been compromised. Let x_m be the number of nodes compromised such that $f_m = f(x_m)$ of the other secure links created during key-setup has been compromised. f_m is a given parameter (see the definition of limited global payoff preceding). Let the average connectivity degree of a single node be d. The adversary thus holds an expected $x_m d$ connections in which the compromised nodes are directly involved. We require that the number of *additional* links compromised elsewhere in the network be less than this number of directly compromised links. There are $\frac{nd}{2}$ total links in the network. Hence, the requirement is that $(\frac{nd}{2} - x_m d) f_m \leq x_m d$. Simplifying,

$$n \leq 2x_m \left(1 + \frac{1}{f_m}\right) \qquad (13.5)$$

Figure 13.3 shows the estimated maximum network sizes for the basic random keys scheme as well as for several parameters of the q-composite keys scheme. We note that the maximum network sizes scale linearly with key ring size m. For example, for $p = 0.33$, $f_m = 0.1$, and $m = 200$, the maximum network size for the 2-composite keys scheme is $1,415$ nodes while the maximum network size for the basic scheme is $1,159$ nodes.

These calculations are our proposed method of estimating the maximum supportable size of a network given that certain security properties hold. Alternative methods may exist that produce different network size estimations.

13.11 MULTIPATH KEY REINFORCEMENT

In this section we present *multipath key reinforcement*, a method to strengthen the security of an established link key by establishing the link key through multiple paths. This method can be applied in conjunction with the basic random key scheme to yield greatly improved resilience against node capture attacks by trading off some network communication overhead. We analyze the resulting scheme and explain why we discourage using multipath key reinforcement in conjunction with a q-composite scheme.

13.11.1 DESCRIPTION OF MULTIPATH KEY REINFORCEMENT

The basic idea behind multipath key reinforcement was first explored by Anderson and Perrig [1]. We assume that initial key-setup has been completed (in the following examples, we assume the basic random key scheme was used for key-setup). There are now many secure links formed through the common keys in the various nodes' key rings. Suppose A has a secure link to B after key-setup. This link is secured using a single key k from the key pool S. k may be residing in the key ring memory of some other nodes elsewhere in the network. If any of those nodes are captured, the security of the link between A and B is jeopardized. To address this, we would like to update the communication key to a random value after key-setup. However, we cannot simply coordinate the key update using the direct link between A and B since if the adversary has been recording all key-setup traffic, it could decrypt the key-update message after it obtained k and still obtain the new communication key.

Our approach is to coordinate the key-update over multiple independent paths. Assume that enough routing information can be exchanged such that A knows all disjoint paths to B created during initial key-setup that are h hops or less. Specifically, $A, N_1, N_2, \ldots, N_i, B$ is a path created during the initial key-setup if and only if each link $(A, N_1), (N_1, N_2), \ldots, (N_{i-1}, N_i), (N_i, B)$ has established a link key during the initial key-setup using the common keys in the nodes' key rings. Let j be the number of such paths that are *disjoint* (do not have any links in common). A then generates j random values $v_1, \ldots, v_j$. Each random value has the same length as the encryption/decryption key. A then routes each random value along a different path to B. When B has received all j keys, then the new link key can be computed by both A and B

as:
$$k' = k \oplus v_1 \oplus v_2 \oplus \ldots \oplus v_j$$

The secrecy of the link key k is protected by all j random values. Unless the adversary successfully manages to eavesdrop on all j paths, they will not know sufficient parts of the link key to reconstruct it.

The more paths we can find between two nodes A and B, the more security multipath key reinforcement provides for the link between A and B. However, for any given path, the probability that the adversary can eavesdrop on the path increases with the length of the path since if any one link on the path is insecure then the entire path is made insecure. Further, it is increasingly expensive in terms of communication overhead to find multiple disjoint paths that are very long. In this paper we will analyze the case where only paths of 2 links (only one intermediate node) are considered. We call this scheme the *2-hop multipath key reinforcement scheme*. This approach has the advantage that path discovery overhead is minimized: for example, A could exchange neighbor lists with B. Once they identify their common neighbors with which both of them share a key, A and B can perform key reinforcement using their secure links through these common neighbors. Furthermore, the paths are naturally disjoint and no further effort needs to be taken to guarantee this property. We will calculate the expected effectiveness of this scheme and evaluate its security properties in simulation.

13.11.2 ESTIMATION OF EXPECTED EFFECTIVENESS OF 2-HOP MULTIPATH KEY REINFORCEMENT

In this section, we first calculate the expected number of common neighbors between two nodes in a random uniform planar deployment of sensors. We then derive a formula for the new expected probability for compromising a given link after multipath key reinforcement has taken place.

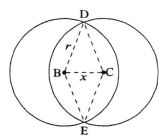

The figure above indicates the parameters to be used in our calculation. B and C denote two communicating sensor nodes. r is the communications range of each sensor node. We assume that each node has the same range for receiving and transmitting. x is the distance between two nodes.

For any given separation x, the area $A(x)$ within both nodes' communication radii is the area of the sectors BDE and CDE minus the area of the rhombus $BDCE$:

$$A(x) = 2r^2 \cos^{-1}\left(\frac{x}{2r}\right) - x\sqrt{r^2 - \frac{x^2}{4}}$$

The probability distribution function of the distance between two nodes within communication radius is given by $F(x) = P(distance < x) = x^2/r^2$. The probability density function is thus $f(x) = F'(x) = 2x/r^2$. The expected area of overlap is thus given by:

$$\int_0^r A(x)f(x)dx$$
$$= \int_0^r \left(2r^2 \cos^{-1}\left(\frac{x}{2r}\right) - x\sqrt{r^2 - \frac{x^2}{4}}\right)\frac{2x}{r^2} dx$$
$$= \left(\pi - \frac{3\sqrt{3}}{4}\right)r^2 = 0.5865\pi r^2$$

We define the term *reinforcing neighbors* of two nodes sharing a secure link as the common neighbors with whom both nodes share a secure link. Since the expected area of overlap is 0.5865 of a single communication radius, the expected number of reinforcing neighbors is thus $0.5865 p^2 n'$ where p is the probability of sharing sufficient keys to communicate, and n' is the number of neighbors of each node. Via Equation 13.2, this can also be expressed as $0.5865\frac{d^2}{n'}$. As an example, for $d = 20$ and $n' = 60$ (i.e. $p = 0.33$), the expected number of reinforcing neighbors is 3.83.

In general, if a link is reinforced by k common neighbors, then the adversary must be able to eavesdrop on that link, as well as at least one link on each of the k 2-hop paths. If the adversary's base probability of compromising a link is b, then the probability of compromising at least one hop on any given 2-hop path is the probability of compromising hop 1 in the path plus the probability of compromising hop 2 in the path minus probability of compromising both hops in the path $= 2b - b^2$. Hence, the final probability of breaking the link is now

$$b' = b(2b - b^2)^k$$

For example, if the adversary has a base 0.1 chance of eavesdropping on a given link before reinforcement, for a link reinforced by 3 neighbors, the chance of eavesdropping after reinforcement improves to 6.86×10^{-4}, or about 1 in 1,458.

From the expected number of reinforcing neighbors we can estimate the expected network communications overhead of the 2-hop multipath reinforce-

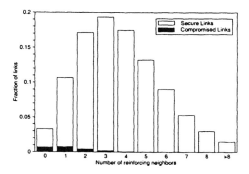

Figure 13.4. Reinforcement and compromise statistics for base compromise probability $b = 0.2$

ment scheme. Each reinforcing neighbor represents an extra 2-hop communication to help reinforce a given 1-hop link. Hence, on average, the total additional communications overhead for key-reinforcement is at least $2 \times 0.5865p^2n'$ times more than the network communications needed for basic key-setup, not including additional communications for common-neighbor discovery. For example, for $p = 0.33$ and $n' = 60$, we can expect to see at least 7.66 times additional network traffic after key-setup is complete. Including common neighbor discovery, we estimate the final scheme to be approximately 10 times more expensive in network communications than the basic scheme in this case. Given that eavesdropping probabilities can be improved from 0.1 to 6.86×10^{-4} (146 times improvement), this may be a good trade-off.

13.11.3 EVALUATION OF MULTIPATH KEY REINFORCEMENT

The effectiveness of 2-hop multipath key reinforcement is evaluated by simulating the random uniform deployment of 10,000 sensor nodes on a square planar field. The probability of any two nodes being able to establish a secure link is set at $p = 0.33$, and the deployment density is set such that the expected number of neighbors of each node was 60. The eavesdropping attack is modeled by iterating over each secure link and marking it as compromised with random chance based on the simulated probability of compromise c. A link is considered completely compromised only if it is compromised and all its reinforcement paths are also compromised.

Figure 13.4 reflects the relative distribution of the number of reinforcing neighbors for each link in the simulation. The results indicated reflect support for our calculated average of 3.83 reinforcing neighbors between any 2 nodes within communication distance. The figure also shows the distribution of reinforced links that were compromised by an adversary with a base 0.2 probability

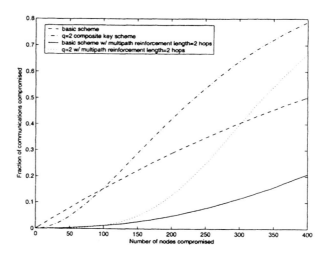

(a) Resistance against node capture

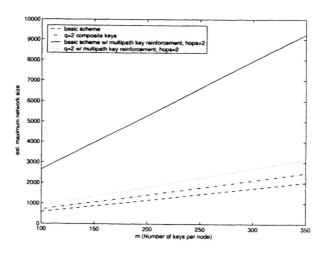

(b) Maximum network sizes

Figure 13.5. Multipath key reinforcement results ($m = 200, p = 0.33$)

of compromising any link prior to reinforcement. In this simulation, links with more than 3 reinforcing neighbors did not suffer significant rates of compromise. The overall rate of compromise was lowered by an order of magnitude, from 0.2 to 0.022.

Figure 13.5a indicates the amount of communications compromised versus the number of nodes compromised, with and without key reinforcement for the various schemes. Successfully implementing multipath key reinforcement on the basic scheme enables it to outperform the q-composite scheme for $q \geq 2$ even when the q-composite scheme is supplemented by key reinforcement. The intuitive reason for this is that multipath key reinforcement acts similarly to the q-composite keys scheme in that it compounds the difficulty of compromising a given link by requiring the adversary possess multiple relevant keys to eavesdrop on a given link. The trade-off for this benefit in the q-composite case is a smaller key pool size; the trade-off for the multipath key reinforcement scheme is increased network overhead. Compounding both the schemes compounds their weaknesses - the smaller key pool size of the q-composite keys scheme undermines the effectiveness of multipath key reinforcement by making it easier to build up a critically large collection of keys.

Figure 13.5b shows the maximum network size of the basic scheme with multipath key reinforcement. The graphs show that multipath key reinforcement gives a significant boost to network size performance when implemented on the basic scheme, but has little effect with the q-composite scheme.

The cost of the improved security due to multipath key reinforcement is an added overhead in neighbor discovery and key establishment traffic. Whether this tradeoff is a good one will depend on the specific application as well as the deployment density characteristics of the sensor network.

While the analysis presented is for using multipath key reinforcement to secure links that have been formed after key-setup, the scheme can also be used to reinforce *path-keys* that are established between nodes that did not share keys during key setup. This will further improve the security of the system.

13.12 PAIRWISE KEY SCHEMES

In the random key pool distribution schemes described above, keys can be issued multiple times out of the key pool, and node-to-node authentication is not possible [9]. In contrast, pairwise key distribution assigns a unique key to each pair of nodes. In this section ,we review several different approaches for pairwise key distribution: the random pairwise key scheme by Chan, Perrig and Song [9], the single-space pairwise key distribution approaches by Blom [5] and Blundo et al. [6], and the multi-space pairwise key scheme by Du et al. [10] and by Liu and Ning [16].

13.12.1 RANDOM PAIRWISE KEY SCHEME

Recall that the size of each node's key rings is m keys, and the probability of any two nodes being able to communicate securely is p. The random pairwise keys scheme proceeds as follows:

i. In the pre-deployment *initialization* phase, a total of $n = \frac{m}{p}$ unique node identities are generated. The actual size of the network may be smaller than n. Unused node identities will be used if additional nodes are added to the network in the future. Each node identity is matched up with m other randomly selected distinct node IDs and a pairwise key is generated for each pair of nodes. The key is stored in both nodes' key rings, along with the ID of the other node that also knows the key.

ii. In the post-deployment *key-setup* phase, each node first broadcasts its node ID to its immediate neighbors. By searching for each other's IDs in their key-rings, the neighboring nodes can tell if they share a common pairwise key for communication. A cryptographic handshake is then performed between neighbor nodes who wish to mutually verify that they do indeed have knowledge of the key.

13.12.2 SINGLE-SPACE PAIRWISE KEY SCHEMES

Both Blom's and the polynomial scheme require a sensor node i to store unique public information U_i and private information V_i. During the bootstrapping phase, nodes exchange public information, and node i could compute its key with node j with $f(V_i, U_j)$. It is guaranteed that $f(V_i, U_j) = f(V_j, U_i)$. Both approaches ensure the λ-secure property: the coalition of no more than λ compromised sensor nodes reveals nothing about the pairwise key between any two non-compromised nodes.

13.12.3 MULTI-SPACE PAIRWISE KEY SCHEMES

Recently, researchers have proposed the idea of multiple key spaces [10, 16] to further enhance the security of single-space approaches. The idea of introducing multiple key spaces can be viewed as the combination of the basic key pool scheme and the above single space approaches. The setup server randomly generates a pool of m key spaces each of which has unique private information. Each sensor node will be assigned k out of the m key spaces. If two neighboring nodes have one or more key spaces in common, they can compute their pairwise secret key using the corresponding single space scheme.

13.13 OTHER RELATED WORK

We first review work in establishing shared keys in mobile computing, then review work in sensor network key establishment.

Tatebayashi, Matsuzaki, and Newman consider key distribution for resource-starved devices in a mobile environment [23]. Leighton and Micali present two mechanisms for key agreement using a predistributed set of symmetric keys [15]. Their first scheme is similar in nature to our q-composite protocols, their keys in the key pool are deterministically selected, such that any two nodes can certainly establish a shared key. Park et al.[19] point out weaknesses and improvements. Beller and Yacobi further develop key agreement and authentication protocols [3]. Boyd and Mathuria survey the previous work on key distribution and authentication for resource-starved devices in mobile environments [7]. The majority of these approaches rely on asymmetric cryptography. Bergstrom, Driscoll, and Kimball consider the problem of secure remote control of resource-starved devices in a home [4].

Stajano and Anderson discuss the issues of bootstrapping security devices [22]. Their solution requires physical contact of the new device with a master device to imprint the trusted and secret information.

Carman, Kruus, and Matt analyze a wide variety of approaches for key agreement and key distribution in sensor networks [8]. They analyze the overhead of these protocols on a variety of hardware platforms.

Wong and Chan propose a key exchange for low-power computing devices [24]. However, their approach assumes an asymmetry in computation power, that is, one of the participants is a more powerful server.

Perrig et al. propose SPINS, a security architecture specifically designed for sensor networks [20]. In SPINS, each sensor node shares a secret key with the base station. To establish a new key, two nodes use the base station as a trusted third party to set up the new key.

We review the related work by Eschenauer and Gligor [11] in Section 13.9. Anderson and Perrig propose a key establishment mechanism for sensor networks based on initially exchanging keys in the clear [1]. Their key infection approach is secure as long as an attacker arrives after key exchange and did not eavesdrop the exchange.

Zhou and Haas propose to secure ad hoc networks using asymmetric cryptography [25]. Kong et al. propose localized public-key infrastructure mechanisms, based on secret sharing and multiparty computation techniques [14]. Such approaches are expensive in terms of computation and communication overhead.

Broadcast encryption by Fiat and Naor [12] is another model for distributing a shared key to a group of receivers. However, this model assumes a single sender, and that the sender knows the key pools of all receivers. Subsequent papers further develop this approach [2, 13, 17].

REFERENCES

[1] Ross Anderson and Adrian Perrig. Key infection: Smart trust for smart dust. Unpublished Manuscript, November 2001.

[2] Dirk Balfanz, Drew Dean, Matt Franklin, Sara Miner, and Jessica Staddon. Self-healing key distribution with revocation. In *Proceedings of the IEEE Symposium on Research in Security and Privacy*, pages 241–257, May 2002.

[3] M. Beller and Y. Yacobi. Fully-fledged two-way public key authentication and key agreement for low-cost terminals. *Electronics Letters*, 29(11):999–1001, May 1993.

[4] Peter Bergstrom, Kevin Driscoll, and John Kimball. Making home automation communications secure. *IEEE Computer*, 34(10):50–56, Oct 2001.

[5] R. Blom. Non-public key distribution. In *Advances in Cryptology: Proceedings of Crypto '82*, pages 231–236, 1982.

[6] C. Blundo, A. De Santis, A. Herzberg, S. Kutten, U. Vaccaro, and M. Yung. Perfectly-secure key distribution for dynamic conferences. In *Advances in Cryptology - Crypto '92*, pages 471–486, 1992.

[7] Colin Boyd and Anish Mathuria. Key establishment protocols for secure mobile communications: A selective survey. In *Australasian Conference on Information Security and Privacy*, pages 344–355, 1998.

[8] David W. Carman, Peter S. Kruus, and Brian J. Matt. Constraints and approaches for distributed sensor network security. *NAI Labs Technical Report #00-010*, September 2000.

[9] Haowen Chan, Adrian Perrig, and Dawn Song. Random key predistribution schemes for sensor networks. In *IEEE Symposium on Security and Privacy*, May 2003.

[10] Wenliang Du, Jing Deng, Yunghsiang S. Han, and Pramod K. Varshney. A pairwise key pre-distribution scheme for wireless sensor networks. In *ACM CCS 2003*, pages 42–51, October 2003.

[11] Laurent Eschenauer and Virgil D. Gligor. A key-management scheme for distributed sensor networks. In *Proceedings of the 9th ACM Conference on Computer and Communication Security*, pages 41–47, November 2002.

[12] Amos Fiat and Moni Naor. Broadcast encryption. In *Advances in Cryptology – CRYPTO '93*, volume 773 of *Lecture Notes in Computer Science*, 1994.

[13] J. Garay, J. Staddon, and A. Wool. Long-lived broadcast encryption. In *Advances in Cryptology — CRYPTO '2000*, pages 333–352, 2000.

[14] Jiejun Kong, Petros Zerfos, Haiyun Luo, Songwu Lu, and Lixia Zhang. Providing robust and ubiquitous security support for mobile ad-hoc networks. In *9th International Conference on Network Protocols (ICNP'01)*, 2001.

[15] T. Leighton and S. Micali. Secret-key agreement without public-key cryptography. In *Advances in Cryptology - Crypto '93*, pages 456–479, 1993.

[16] Donggang Liu and Peng Ning. Establishing pairwise keys in distributed sensor networks. In *ACM CCS 2003*, pages 52–61, October 2003.

[17] M. Luby and J. Staddon. Combinatorial bounds for broadcast encryption. In *Advances in Cryptology — EUROCRYPT '98*, pages 512–526, 1998.

[18] R. Merkle. Secure communication over insecure channels. *Communications of the ACM*, 21(4):294–299, 1978.

[19] C. Park, K. Kurosawa, T. Okamoto, and S. Tsujii. On key distribution and authentication in mobile radio networks. In *Advances in Cryptology - EuroCrypt '93*, pages 461–465, 1993. Lecture Notes in Computer Science Volume 765.

[20] Adrian Perrig, Robert Szewczyk, Victor Wen, David Culler, and J. D. Tygar. SPINS: Security protocols for sensor networks. In *Seventh Annual ACM International Conference on Mobile Computing and Networks (MobiCom 2001)*, July 2001.

[21] J. Spencer. *The Strange Logic of Random Graphs*. Number 22 in Algorithms and Combinatorics. 2000.

[22] Frank Stajano and Ross Anderson. The resurrecting duckling: Security issues for ad-hoc wireless networks. In *Security Protocols, 7th International Workshop*, 1999.

[23] M. Tatebayashi, N. Matsuzaki, and D. B. Jr. Newman. Key distribution protocol for digital mobile communication systems. In *Advances in Cryptology - Crypto '89*, pages 324–334, 1989. Lecture Notes in Computer Science Volume 435.

[24] Duncan S. Wong and Agnes H. Chan. Efficient and mutually authenticated key exchange for low power computing devices. In *Advances in Cryptology — ASIACRYPT '2001*, 2001.

[25] Lidong Zhou and Zygmunt J. Haas. Securing ad hoc networks. *IEEE Network Magazine*, 13(6):24–30, November/December 1999.

Chapter 14

SECURITY IN SENSOR NETWORKS: WATERMARKING TECHNIQUES

Jennifer L. Wong
University of California, Los Angeles
Los Angeles, CA 90095
jwong@cs.ucla.edu

Jessica Feng
University of California, Los Angeles
Los Angeles, CA 90095
jessicaf@cs.ucla.edu

Darko Kirovski
Microsoft Research
Redmond, WA 98052
darkok@microsoft.com

Miodrag Potkonjak
University of California, Los Angeles
Los Angeles, CA 90095
miodrag@cs.ucla.edu

Abstract   The actual deployment of the majority of envisioned applications for sensor networks is crucially dependent on resolving associated security, privacy, and digital rights management (DRM) issues. Although cryptography, security, and DRM have been active research topics for the last several decades, wireless sensor networks (WSN) pose a new system of conceptual, technical, and optimization challenges.

In this Chapter we survey two areas related to security in WSN. First, we briefly survey techniques for the protection of the routing infrastructure at the network level for mobile multi-hop (ad-hoc) networks. Secondly, we discuss the first-known watermarking technique for authentication of sensor network data and information. We conclude the Chapter by providing a short discussion of future research and development directions in security and privacy in sensor networks.

Keywords: Wireless sensor networks, Security, Privacy, Digital Rights Management

14.1 INTRODUCTION

Wireless ad-hoc sensor networks (WSN) are distributed embedded systems where each unit is equipped with a certain amount of computation, communication, storage, and sensing resources. In addition each node may have control over one or more actuators and input/output devices such as displays. A variety of applications for sensor networks are envisioned, starting from nano-scale device networks to interplanetary scale distributed systems. In many senses, WSN are a unique type of systems which have unique technical and operational challenges. Among these, security and privacy are most often mentioned as the key prerequisite for actual deployment of sensor networks.

There are at least three major reasons why security and privacy in WSN is such an important topic. The first one is that sensor networks are intrinsically more susceptible to attacks. They are often deployed in uncontrolled and sometimes even hostile environments. Wireless communication on a large scale can be easily observed and interfered with. WSN nodes are both complex component systems with numerous weak points from a security point of view. In addition, they are severely constrained in terms of energy and therefore extensive on-line security checking is not viable. Finally, sensors can be manipulated even without interfering with the electronic subsystem of the node and actuators can pose strong safety and hazard concerns.

The second argument that emphasizes the role of security in WSN is the importance of protecting typical applications. WSN can not only have data about one or more users, but can also contain a great deal of information about their past and even future actions. In addition, they may contain significant amounts of information about a users physiological and even psychological profiles. Furthermore, once the sensors are equipped with actuators both the sensors and the environment can be impacted in a variety of ways.

The third reason for security in WSN is, in a sense, the most scientific and engineering based reason. WSN require new concepts and a new way of thinking with respect to security, privacy, digital rights management, and usage measurement. The Internet was a great facilitator of computer and communication security on a large scale. Note that the Internet itself created opportunities for new types of attacks such as denial of service (DoS) and intrusion detection. It also created new conceptual techniques on how to defend the Internet infrastructure. For example, Honeypots are now widely used to obtain information about the behavior and thinking of an attacker [33]. It is easy to see that WSN will further accentuate these trends. For example, denial of sleep attacks will be brought to a new level of importance.

In this chapter we present two case studies: how to leverage on mobility to provide security, and how to develop and evaluate watermarking schemes for the protection of data and information in sensor networks. We conclude the chapter with a section on future directions where we identify eight security dimensions that we expect to be of high importance for both research and practical deployment of WSN.

14.2 CASE STUDY: MOBILITY AND SECURITY

In this Section, we discuss the interplay between mobility and security. More specifically, we focus on security and mobility with respect to multi-hop wireless networks. This area of research is still in the very early phases of its development and only a few research results has been reported. We expect very rapid growth in this direction in the near future.

It is expected that a large percentage of wireless networks will be mobile. There are two main reasons for this prediction. The first is that in many applications one can achieve significantly higher performance if mobility is provided and exploited. For example, sensors nodes may move closer to the phenomenon or event of interests. The second reason is even more compelling: sensor networks associated with individual users, cars, trains, airplanes and other transportation vehicles are intrinsically mobile.

It is not clear whether mobility makes security in wireless sensor networks easier or more difficult to be achieved. From one point of view, it makes it easier because one can leverage on conducting specific security tasks on the nodes of interest which are in favorable locations. From the other point of view, it makes it more difficult due to the dynamic structure of the topology and potential introduction and departure of any given node.

Until recently, mobility received relatively little attention in wireless ad-hoc networks. The main reason for this is that the majority of standard tasks in wireless ad-hoc networks are easier to address in the static scenario. Even more importantly, there experimental data that would enable realistic modeling of mobility does not exist. Essentially all current models are of random statistical nature [7]. Notable exception include [27, 32].

While a number of notable research results have been reported on security in mobile ad hoc networks [8, 18, 19, 24, 30], we will focus our attention on the first paper to address using mobility to assist in the security of mobile networks [9]. Capkun et al. consider a self-organized mobile wireless network with no security infrastructure. Therefore, there is no central authority, no centralized trusted party, and no other centralized security service provider. Due to this fact, the approach can be applied at any network layer, and will allow for the addition of new nodes into an existing and operating network. Each node in the network is given the same role in the security exchange, and there for there is no single point of failure for the system (as there is with a central security certification authority).

The focus of the work is on one of the most critical phases in establishing secure communication: exchange of cryptographical keys. Their underlying and main assumption is that when nodes are in close vicinity of each other they can communicate using a secure side channel, such as infrared or wired communication. Under this assumption, no middle man attack that would alter any communication between them at this point is possible. However, their scheme does not require that secrecy of the initial message is guaranteed, due to the fact that they are focusing on public key cryptography schemes. However, their scheme is general and can be applied to secret key schemes too. An additional assumption in their scheme is that each node can generate cryptographic keys and verify signatures.

Under these assumptions they assume that two types of security associations can be created between nodes in the network. The first association is a direct association. In this case, when nodes come in contact with each other they can physically verify each other. At this point, the nodes both consciously and simultaneously exchange cryptographic keys. The second type of association is an indirect association, which they introduce in order to expedite the key establishment throughout the network. An indirect association is established through a "friend". Two nodes are friends if they trust each other to always provide correct information about themselves and about other nodes that they have encountered/established associations with and they have have already established a security association between each other. A friend can interchange security information (i.e. public keys) between two nodes if it has a security association with both nodes that want to establish a security association. Through this process, all nodes do not have to come into contact directly with each other, but only with a node who has already established an association with another node. This process in not transitive beyond a chain that consists of more than one friend. Their scheme protects against attacks in the form of eavesdropping on communications, manipulation of messages, and nodes with misrepresented identity.

They evaluated their approach under the random walk mobility approach [7] and demonstrated that almost all nodes in the network can establish security associations in a relatively short period of time. It is important to note that the mobility model crucially impacts this conclusion and that in more realistic cases where nodes have limited mobility ranges this will not be the case. We believe that inclusion of Internet gateway points or bases stations would greatly enhance the applicability of the approach.

Specifically, for WSN this type of approach has both advantages and limitations. No central authority is needed, however each node in the network must be able to generate and verify public keys. The use of public and private key cryptography in sensor networks is questionable due to heavy computation requirements. The notion of using mobile nodes in conjunction with static nodes in sensor networks to establish secure relationships between static nodes would be a possible approach. However, these types of relationships may not be applicable for WSN deployed in hostile environments. In the very least, this work and other security approaches for mobile ad hoc networks have not only potential for direct application in WSN but also provide foundations for new techniques.

14.3 CASE STUDY: REAL-TIME WATERMARKING

One of the major security issues in the Internet is digital right management (DRM). It is easy to see that DRM will also play a major role in wireless sensor networks. In addition, data authentication will be exceptionally important. To address these problems, Feng et al [14] have developed the first watermarking techniques for cryptologically embedding an authorship signature into data and information acquired by a WSN.

14.3.1 RELATED WORK: WATERMARKING

The notion of intellectual property protection and specifically watermarking has been widely studied for items such as text [3], audio/video [37], and circuit designs. Specifically, watermarking techniques have been proposed for two domains: static artifacts and functional artifacts.

Static artifacts [34, 17] are artifacts that consist of only syntactic components which are not altered during their use, ie. images [36] and audio [12, 23]. Watermarks can also be placed in graphical objects such as 3D graphics [28] and animation [15]. The essential property of all watermarking techniques for static artifacts is that they leverage the imperfection of human perception. The main objectives of watermarking techniques for static artifacts include requirements for global placement of the watermark in the artifact, resiliency against removal, and suitability for rapid detection.

Watermarking for functional artifacts, such as software and integrated circuits design have also been proposed. The common denominator for functional artifacts is that they must fully preserve their functional specifications and therefore can not leverage on the principles for watermarking static artifacts. Functional artifacts can be specified and therefore watermarked at several levels of abstraction such as system level designs, FPGA designs [25], at the behavioral and logic synthesis levels, and the physical design level [20, 21]. These approaches leverage on the fact that for a given optimization problem, a large number of similar quality solutions exist which can be selected from in order to have certain characteristics which match a designer's signature. More complex watermarking protocols, such as multiple watermarks [25], fragile watermarks [16], publicly detectable watermarks [31] and software watermarking [29], have also been proposed. Techniques have also been developed for watermarking of DSP algorithms, sequential circuits, sequential functions [10], and analog designs.

Additionally, other techniques for intellectual property protection such as fingerprinting [2, 6], obfuscation [5], reverse engineering [4], and forensic engineering [22] have been proposed.

In sensor networks, watermarking and other intellectual property protection techniques can be applied at a variety of levels. The design of the sensor nodes and the software used in the network can be protected using functional techniques. Additionally, both static and functional watermarking can be applied on the data collected from the network depending on the types of sensors and actuators deployed (i.e. video, audio, measured data). In the remainder of this section, we survey work which proposes the first watermarking technique for the protection of data collected in a sensor network.

14.3.2 REAL-TIME WATERMARKING

Real-time watermarking aims to authenticate data which is collected by a sensor network. The first watermarking technique for cryptologically watermarking data and information acquired by a WSN has been developed by Feng et al [14].

The key idea of their technique is to impose additional constraints to the system during the sensing data acquisition and/or sensor data processing phases. Constraints that correspond to the encrypted embedded signature are selected in such a way that they provide favorable tradeoffs between the accuracy of the sensing process and the strength of the proof of authorship. The first set of techniques embeds the signature into the process of sensing data. The crucial idea is to modulate by imposing additional constraints on the parameters which define the sensor relationship with the physical world. Options for these parameters include the location and orientation on sensor, time management (e.g. frequency and phase of intervals between consecutive data capturing), resolution, and intentional addition of obstacles and use of actuators. In particular, an attractive alternative is to impose constraints on intrinsic properties (e.g. sensitivity, compression laws) of a particular sensor, therefore the measured data will have certain unique characteristics that are strongly correlated with the signature of the author/owner.

The second technique is to embed a signature during data processing, either in the sensor or control data. There are at least three degrees of freedom that can be exploited: error minimization procedures, physical world model building, and solving computationally intractable problems. In the first scenario, there are usually a large number of solutions that have similar levels of error. The task is to choose one that maintains the maximal consistency in measured data and also contains a strong strength of the signature. Typical examples of this type of tasks are location discovery and tracking. In the second scenario, they add additional constraints during the model building of the physical world.

In the final scenario, they are dealing with NP-complete problems, and therefore it is impossible to find the provably optimal solution. Therefore, the goal is to find a high quality solution that also has convincing strength of the signature.

Probably the best way to explain the watermarking approach for sensor networks is to demonstrate its essential features using a simple, yet an illustrative example. For this purpose the authors demonstrate how a watermark can be embedded during the atomic trilateration process. Atomic trilateration is a widely used basic algorithmic block for location discovery that can be formulated on the example shown in Figure 14.1 in the following way.

Problem: There are four sensors: A, B, C, and D. Sensors A, B, and C know their locations in terms of x and y coordinates. The distances between themselves and node D are measured with a certain level of accuracy and are reported by A, B, and C.

Goal: The objective is to discover the location of sensor D in term of its x and y coordinates.

The problem can be stated as a system of three nonlinear equations that contain nine known values and two unknown variables as stated bellow.
Known values: $(A_x, A_y), (B_x, B_y), (C_x, C_y), M_{AD}, M_{BD}, M_{CD}$ where $(A_x, A_y), (B_x, B_y), (C_x, C_y)$ are the x and y coordinates of sensor node A, B and C respectively. M_{AD}, M_{BD}, M_{CD} are the measured distances from A to D, B to D and C to D respectively.
Unknown variables: (D_x, D_y); that are components of the location of sensor node D that it suppose to conclude from the measured distances from all other three nodes to itself.

The key observation is that all distance measurements are noisy. Therefore, the equations can be solved in such a way that all of them are simultaneously satisfied. Instead, the goal is to assign the values to unknown variables in such a way that the solution to all the equations is maximally consistent. Maximal consistency, of course, can be defined in an infinite number of ways. For example, the following three measures are often advocated:

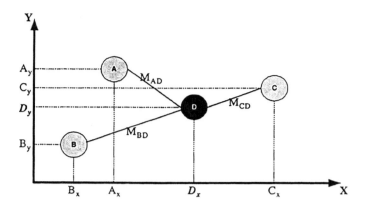

Figure 14.1. Atomic Trilateration.

$$L_1 = |M_{AD} - E_{AD}| + |M_{BD} - E_{BD}| + |M_{CD} - E_{CD}| \quad (14.1)$$
$$L_2 = \sqrt{(M_{AD} - E_{AD})^2 + (M_{BD} - E_{BD})^2 + (M_{CD} - E_{CD})^2} \quad (14.2)$$
$$L_\infty = max(|\frac{(M_{AD} - E_{AD})}{E_{AD}}|, |\frac{(M_{BD} - E_{BD})}{E_{BD}}|, |\frac{(M_{CD} - E_{CD})}{E_{CD}|}) \quad (14.3)$$

where

$$E_{AD} = \sqrt{(D_x - A_x)^2 + (D_y - A_y)^2} \quad (14.4)$$

$$E_{BD} = \sqrt{(D_x - B_x)^2 + (D_y - B_y)^2} \quad (14.5)$$

$$E_{CD} = \sqrt{(D_x - C_x)^2 + (D_y - C_y)^2} \quad (14.6)$$

The first measure, L_1, combines the errors in a linear way and asks for their simultaneous minimization. The second measure, L_2, is widely used and specifies the errors as linear combination of quadratic values. The intuition is that one will obtain a solution that will have not just a relatively low linear sum, but also will minimize, to some extent, the maximal error. The third measure L_∞ aims to reduce the maximal error among all three measurements. E_{AD}, E_{BD}, E_{CD} are the expected distances from A, B, C to D. Essentially, the goal is to minimize the differences between expected distances (E's) and measured distances (M's). The expected distances are written in terms of the location of node D. Thus, by minimizing the distances, the closest estimate of the real correct location of node D is determined. There are many ways to solve this small and simple system of equations. For example, one can use the conjugate gradient method or a multi-resolution grid to obtain the solution according to the selected measure of quality. If they just solve the system of equations, they will have the requested location information, but will not have the proof that they conducted measurements and solve the system. However, if they impose additional constraints on the system of equations or on selected objective function, they will have both the high quality solution and the strong proof of the ownership.

One potential watermarking alternative process, where they modify the objective function, is illustrated using the following example. Suppose that "000101001101110001" is the binary string/signature that they want to embed. One option is to embed the signature by assigning weight factors to each term of the objective function according to the binary string. The binary string can be partitioned into sections (in this case, three sections), then converted to decimal numbers and used to assign weight factors. The process can be illustrated as in the following figure:

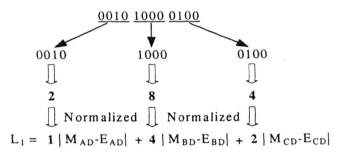

Figure 14.2. Embedding watermarks by assigning weight factors to the objective function during atomic trilateration.

14.3.3 GENERIC PROCEDURE

There exist numerous types of sensor networks and they can be used for many different purposes. Their goal is to watermark all data provided by a sensor network generically regardless of the type of data the network is collecting or what the purpose of the network is.

There exist two types of data being produced by a sensor network: raw sensor data and processed application data. The first type, sensor data, is the original data the sensor network captures or measures. It may or may not be what the user of the network desires. However, the second type, processed data, is the output of the network to the user. The distinction of these two types of data provides insight into where watermarking can take place: i) during the process of sensing data (original data capturing); ii) during the process of processing the original data. Therefore, they call these two processes watermarking in sensing data and watermarking in processing data.

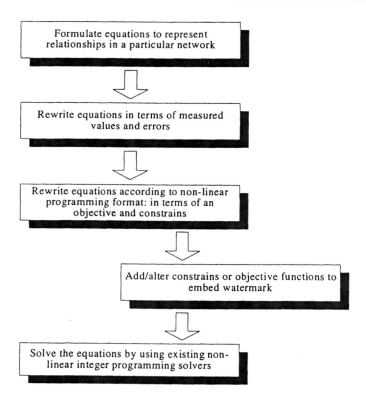

Figure 14.3. General procedure for embedding a watermark.

An important question to ask is how is the original raw data being processed in order to generate the processed application data? In this case, Feng and Potkonjak enquired the technique of non-linear programming. The general procedure can be summarized as Figure 14.3.

They first represent all the relationships that exist in the network using equations. Since everything is measured there always exists some degree of error. Realizing this, they replace the variables with the summation of a reasonable estimate and some error value. Their next goal is to minimize the errors in the equations, and achieve the closest possible estimates to the true values. This can be achieved by using effective non-linear programming solvers.

In order to illustrate this process, consider the example of navigation shown in Figure 14.4

Problem: A sensor node is moving over a period of time. At each point of time, atomic trilateration can be performed to determine its location.

Goal: The trajectory motion of a particular node over a period of time in terms of coordinates at each point of time.

Security in Sensor Networks: Watermarking Techniques

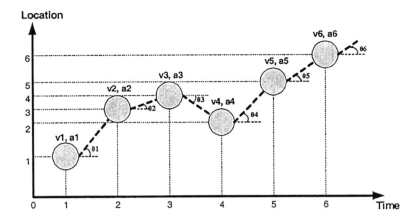

Figure 14.4. Trajectory process.

Known (Measured) variables			
$V_{obj,0}$	$V_{obj,1}$	$V_{obj,2}$	velocity
$a_{obj,0}$	$a_{obj,1}$	$a_{obj,2}$	acceleration
	Δt		time interval
$d_{obj,a}$	$d_{obj,c}$	$d_{obj,d}$	measured distance
(x_a, y_a)	(x_c, y_c)	(x_d, y_d)	3 known sensors
$d_{obj,f}$	$d_{obj,g}$	$d_{obj,h}$	measured distance
(x_f, y_f)	(x_g, y_g)	(x_h, y_h)	3 known sensors
$d_{obj,I}$	$d_{obj,j}$	$d_{obj,k}$	measured distance
(x_i, y_i)	(x_j, y_j)	(x_k, y_k)	3 known sensors
Unknown variables			
$(x_{obj,0}, y_{obj,0})$			coordinates of object at time 0
$(x_{obj,1}, y_{obj,1})$			coordinates of object at time 1
$(x_{obj,2}, y_{obj,2})$			coordinates of object at time 2

Table 14.1. Known and unknown variable at time 2 for Figure .

Consider the case where the time is 2, they have the known and unknown variables which are shown in Table 14.1.

Now, this trajectory motion can be described by the following

$d_{obj,a} = \sqrt{(x_a - x_{obj,0}) + (y_a - y_{obj,0})}$ 9 equations

$d_{t_0 \to t_1} = (V_{obj,0})\Delta t + \frac{a_{obj,0}}{2}(\Delta t)^2$ 2 equations

$V_{obj,1} = V_{obj,0} + (a_{obj,0})\Delta t$ 2 equations

$x_{obj,1} = (d_{t_0 \to t_1})cos(\alpha_{obj,0}) + x_{obj,0}$ 4 equations

$$y_{obj,1} = (d_{t_0 \to t_1})sin(\alpha_{obj,0}) + y_{obj,0} \qquad \text{4 equations}$$

Feng et al. incorporate errors to each variable:

$$\varepsilon_1 \Leftrightarrow x_{obj,0}$$
$$\varepsilon_2 \Leftrightarrow y_{obj,0}$$
$$\varepsilon_3 \Leftrightarrow x_{obj,1}$$
$$\varepsilon_4 \Leftrightarrow y_{obj,1}$$
$$\varepsilon_5 \Leftrightarrow x_{obj,2}$$
$$\varepsilon_6 \Leftrightarrow y_{obj,2}$$

Now, they can rewrite the system of equations in terms of objective function and constraints:

$$\text{OF: MIN}(|\varepsilon_1| + |\varepsilon_2| + |\varepsilon_3| + |\varepsilon_4| + |\varepsilon_5| + |\varepsilon_6|)$$

such that:

$$d_{obj,a} = \sqrt{(x_a - x_{obj,0}) + (y_a - y_{obj,0})} \qquad \text{9 equations}$$
$$d_{t_0 \to t_1} = (V_{obj,0})\Delta t + \frac{a_{obj,0}}{2}(\Delta t) \qquad \text{2 equations}$$
$$V_{obj,1} = V_{obj,0} + (a_{obj,0})\Delta t \qquad \text{2 equations}$$
$$x_{obj,1} = (d_{t_0 \to t_1})cos(\alpha_{obj,0}) + x_{obj,0} \qquad \text{4 equations}$$
$$y_{obj,1} = (d_{t_0 \to t_1})sin(\alpha_{obj,0}) + y_{obj,0} \qquad \text{4 equations}$$
$$x_{obj,0} = Ex_{obj,0} + \varepsilon_1$$
$$y_{obj,0} = Ey_{obj,0} + \varepsilon_2$$
$$x_{obj,1} = Ex_{obj,1} + \varepsilon_3$$
$$y_{obj,1} = Ey_{obj,1} + \varepsilon_4$$
$$x_{obj,2} = Ex_{obj,2} + \varepsilon_5$$
$$y_{obj,2} = Ey_{obj,2} + \varepsilon_6$$

There are a number of methods that can be used to solve problems posed as non-linear programming problem in the form of objective function and constraint. The most popular options includes feasible direction, active set, gradient projection, penalty, barrier, augmented lagrangians, cutting plane, direct, and quasi-Newton methods. The standard nonlinear programming references include [26].

The watermarking procedure is a self-contained block that is embedded in the overall multi-model sensor fusion process, as shown in Figure 14.3. The watermarking procedure can be conducted in many ways. For example, one can augment or alter the objective function with new components that correspond to the signature. Or one can superimpose additional constraints that correspond to a pseudorandom binary string that correspond to the signature. The advantage of the former technique is that it usually provides rather low overhead in terms of the solution quality. The advantage of the later technique is that it usually provides exceptionally strong proof of the authorship. In all

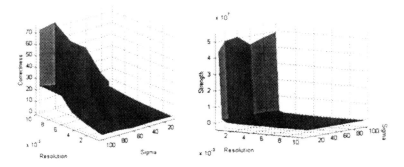

Figure 14.5. Correctness & Strength of authorship of the watermarking scheme given various Resolution and Sigma.

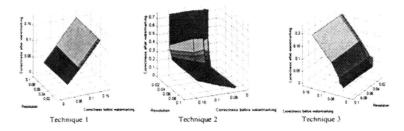

Figure 14.6. Comparison of correctness based on changes on resolution: before vs. after embedding watermarks.

cases, exact mapping of the pseudorandom string onto the constraints or objective function can be conducted in many ways. Three specific instances are presented in their sxperimental results.

The objective is to demonstrate the effectiveness of the approach on very small example (where it is most difficult to hide information) by statistically analyzing the relationships between correctness, strength of authorship, measurement errors, and resolution used for measurements and computations in terms of bits. Correctness is defined as the normalized difference in errors from the optimal solution between the watermarked solution and the solution obtained without watermarked. Strength of authorship is defined as 1 out of the all possible solutions that have at least the same quality as the watermarked solution.

The simulation process was conducted in the following way. They first generated the coordinates of three points according to the uniform distribution on the interval [0.0, 1.0]. For comparison and evaluation purposes later on, they also generate the coordinates of the point that they are trying to determine its location. After that, they calculated the exact distances between the forth point

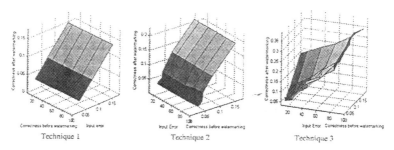

Figure 14.7. Comparison of correctness based on changes on sigma: before vs. after embedding watermarks.

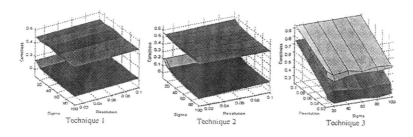

Figure 14.8. Comparison of correctness: 2-D vs. 3-D.

and the three beacon points. Furthermore, they add a small error value to the correct distances in order to simulate the estimated/measured distances. These small changes are randomly generated according to the Gaussian distribution (0, 1).

They consider three specific watermarking schemes. The first one just alters the least significant bit according to the signature. It is well known that this technique is not adequate for watermarking. They used it solely to provide basis for comparison for two other techniques. The second technique alters the components of the objective function according to the user's signature. The final technique, finds among all solutions that differ at most k% (they used value k = 5 in their experiments) on terms of estimated error from the non-watermarked solution, one that has the smallest Hamming distance from the signature stream.

From Figures 14.5-14.9, it is easy to see that two last techniques perform well, in particular when they consider 3D trilateration. The following series of figures show the comparisons of embedding signature in 2-D vs. 3-D by applying three different watermarking techniques. As they can observe from the figures, spreading the signature into more places (i.e. embedding watermarks in 3-D) produces a more accurate and stronger solution.

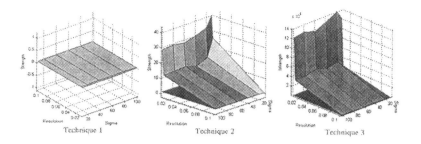

Figure 14.9. Comparison of strength of authorship: 2-D vs. 3-D

14.4 FUTURE DIRECTIONS

In this section, we briefly discuss the research directions that we perceive as the most challenging and most promising. We classify these directions in eight broad categories: (i) individual nodes, (ii) network infrastructure, (iii) sensor data and information, (iv) applications at the semantic level, (v) denial of service techniques, (vi) mobility, (vii) actuators, and ($viii$) theoretical foundations.

As we previously mentioned, wireless sensor networks are highly susceptible to security attacks due to their deployment, their hardware, and their resource constraints. Due to these factors, we can expect that fault inducing attacks and power consumption attacks will be conducted much more often on sensor network nodes. In addition, we expect that sensor and actuator related attacks will attract a significant amount of attention. As of now, no work has been done to address these security issues.

Similarly to the case of individual nodes, wireless sensor network infrastructures are potentially more susceptible to attacks than traditional networks. The operational state of the nodes are also conceptually very different because many nodes will often go into sleep mode and many nodes will be rarely active. Although significant progress has already been made on securing basic network protocols, it is clear that we need additional work to produce techniques for protecting canonical tasks in wireless senor networks such as routing, broadcast, multicast, and data aggregation. It will also be important to develop techniques which ensure that at least some nodes in each geographic area are operational. From the system point of view, we expect to see the development of firewalls specifically designed for the needs of WSN. For example, nodes in the network could be grouped in clusters and access to each cluster will be available through only a single node. Low power requirements may insist that nodes interchangeably serve as the firewall clusterhead.

WSN are designed and deployed because of sensors and therefore the primary objects of protection should be the sensor data and information. Very little work has been reported on these topics. A number of security and privacy issues need to be addressed including how to ensure the integrity of sensor data, how to provide mechanisms for authentication and access control, and how to efficiently, in terms of energy and storage, measure the usage of each sensor node by each user query.

In a sense, conceptually the most novel techniques will be developed for securing applications at the semantic level.We expect that most users will be most concerned with the privacy of their actions and information about their physiological state and the environment that surrounds them. Therefore, there is an urgent need to develop techniques that ensure privacy of subject and objects in sensor networks. One potential starting point for these efforts could be work done by the database community [1, 13, 35].

Recently, with the proliferation of the Internet, denial of service (DoS) techniques have gathered a great deal of attention [11, 38]. A number of static and dynamic defense mechanisms have been proposed. Due to the unique nature of WSN, we expect that denial of service attacks will be very popular. In addition, node sensitivity to energy consumption will further facilitate the effectiveness of denial of service attacks. One way to better learn about the most dangerous attacks is the development of controlled Honeypots of WSN that would register the exact sequence of steps taken by attackers and therefore eventually facility the development of better defense techniques [33]. Also we expect that a number of intrusion detection techniques for WSN will be soon developed.

Numerous WSN will be mobile, and as we already stated, mobility makes security and privacy significantly more difficult from one point of view and possibly easier from another point of view. Research in mobile WSN has just started and there is a need to develop tractable yet realistic mobility models. Note that in some mobile scenarios, restrictions on the size and power of the energy supply will not be as strict as in tradition wireless networks. Furthermore, note that mobility itself impacts essentially all network infrastructure tasks such as routing and data aggregation as well as how data is collected and processed.

Lastly, actuators will close the loop between the physical and information world. They have the potential to greatly improve the quality of individual life and industrial and economic processes. However, it is apparent that once the control of actuators is compromised, an attacker will be positioned to induce not just intellectual but also direct physical harm and damage. To the best of our knowledge, security of an actuator is a topic which still needs to be addressed. We expect that authentication and control techniques based on secret sharing will play an important role. Finally, in addition to outlining a great number of security techniques for sensor networks their exists a clear need to develop a sound theoretical foundation of the field. This is particularly true with emerging misinformation and privacy research.

14.5 CONCLUSION

In this Chapter we discussed security issues in wireless sensor networks at the network layer, specifically mobility assisted security. Additionally, we survey a watermarking technique for digital right management of data and information from sensor networks. Furthermore, we summarize some of most promising pending research directions related to security and privacy in sensor networks.

REFERENCES

[1] R. Agrawal and R. Srikant. Pricacy-preserving data minimg In *International Conference on Management of Data*, pages 429–450, 2000.

[2] D. Boneh and J. Shaw. Collusion-secure fingerprinting for digital data In *Advances in Cryptology*, pages 452–465, 1995.

[3] J. T. Brassil, S. Low, and N. F. Maxemchuk. Copyright protection for the electronic distribution of text documents In *Proceedings of the IEEE*, volume 87, pages 1181–1196, 1999.

[4] P.T. Breuer and K.C. Lano. Creating specifications from code: Reverse engineering techniques *Journal of Software Maintenance: Research and Practice*, 3:145–162, 1991.

[5] C. Thomborson C. Collberg and. Watermarking, tamper-proofing, and obfuscation - tools for software protection *Transactions on Software Engineering*, 28(2):735–746, 2002.

[6] A.E. Caldwell, H. Choi, A.B. Kahng, S. Mantik, M. Potkonjak, G. Qu, and J.L. Wong. Effective iterative techniques for fingerprinting design IP In *Design Automation Conference*, pages 843–848, 1999.

[7] L.J. Camp. Drm: doesn't really mean digital copyright management In *ACM Computer and Communications Security*, pages 78–87, 2002.

[8] S. Capkun, L. Buttyan, and J. Hubaux. Self-organized public-key management for mobile ad hoc networks In *ACM International Workshop on Wireless Security*, pages 52–64, 2002.

[9] S. Capkun, J.P. Hubaux, and L. Buttyan. Mobility helps security in ad hoc networks In *ACM Symposium on Mobile Ad Hoc Networking and Computing*, pages 46–56, 2003.

[10] R. Chapman and T. Durrani. Ip protection of dsp algorithms for system on chip implementation *IEEE Transactions on Signal Processing*, 48(3):854–861, 2000.

[11] S. Cheung and K. N. Levitt. Protecting routing infrastructures from denial of service using cooperative intrusion detection In *New Security Paradigms Workshop*, pages 94–106, 1997.

[12] I. Cox, J. Killian, T. Leighton, and T. Shamoon. Secure spread spectrum watermarking for images In *IEEE Int. Conf. on Image Processing*, pages 243–246, 1996.

[13] D. Dobkin, A. Jones, and R. Lipton. Secure databases: Protection against user influence *Transactions on Database Systems*, 4(1):97–106, 1979.

[14] J. Feng and M. Potkonjak. Real-time watermarking techniques for sensor networks In *SPIE Security and Watermarking of Multimedia Contents*, pages 391–402, 2003.

[15] C. Fornaro and A. Sanna. Public key watermarking for authentication of csg models *Computer Aided Design*, 32(12):727–735, 2000.

[16] J. Fridrich, M. Goljan, and A.C. Baldoza. New fragile authentication watermark for images In *International Conference on Image Processing*, pages 446–449, 2000.

[17] F. Hartung and M. Kutter. Multimedia watermarking techniques In *Proceedings of the IEEE*, volume 87, pages 1079–1107, 1987.

[18] Y.-C. Hu, D. B. Johnson, and A. Perrig. Sead: Secure efficient distance vector routing for mobile wireless ad hoc networks In *IEEE Workshop on Mobile Computing Systems and Applications*, pages 3–13, 2002.

[19] H. Huang and S. F.x Wu. An approach to certificate path discovery in mobile ad hoc networks In *ACM Workshop on Security of Ad Hoc and Sensor Networks*, 2003.

[20] A. B. Kahng, S. Mantik, I. L. Markov, M. Potkonjak, P. Tucker, H. Wang, and G. Wolfe. Robust IP watermarking methodologies for physical design In *Design Automation Conference*, pages 782–787, 1998.

[21] D. Kirovski, Y.-Y. Hwang, M. Potkonjak, and J. Cong. Intellectual property protection by watermarking combinational logic synthesis solutions In *International Conference on Computer-Aided Design*, pages 194–198, 1998.

[22] D. Kirovski, D. Liu, J. L. Wong, and M. Potkonjak. Forensic engineering techniques for VLSI CAD tools In *Design Automation Conference*, pages 581–586, 2000.

[23] D. Kirovski and H.S. Malvar. Robust spread-spectrum watermarking In *International Conference on Acoustics, Speech, and Signal Processing*, pages 1345–1348, 2001.

[24] J. Kong, P. Zerfos, H. Luo, S. Lu, and L. Zhang. Providing robust and ubiquitous security support for mobile ad-hoc networks In *IEEE International Conference on Network Protocols*, pages 251–260, 2001.

[25] J. Lach, W.H. Mangione-Smith, and M. Potkonjak. Robust FPGA intellectual property protection through multiple small watermarks In *Design Automation Conference*, pages 831–836, 1999.

[26] D. Luenberger. *Linear and Nonlinear Programming*. Addison Wesley, 1984.

[27] J.G. Markoulidakis, G. L. Lyberopoulos, D. F. Tsirkas, and E. D. Sykas. Mobility modeling in third-generation mobile telecommunication systems In *IEEE Personal Communications*, pages 41–56, 1997.

[28] R. Ohbuchi, H. Masuda, and M. Aono. Watermarking three-dimensional polygonal models In *ACM International Multimedia Conference*, pages 261–272, 1997.

[29] J. Palsberg, S. Krishnaswamy, M. Kwon, D. Ma, Q. Shao, and Y. Zhang. Experience with software watermarking In *Annual Computer Security Applications Conference*, pages 308–316, 2000.

[30] P. Papadimitratos and Z.J. Haas. Securing mobile ad hoc networks In *Handbook of Ad Hoc Wireless Networks*. CRC Press, 2002.

[31] G. Qu. Publicaly detectable techniques for the protection of virtual components In *Design Automation Conference*, pages 474–479, 2001.

[32] J. Scourias and T. Kunz. An activity-based mobility model and location management simulation framework *Workshop on Modeling and Simulation of Wireless and Mobile Systems*, pages 61–68, 1999.

[33] L. Spitzner. *Honeypots: Tracking Hackers*. Addison-Wesley., 2002.

[34] M. Wagner. Robust watermarking of polygonal meshes In *Geometric Modeling and Processing*, pages 201–208, 2000.

[35] S. Warner. Randomized response: A survey technique for eliminating evasice answer bias *Am. Stat. Assoc*, 60(309):62–69, 1965.

[36] R. Wolfgang and E. Delp. A watermark for digital images In *International Conference on Images Processing*, pages 219–222, 1996.

[37] R. Wolfgang, C.I. Podilchuk, and E. Delp. Perceptual watermarks for digital images and video In *International Conference on Security and Watermarking of Multimedia Contents*, volume 3657, pages 40–51, 1996.

[38] A. D. Wood and J. A. Stankovic. Denial of service in sensor networks *Computer*, 35(10):54–62, 2002.

LOCALIZATION AND MANAGEMENT

Chapter 15

LOCALIZATION IN SENSOR NETWORKS

Andreas Savvides
Electrical Engineering Department
Yale University
andreas.savvides@yale.edu

Mani Srivastava
Electrical Engineering Department
University of California, Los Angeles
mbs@ee.ucla.edu

Lewis Girod and Deborah Estrin
Computer Science Department
University of California, Los Angeles
{girod,destrin}@lecs.cs.ucla.edu

Abstract The development of large scale distributed sensor systems is a significant scientific and engineering challenge, but they show great promise for a wide range of applications. The capability to sense and integrate spatial information with other elements of a sensor application is critical to exploring the full potential of these systems. In this article we discuss the range of application requirements, introduce a taxonomy of localization mechanisms, and briefly discuss the current state of the art in ranging and positioning technologies. We then introduce two case studies that illustrate the range of localization applications.

Keywords: Wireless Sensor Network, Localization, Position Estimation, Ranging

15.1 INTRODUCTION

The development of large scale distributed sensor systems is a significant scientific and engineering challenge, but they show great promise for a wide range of applications. By placing sensors close to the phenomena they sense, these systems can yield increased signal quality, or equivalent signal quality at reduced cost. By reducing deployment overhead, whether in terms of unit cost or installation time, they enable the sensors to be placed in greater numbers.

Relative to other types of distributed systems, distributed sensor systems introduce an interesting new twist: they are coupled to the physical world, and their spatial relationship to other objects in the world is typically an important factor in the task they perform. The term **localization** refers to the collection of techniques and mechanisms that measure these spatial relationships.

When raw sensor data is combined with spatial information, the value of the data and the capability of the system that collects it increases substantially. For example, a collection of temperature readings without location information is at best only useful to compute simple statistics such as the average temperature. At worst, analysis of the data might yield incorrect conclusions if inaccurate assumptions are made about the distribution of physical sampling. By combining the data with location information, the resulting temperature map can be analyzed much more effectively. For instance, statistics can be computed in terms of spatial sampling rather than the count of sensor readings, and the confidence of the results can be assessed more meaningfully. Location also opens up entirely new application possibilities: a model for heat transfer can be applied to filter out noise and pinpoint the location of heat sources.

This simple example is intended to illustrate a more general point. As anyone who has worked with distributed sensor systems is painfully aware, there is a high cost in moving from a centralized, wired application to a large-scale, distributed, wireless application. The application is certain to grow in complexity; new techniques must be developed, new protocols deployed, and the application must be resilient to the whims of nature. In addition, there are the more mundane details of dealing with large numbers of independent parts, each of which needs the right software version and fresh batteries, and each of which can independently fail. But what makes all of this effort worthwhile is the ability to deploy applications that collect data that could never be collected before, and for this we need localization.

In this article, we will first present a range of application requirements in Section 15.2. In Section 15.3, we will introduce a taxonomy of localization mechanisms to satisfy those requirements. In Section 15.4, we will summarize the state of the art in available ranging and positioning technologies. Section 15.5 discusses node localization in multihop networks and section 15.6 examines how error behaves with respect to different network parameters. Section 15.7 concludes the chapter with a description of an example ad-hoc localization system.

15.2 APPLICATION REQUIREMENTS

The field of networked sensor systems encompasses a very broad array of applications, with a broad range of requirements. Often different application requirements can motivate very different systems. While these differences are sometimes tunable parameters, often they are significant structural choices. For example, adding a low-power requirement rules out many possible designs from the start.

To introduce this variety, we first present two points in the application space, and then enumerate a set of requirements axes that characterize the requirements space of localization systems.

15.2.1 PASSIVE HABITAT MONITORING

Scientists in numerous disciplines are interested in methods for tracking the movements and population counts of animals in their natural habitat. While there are various techniques currently employed (e.g. rings on birds' claws, "drop buckets" for small animals on the ground), these techniques do not scale well in terms of the time required of experimenters. One of the open challenges in this field is to develop an automated system that can build a record of the passage and habits of a particular species of animal, without disturbing it in its natural habitat.

One possible solution might be built around a passive source localization and species identification system. Such a system would detect and count animals by localizing the sounds they make, then training a camera system on them to aid in counting.

Sensor nodes equipped with microphones would be distributed through the target environment. When an acoustic source is detected by a node, it communicates with nearby nodes to try to estimate the location of the source by comparing the times of arrival of the signals. Analysis techniques such as beam-forming [19] might apply to this application, along with species recognition techniques to filter out acoustic sources not relevant to the task.

From this application we can derive a number of requirements:

- **Outdoor Operation.** The system must be able to operate outdoors, in various weather conditions.

- **Power Efficiency.** Power may be limited, whether by battery lifetime or by the feasibility of providing sufficient solar collectors.

- **Non-cooperative Target, Passive Infrastructure.** The animal does not emit signals designed to be detected by the system (i.e. non-cooperative), and the system does not emit signals to aid in localization.

- **Accuracy.** The system must be accurate enough to be able to produce a reliable count, and to accurately focus a camera on suspected locations.

- **Availability of Infrastructure.** In some cases GPS may be available to localize the sensors themselves. However, in many cases sensors will need to be placed under canopies where GPS signals are unavailable. In these cases, if surveying the sensors is inconvenient, the sensors will need to self-localize. The self-localization system may have a different set of requirements.

15.2.2 SMART ENVIRONMENTS

Smart environments are a second class of applications where location awareness is a key component. Smart environments are deeply instrumented systems with very demanding localization requirements. These systems need localization for two different purposes. First, rapid installation and self-configuration of a set of infrastructure "beacons" is required to reduce installation cost and increase flexibility. Second, very fine-grained localization and tracking of the system components is required during normal system operation.

The operation requirements can also be assessed along similar axes:

- **Indoor Operation.** The system must operate indoors. While the weather indoors is generally predictable, there are typically many reflectors that cause multipath interference for both RF and acoustic signals. If the environment is an office environment, acoustic signals should be outside of the range of human hearing. The system will need to operate in the presence of obstacles.

- **Power Requirements.** Some of the infrastructure components may be close to power sources, but a large number of the system components should be untethered and free to move around the space. This implies that the majority of the system components should be as low power as possible.

- **Cooperative Target, Active Infrastructure.** Because the target badge localizes itself relative to the beacons and reports its location, we consider it to be cooperative. The beacon infrastructure is active because it emits periodic signals that the badge receives.

- **Accuracy.** Fine-grained localization of people and objects with 10cm accuracy may be required by many systems.

- **Availability of Infrastructure.** Infrastructure may be present, but infrastructure installation usually becomes a dominant cost factor. Ideally the infrastructure should be self-configuring in a way that reduces installation cost.

15.2.3 AXES OF APPLICATION REQUIREMENTS

These points in the application space demonstrate a broad spectrum of requirements that seem to fall along nine independent axes:

- **Granularity and Scale of Measurements.** What is the smallest and largest measurable distances? For instance, local coordinate systems for a sensor network might scale from centimeters to hundreds of meters, whereas GPS coordinates have a global scale and a granularity on the order of meters.

- **Accuracy and Precision.** How close is the answer to ground truth (accuracy), and how consistent are the answers (precision)?

- **Relation to Established Coordinate System.** Are absolute positions needed, or is a relative coordinate system sufficient? Do locations need to be related to a global coordinate system such as GPS, or an application specific coordinate system such as a forest topography or building floorplan?

- **Dynamics.** Are the elements being localized fixed in place or mobile? Can a static infrastructure be assumed? What refresh rate is needed? Is motion estimation required?

- **Cost.** Node hardware cost, in terms of both power consumption and monetary cost; Latency of localization mechanism; Cost of installing infrastructure (if needed), in terms of power, money, and labor.

- **Form Factor.** How large can a node be? If the node has multiple sensors separated by a baseline, what kind of baseline is required for sensors to work effectively?

- **Communications Requirements.** What kind of coordination is required among nodes? What assumptions does the system make about being able to send or receive messages at any time? What kind of time synchronization is needed? Does the algorithm rely on the existence of a cluster head or a "microserver"?

- **Environment.** How sensitive is a given technique to environmental influences, and in what range of environments does it work? For instance, indoors (multipath), outdoors (weather variations), underwater, or on Mars?

- **Target Cooperation, System Passivity.** Does the target play a cooperative role in the system? Can the system emit signals into the environment without interfering with the task at hand?

Given such a complex requirement space, it seems that few system designs, techniques or technologies will fit that space uniformly. This complicates efforts to develop reusable designs and components, especially in the early stages of a research program where the canonical applications are not well understood. It also means that the importance of a new technique or system must be evaluated in an appropriate context. In order to equitably compare different mechanisms and systems, we need a taxonomy of mechanisms and system structures to provide context for our comparisons.

15.3 TAXONOMY OF LOCALIZATION MECHANISMS

The capsular conclusion of the last section was simple: localization systems will differ not only in details of algorithms and protocols, but also fundamentally in the structure of their system and in the assumptions they make. The challenge of this section is therefore to construct a taxonomy of general system structures that capture the breadth and depth of the solution space. Having done this, hopefully we can better classify localization systems and components for the purposes of comparison and contrast.

15.3.1 CLASSIFYING OUR EXAMPLES

Returning to our examples of the previous section, we can motivate a taxonomy by summarizing their important structural features. Perhaps the distinction that stands out the most among our example applications has to do with "cooperative" targets. In the case of an animal localizer, we can't assume that the animal is acting with the intent of being localized. In fact, under some circumstances, targets may intentionally avoid detection. This property is in sharp contrast with the case of locating badges. While a small child might attempt to subvert the localization system, we can assume that the badges themselves will be cooperative in whatever ways will simplify the implementation of the system.

This distinction has numerous repercussions on the overall design of the system. First, while in the non-cooperative case the animal detector must be constantly vigilant, at considerable energy cost, the cooperative smart badge can take advantage of a simple mechanical motion detector (e.g. a jiggle switch) to implement a simple zero-power wakeup mechanism.

Second, while the characteristic sounds of animal calls and motions may be detectable, the detection process is more complex, and ultimately more failure-prone than the detection of a synthetic ranging signal. Not only will the processing be more expensive, both in terms of processing and communications costs, the false positive rate will tend to be higher. In contrast, a cooperative system has the luxury of designing its signals to be detected efficiently and accurately, e.g. low-autocorrelation codes and a simple matched filter detector at each receiver.

Third, a non-cooperative system can only use Time Difference of Arrival (TDoA) techniques to estimate position, because the true time of the signal emission is not known. In contrast, a cooperative system can sometimes use an out-of-band synchronization protocol to establish a consistent timebase, and then provide receivers with the send time so that they can measure Time of Flight (ToF).

These differences result in important structural differences among our example systems. We place the habitat monitoring application in the "Passive Target Localization" category. The smart environment application can be broken into two phases, that fall into different categories: a "bootstrapping" phase in which the infrastructure of beacons self-organizes into a coordinate system, and a "service" phase in which the badges localize themselves with respect to the beacon infrastructure. The bootstrapping phase fits into the "Cooperative Target" category, while the service phase fits into "Cooperative Infrastructure".

15.3.2 A TAXONOMY OF LOCALIZATION SYSTEMS

We have discussed examples representing three categories of localization system. Extending this process, we can identify six categories of localization system, that appear to cover all the systems of which we are currently aware. These six categories are partitioned into "active" and "passive".

Active Localization. Active localization techniques emit signals into the environment that are used to measure range to the target. These signals may be emitted by infrastructure components or by targets. Within the category of active localization there are three subcategories:

- **Non-cooperative.** In an active, non-cooperative system, system elements emit ranging signals, which are distorted or reflected in flight by passive elements. The system elements then receive the signals and analyze them to deduce their location relative to passive elements of the environment. Examples include radar systems and reflective sonar systems often used in robotics.

- **Cooperative Target.** In a cooperative target system, the targets emit a signal with known characteristics, and other elements of the system detect the signals and use information about the signal arrivals to deduce the target's location. Often a cooperative target system also involves some synchronization mechanism to readily compute signal ToF. This category includes both infrastructure-less systems and systems that localize with respect to infrastructure receivers. Infrastructure-based systems include the ORL Active Bat and the service phase of the GALORE localization system [7]. Infrastructure-less systems include the bootstrapping phases of both the GALORE and Smart Kindergarten systems.

- **Cooperative Infrastructure.** In a cooperative infrastructure system, elements of the infrastructure emit signals that targets can receive. The infrastructure itself is assumed to be carefully configured and synchronized to simplify the processing done by the target. Another property of this system structure is that receivers can compute their own location passively, without requiring any interaction with the infrastructure. Examples of this type of system include GPS and the MIT Cricket system [14], and the service phase of the Smart Kindergarten system.

Passive Localization. Passive localization techniques differ from active ones in that they discover ranges and locations by passively monitoring existing signals in a particular channel. The term "passive" does not imply that they emit *no* signals, only that the signals they emit are outside the channel that is primarily analysed for time-of-flight measurement. For example, a technique that uses RF signals for synchronization and coordination, but measures range by TDoA of ambient acoustic signals would still be considered passive.

- **Blind Source Localization.** In a blind source localization system, a signal source is localized without any *a priori* knowledge of the type of signal emitted. Typically this is done by "blind beam-forming", which effectively cross-correlates the signals from different receivers. These techniques generally only work so long as the signals being compared are "coherent", which in practice often limits the spacing of receivers because of signal distortion induced by the environment. Coherent combining techniques can generally localize the most prominent source within the convex hull of a sensor laydown, or alternatively can compute a bearing angle to a distant source, but not a range or location. This work is described by Yao et. al.[19].

- **Passive Target Localization.** Similar to blind localization, a passive target localization system is usually based on coherent combination of signals, with the added assumption of some knowledge of the source. By assuming a model for the signals generated by the source, filtering can be applied to improve the performance of the algorithms and to reduce the computational and communications requirements. Examples include our previous example of habitat monitoring, UCLA work on beamforming [20], and some E911 cell phone location proposals.

- **Passive Self-localization.** In passive self-localization, existing beacon signals from known infrastructure elements are used by a target to passively deduce its own location. Most commonly, properties of RF signals from base stations are used to deduce location of a mobile unit. Examples include RADAR [1], which measured RSSI to different 802.11 access points, and the work of Bulusu et. al.[2], which measured RSSI to Ricochet transmitters.

Cross-cutting Issues. As a rule, active and cooperative techniques tend to be more accurate, more efficient, and generally more effective. Because cooperative techniques can design both the receiver and transmitter, the designs can be optimized for performance much more effectively. Cooperative systems can also synchronize explicitly, improving the performance of ranging based on signal propagation time. However, applications such as habitat monitoring can only be addressed using passive techniques. Although passive techniques are attractive because they can leverage existing signaling, they often perform poorly when the signaling is not designed with ranging or localization in mind.

Another aspect of sensor network localization that cuts across these categories is an ability to support ad-hoc deployment and operation. In an ad-hoc setup, there is no guarantee that all the sensor nodes will be in communication and sensing range to each other, nor that the sensing and communications properties will remain constant over time. Thus, regardless of category, systems that can operate in an ad-hoc fashion must collaborate across the sensor nodes, must operate within a multihop network, and must react to system dynamics. These issues will be addressed further in Section 15.5.

15.4 RANGING TECHNOLOGIES

When designing a localization system, an important factor in the design are the mechanisms used to measure physical distances and angles. Typically for cooperative systems this will involve some kind of emitter and detector pair. The selection of these elements has a significant impact on how well the final system will fit the application requirements. In this section, we will discuss the relative merits of three types of ranging mechanism, based on visible light, radio signals, and acoustic signals.

15.4.1 RANGING USING RF

RF ranging generally follows one of two approaches: distance measured based on received signal strength, and distance measured based on the ToF of the radio signal.

RF RSS. Received Signal Strength (RSS) is roughly a measure of the amplitude of a detected radio signal at a receiver. If we assume a model for path loss as a function of distance, the received signal strength should generally decrease as a function of distance. The path loss model is highly dependent on environmental factors: in open space the model is $1/R^2$; near the ground, the model is closer to $1/R^4$. Under some conditions (e.g. waveguides, corridors, etc.), path loss can actually be lower than in free space, e.g. $1/R^{1.5}$. Because the path loss model is dependent on details of the environment, automatically choosing a valid model can be difficult.

In practice, the behavior of RSS is dependent on a number of factors, not the least of which is simply whether the RSS estimator is well designed. Some radios, such as the RFM radio used in the Berkeley mote, just sample the baseband voltage to estimate RSS, which is a very crude measurement; other radios have a more capable measurement circuit. Another important factor is the frequency range used by the radio. Multipath fading is a change in RSS caused by the constructive or destructive interference of reflected paths. Multipath fading is dependent on the environment and is in many cases dependent on the frequency of the signals being transmitted. A radio that uses just one frequency will be more susceptible to multipath fades that will cause substantial error in a distance estimate based on RSS. If a variety of frequencies are used and appropriate filtering is applied, the effects of frequency dependent multipath fading may be removed from the RSS estimate, although frequency independent multipath such as ground reflection will still be present. In general, the effectiveness of RSS estimation varies from system to system and cannot be implemented without hooks into the internals of the radio hardware.

Another difficulty with using RSS is that the transmit power at the sender may not be accurately known. In many cases this is a function of specific component values on a given board, or a function of battery voltage. Without knowing the original transmit power it it may not be possible to correctly estimate the path loss.

RF Time of Flight. Measuring the time of flight of radio signals is another possible solution. Because the ToF of a radio signal is not very dependent on the environment, ToF approaches can be much more precise than approaches based on measuring RSS. The two main challenges in implementing an RF ToF scheme are: (1) synchronization must use signals also traveling at the speed of light, and (2) to achieve high precision ToF measurements require high frequency RF signals and fast, accurate clocks.

The timing and synchronization issues are the central problems with RF ToF ranging. The synchonization problem is simplified for Infrastructure based systems where elements of the infrastructure can be synchronized by some out-of-band mechanism, or by taking into account knowledge of their exact locations. However, this does not eliminate the need for accurate clocks, which tend to be expensive in terms of power. For example, GPS satellites carry atomic clocks for timing, but these clocks are continually adjusted to account for relativistic effects as they orbit the earth, and the trajectory of the satellites is carefully measured. In more down-to-earth implementations, in most RF ToF ranging implemenations, the infrstructure elements are connected by carefully measured cables to achieve synchronization.

However, for ad-hoc deployments, the lack of a common timebase means that "round trip" messages must be used in order to compare send and receive times within the same timebase. The time spent "waiting" at the remote transponder must then be measured and subtracted. This requires that the two systems' clocks be running at close to the same rate, or that the turnaround time be a fixed constant. Getting these details to work correctly, given the fact that all the timing must be very precise (30 cm of error per ns), can be quite challenging from a hardware perspective.

Probably the best path toward an ad-hoc RF ToF solution is to leverage the hardware of a sufficiently advanced radio system, such as 802.11 or an ultra-wide band (UWB) receiver. Because these systems operate at high bit rates and must match clock rates in order to inter-operate, there is a better chance that it is possible to exploit their features to implement ranging. Some 802.11 chipsets have ranging support, although at press time we do not have any references to implementations that use them successfully. UWB ranging solutions have been advertised, however because of licensing restrictions and other issues there have been no documented implementations in the sensor networks space.

Bearing estimates for incoming RF signals require similar types of designs. By implementing a radio with an array of antennae, the signals from those independent reception points can be compared to estimate direction of arrival (DoA). While these techniques are commonplace in radar systems and commercial wireless systems, currently this kind of feature is not available on small, low-power platforms. However, similar designs to those that enable ranging might someday also enable DoA estimation.

15.4.2 RANGING USING ACOUSTICS

Acoustic ranging is probably the most developed ranging technology in use in sensor networks. There are a number of factors that make acoustics attractive, given currently available COTS components. Acoustic transducers are easy to interface, and simple, inexpensive detector chipsets are available for ultrasound. However the key advantage to using acoustics is that timing and synchronization is much easier to implement. A 32 KHz clock is sufficient to achieve ranging accuracy to 1 cm, and synchronization between sender and receiver can be implemented using most radio modules without modification.

In terms of power, acoustics performs quite well, even near the ground. Whereas RF communication suffers r^{-4} path loss near the ground because ground reflections are phase-shifted by 180 degrees, this is not the case for acoustic waves. Acoustic path loss near the ground under good conditions is much closer to r^2. Outdoors, acoustics is susceptible to interference from weather conditions, such as wind that causes noise, and convective updrafts that carry signals up and away from the ground.

However, acoustics has a few disadvantages as well. First, acoustic emitters tend to be physically large, especially if they emit low frequencies. The other main disadvantage is that acoustic signals are stopped by solid obstructions. However, for some applications this can be advantageous, such as the case of an asset tracking system which only needs to know which room the asset is in.

When using acoustics, a wide band of frequencies are available for use. Some systems are based on ultrasound frequencies (typically 40 KHz to 1 MHz), while others are based on audible frequencies (100 Hz to 20 MHz). Some systems use tuned piezo emitters at specific frequencies, while others use wide-band acoustic signals. The choice of frequency depends on the application (e.g. is audible sound acceptable), as well as the environment.

Experience with 40 KHz ultrasound systems outdoors indicates a typical range of about 10 meters at a voltage of 3 volts, and about 16 meters at 16 volts. The type of emitter used also has a significant effect on the performance of the system. Many ultrasound emitters are directional, substantially increasing their output in a conical beam. This can be disadvantageous from a packaging perspective, as it may require many emitters and receivers in order to support ad-hoc deployment.

Audible acoustics can be very effective outdoors, because of the wide diversity of wavelengths possible. A wide-band signal will be more robust to environmental interference, because of the process gain in the detection process. A wide-band signal is also less susceptible to narrowband sources of noise, as well as absorption and scattering of specific frequencies.

Under ideal weather conditions, audible ranging systems have been shown to achieve ranges as large as 100m for power levels of 1/4 Watt. High power emitters such as heavy vehicles are detectable at ranges of 10's of kilometers. Acoustic range is longest at night when the air is still and cool. The worst conditions for acoustics are warm, sunny afternoons, when heated air near the ground rises and deflects signals up and away from other ground-based receivers. Under these conditions, the same acoustic system might achieve only 10m range.

Errors in line-of-sight (LoS) acoustic ranges tend in general to be independent of distance, up to the limit of the signal detector. However, when obstructions or clutter are present, severe attenuation can be observed, as well as radical outliers when the LoS path is completely blocked and a reflected path is detected. When designing positioning algorithms around an acoustic ranging system it is important to take these issues into account.

Bearing estimates for acoustic signals can often be implemented without much difficulty using simple hardware and software solutions. If the baseline between sensors is known with sufficient accuracy, a bearing estimate can be derived from the time difference of arrivals. There are several examples of implemented systems that measure DoA using acoustics, the MIT software compass[21], and beam-forming systems[20].

15.5 POSITIONING IN MULTIHOP SENSOR NETWORKS

The importance and plethora of applications in multihop sensor networks, motivated the development of diverse set of positioning algorithms. The ability to operate in a multihop regime allows nodes with short-range signal transmissions to collaboratively localize themselves across larger areas. These properties make multihop ad-hoc localization an appealing choice in ad-hoc deployed sensor networks, rapidly installable infrastructures and fine-grained localization in indoor settings where the multihop and ad-hoc nature of the system can compensate for the presence of obstacles and many other settings where other infrastructure based technologies such as GPS cannot operate.

15.5.1 LOCALIZATION CHALLENGES IN MULTIHOP AD-HOC SENSOR NETWORKS

Despite the attractiveness of ad-hoc multihop localization, the application requirements need to be carefully reviewed before any design choices are made. Unfortunately the flexibility promised by such localization systems is also coupled with large set of challenges and trade-offs that have so far inhibited their widespread deployment. Some of these challenges are listed here.

Physical Layer Challenges. As described in the previous section, measurements are noisy and can fluctuate with changes in the surrounding environment.

Algorithm Design Challenges. The algorithm designer needs concurrently consider multiple issues when designing such systems.

- **Noisy measurements** call for the use of optimization techniques that minimize the error in position estimates. Despite the well-established body of knowledge in optimization techniques, the use of any optimization algorithm is only as good as the validity of the assumptions on the underlying measurement error distribution in the actual deployment scenario.

- **Computation and communication trade-offs.** Cost and energy limitations force designers to consider the development of lightweight distributed algorithms that can operate on low cost resource constrained nodes, where the computation is performed inside the network.

- **Problem setup.** A large variety of problem setups has appeared in the literature. Some approaches consider the use of a small percentage of location aware anchor nodes spread randomly distributed inside the network. Some other approaches, suggest that one should ensure that enough anchor nodes are placed on the network perimeter, while some others advocate anchor free setups. In addition to the detup decision, the type of measurements used in each case, vary across different solutions, some try to infer locations based on mere connectivity information while others, use angular and/or distance measurements.

- **Error behavior and scalability.** Perhaps the most overlooked aspect of multihop localization in currently proposed solutions is understanding how the network parameters affect the resulting position error behavior and scalability. Network topology and geometry between nodes, network density, ranging accuracy, anchor node concentration and uncertainty in anchor node locations, affect the quality of location estimates; therefore their behavior needs to be formally understood.

System Integration Challenges. All the previously discussed requirements imply a non-trivial system integration effort. Many off-the-shelf measurement technologies are not directly suitable for use in sensor networks, so customized hardware and software often needs to be developed to make a functional system.

15.5.2 OVERVIEW OF MULTIHOP LOCALIZATION METHODS

Despite the numerous proposals, very few ad-hoc localization systems have been built and evaluated in practice. Furthermore, the side-by-side comparison of different approaches is a non-trivial task due to the differences in problem setup and underlying assumptions. In the remainder of this section we highlight some of the recently proposed approaches by broadly classifying them as connectivity based and measurement based approaches. Later on we also comment on some of the trends associated on position error based Cramér Rao bound analysis.

15.5.3 RADIO CONNECTIVITY BASED APPROACHES

Connectivity-based approaches try to leverage radio connectivity to infer node locations. Although radio connectivity alone cannot provide fine-grained localization, it can provide a good indication of proximity that is useful in supporting other network level tasks such as geographic routing. The GPS-less low cost localization system described in [2] is an example of a connectivity based system. In this system, a set of pre-deployed, location aware reference nodes transmit spatially overlapped beacon signals. Other nodes with unknown locations can localize themselves at the centroid of the reference nodes from which they can receive beacon signals. The best results are obtained when the nodes are arranged in a mesh pattern.

The convex position estimation approach proposed by Doherty et. al. in [4] also localizes nodes using radio connectivity. In this case the localization is formulated as a linear or semi definite program that is solved at a central location. This approach also requires a set of nodes with known locations to act as beacons. With careful placement of the beacon nodes on the perimeter of the network the authors have shown that node locations between 0.64 and 0.72R (where R is the radio transmission range) are possible at density of 5.6 neighbors per node.

A more recent proposal based on multidimensional scaling (MDS) can solve for the relative position of the nodes with respect to each other without requiring any beacon nodes [17]. This is done by using a classical MDS formulation that takes node connectivity information as inputs and creates a two dimensional relative map of the nodes that preserves the neighborhood relationships. The connectivity only approach uses hop distances between nodes to initialize a distance matrix. The same MDS formulation can take more accurate internode distances to construct more accurate maps.

15.5.4 MEASUREMENT BASED APPROACHES

Measurement based approaches build upon a wide range of measurement technologies. While different approaches focus on specific ranging systems, a large source of disparity in measurement-based algorithms stems from different assumptions about measurement error distribution. Some systems assume additive Gaussian noise, while others assume that measurement error is proportional to distance. Furthermore, some algorithms require a set of initial anchor nodes, whereas others perform relative localization and use anchors only at the end of the localization process to translate the derived relative coordinate system to an absolute coordinate system. Because of these reasons, in this section we do not attempt a direct comparison of existing approaches, instead highlighting the key features of each approach. We begin with anchor free approaches, followed by approaches that use anchor nodes.

An example algorithm that does not require anchor nodes is described in [3]. This relative localization system is based on radio ToF measurements and uses geometric relationships to estimate node positions. First, all the nodes compute their locations with respect to their neighbors. The resulting local coordinate systems are then aligned and merged into a global coordinate system using a simple set of geometric relationships. The position estimates acquired by this method are not very accurate due to the noisy ToF measurements and error propagation. Despite this loss of precision, the location estimates are still adequate to help with network level tasks such as geo-routing.

Another notable anchor-free localization method has been developed by [9]. In this work Moses et. al. have shown that sensor node positions and orientations can be estimated using signals from acoustic sources with unknown locations. Each acoustic source generates a known acoustic signal that is detected by the sensor nodes. The sensor nodes in turn measure the ToA and DoA of the signal and propagate this information to a central information-processing center (CIP). The CIP fuses the information using Maximum Likelihood estimation to obtain the location and orientation of the sensor nodes. The authors also consider cases where partial measurements (i.e either ToA or DoA) are available.

The localization system developed as part of the GALORE project at UCLA is another example of an anchor-free ad-hoc positioning system[7]. This system is composed of standard, unmodified iPAQs and Berkeley Motes with acoustic daughter cards. The system operates in two phases: a self configuration phase in which the iPAQs collectively construct a relative 3-D coordinate system, and a service phase in which the iPAQs can localize a mote and report its location back. In this system, iPAQs are acoustic emitters and receivers, and motes are acoustic emitters only. A time synchronization service component maintains time conversions between all adjacent components of the system, and ranges from one node to another are computed by measuring the time of flight of acoustic signals. The positioning algorithm is a centralized algorithm based on relaxation of a spring model in which range measurements map to spring "lengths". A novel element of the spring algorithm is that the spring constants are non-linear: the springs are modeled as easier to compress than to stretch. This has the effect of favoring short ranges over long ranges, which is more consistent with the errors encountered with acoustics, where excess path measurement is more likely than a "short" range.

The Ad-Hoc Positioning System proposed by Nicolescu and Nath in [11] estimates the locations in an ad-hoc network by considering distances to a set of landmarks. This study explores three alternative propagation methods: *DV-hop*, *DV-distance*, and *Euclidean*. In the DV-hop method, landmarks propagate their location information inside the network. Each node forwards the landmark information to its neighbors and maintains a table with the landmark ID, location, and hop distance. When a landmark receives one of the propagated packets with the position of a different landmark, it uses that information to calculate the average hop-distance between the two landmarks. The computed average hop distance is broadcasted back into the network as a correction to previously known hop distances. The nodes that receive this message use the average hop distances to each of the landmarks to estimate their distances to the landmarks. This information is then used to triangulate the node location. The corrections are propagated in the network using controlled flooding. Each node will forward a correction from a certain landmark only once in an effort to ensure nodes will receive only one correction from the closest landmark. This policy tries to account for anisotropies in the network.

The DV-distance approach is similar to DV-hop but uses radio received signal strength measurements to measure distances. Although this approach gives finer level granularity, it is also the most sensitive to measurement error since the received signal strength is greatly influenced by the surrounding environment and therefore not always consistent.

The Euclidean propagation method uses the true distance measurement to a landmark. In this case, nodes that have at least two distance measurements to nodes that have distance estimates to a landmark can use simple trigonometric relationships to estimate their locations. The reported simulation results indicate that the DV-hop propagation method is the most accurate of the three and determines the positions of nodes within one-third of the radio range in dense networks.

Another approach described in [15] uses an algorithm similar to DV-Hop called Hop-TERRAIN in combination with a least squares refinement. The Hop-TERRAIN finds the number of hops to each anchor node and uses the anchor positions to estimate the average hop lengths. The average hop lengths are broadcasted back into the network and are used by nodes with unknown positions to compute rough estimates of their locations. Each node with unknown location that receives a message with the average hop length, estimates its distance to each anchor by multiplying the average hop distance with the number of hops to each anchor. Once a node knows the distance to each anchor, it estimates its location using triangulation. In the refinement phase, each node uses the more accurate distance measurements to its neighbors to obtain a more accurate position estimate using least squares refinement.

The collaborative multilateration approach described in [16] uses a three-phase process to estimate node locations. During the first phase, the nodes compute a set of initial estimates by forming a set of bounding boxes around the nodes. The nodes then organize themselves into over-constrained groups in which their positions are further refined using least squares. The refinement phase is presented in two computation models centralized and distributed. The centralized computation model requires global information over the entire network. The distributed computation model is an approximation of the centralized model in which each node is responsible to compute its own location by communicating with its one-hop neighbors. The key attribute that makes the distributed collaborative multilateration possible is its *in-sequence* execution within an over-constrained set of nodes. In distributed collaborative multilateration, each node executes a multilateration using the initial position estimates of its one-hop neighbors and the corresponding distance measurements. The consistent multilateration sequence helps to form a global gradient that allows each node to compute its own position estimate locally by following a gradient with respect to the global constraints.

In addition to the distance-based approaches, some work has also proposed systems using angular measurements. The Angle-of-Arrival system described in [12] is an example of a system that uses angle measurements in a multihop setup to determine node locations.

15.6 NETWORK SETUP ERROR TRENDS

In addition to the error incurred due to noisy measurements, the error in position estimates also depends on network setup parameters such as network size, beacon node concentration and uncertainties in the beacon locations as well error propagation when measurement information is used across multiple hops. In this section we outline the behavior of these effects using results from Cramér Rao bound (CRB) simulations. CRB is a classical result from statistics that give a lower bound on the error covariance matrix of any unbiased estimator. In our discussion, CRB is used as a tool for analyzing the error behavior in multihop localization systems that use angle and distance measurements with Gaussian measurement error. The details on the actual bound derivation can be found in [6]. A close examination of this error behavior can provide valuable insight for the design and deployment of multihop localization systems.

The first notable trend relates to the behavior of localization error with respect to network density. Intuitively, localization accuracy expected to increase with increasing network connectivity. From the CRB simulations results shown in figures 15.1 and 15.2 one can observe that localization accuracy improves asymptotically with network density. Initially, there is a rapid improvement at densities between 6 and 10 neighbors per node. Later on, as the number of neighbors per node increases, the improvement becomes more gradual. Figures 15.1 and 15.2 show the corresponding curves for the cases when distance and angular measurements are used. The y-axis shows the RMS location error normalized by the measurement covariance σ of the measurement technology used.

Localization in Sensor Networks

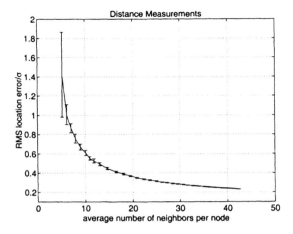

Figure 15.1. Density trend when distance measurements are used

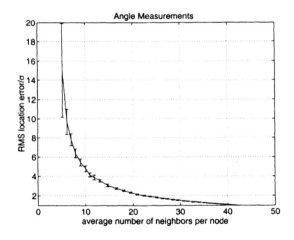

Figure 15.2. Density trend when angle measurements are used

Another useful observation is that in angle-only measurements the location error is approximately one order of magnitude more than the error when distance measurements are used. Furthermore, when angle measurements are used, the error increases proportionately with range. In Figure 15.2 the error when angle-only measurements are used decreases faster than the distance measurement case. This is because to increase density, the area of the sensor field was reduced. As a side effect, the distances between nodes have also been reduced, thus reducing the tangential error in the measurement. The opposite effect would take place if the detection range of the nodes were increased, to increase density.

Another important trend to evaluate is error propagation when measurement information is used over multiple hops. Figure 15.3 shows how the error propagates in an idealized hexagonal placement scenario, where all the nodes with unknown locations have exactly six evenly spaced neighbors. The error in both distance and angular measurements has the same trend. Error propagation is sub-linear with the number of hops. Furthermore, error propagates faster when distance measurements are used than when angle measurements are used.

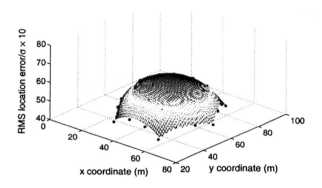

Figure 15.3. Density trend when angle measurements are used

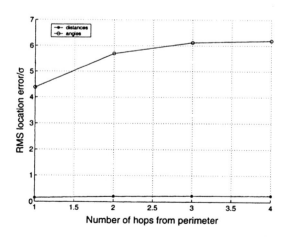

Figure 15.4. Density trend when angle measurements are used

15.7 CASE STUDY: SELF CONFIGURING BEACONS FOR THE SMART KINDERGARTEN

The Smart Kindergarten project at UCLA [18] has developed a deeply instrumented system to study child development in early childhood education. Fine-grained localization is a key component of this system and it is realized using two different sensor node platforms (see figure 15.5). The first one is a wearable tag, called the iBadge [13]. This is attached to a vest worn by the student, or mounted on top of a special cap worn by the student. The localization of these iBadges is made possible by a second type of sensor node, the Medusa MK-2. The Medusa MK-2 nodes are a set of self-configuring beacon nodes deployed on the ceiling to act as anchors for localizing and tracking the iBadges. These nodes are equipped with four pairs of 40KHz ceramic ultrasonic transceivers (4 transmitters and 4 receivers) capable of omni directional transmission and reception of ultrasonic signals over a hemispherical dome. Once deployed on the classroom ceiling, the self-configuring beacons first execute a bootstrapping phase, before entering their beaconing mode. During the bootstrapping phase, each beacon measures distances to its neighboring beacons using its 40KHz ultrasonic ranging system. The distance measurements are forwarded to a central processing station that computes a relative coordinate system and notifies each beacon node of its coordinates.

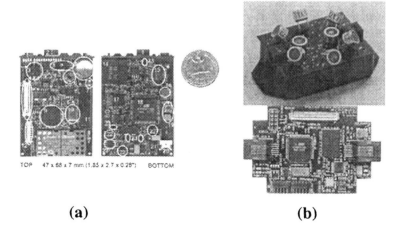

(a)   (b)

Figure 15.5. Smart Kindergarten Sensor Nodes (a) iBadge - the wearable node, (b) Medusa MK-2 - the self-configuring ceiling beacon node

Once all the beacon nodes are initialized with their locations, they enter a service mode. When in service mode, the beacons coordinate with each other to broadcast a sequence of radio and ultrasound signals. These signals are concurrently detected by multiple iBadges in the room. The iBadges use the broadcasted signals to measure distances to each other by timing the difference in the time of detection of the radio and ultrasound signals. With this information, the iBadges can compute and track their location using the onboard, DSP processor, or alternatively they can propagate the raw data back to the central processing station that tracks the iBadges using more powerful tracking algorithms.

15.8 CONCLUSIONS

Despite the recent research efforts, and the diversity of proposed solutions, several research challenges remain to be addressed. In most proposed approaches, the lack of experimental data has so far prevented the evaluation of many localization algorithms under realistic conditions. Furthermore, the multidimensional nature of the problem suggests that the consideration of multiple measurement modalities would improve the robustness of localization systems. Although such schemes have been frequently stated in the literature, the details of fusing information at the measurement level have not been fully explored.

From a theoretical viewpoint, the fundamental characterization of error behavior under different measurement error distributions is still a topic of active research. In this chapter we have described a subset of the initial results characterizing error behavior. Further studies are needed to understand the effects of error propagation under different measurement distributions. Finally we note that node localization is an application specific problem for which a *one size fits all* solution is unlikely to exist for all applications. Since each application is likely to have its own requirements in terms of accuracy, latency and power consumption, node localization needs to be explored further in the context of each target application.

REFERENCES

[1] P. Bahl, V. Padmanabhan, "An In-Building RF-based User Location and Tracking System", Proceedings of INFOCOM 2000 Tel Aviv, Israel, March 2000, p775-84, vol 2

[2] N. Bulusu, J. Heidemann and D. Estrin, *GPS-less low cost outdoor localization for very small devices*, IEEE Personal Communications Magazine, Special Issue on Networking the Physical World, August 2000

[3] S. Capkun, M Hamdi and J. P. Hubaux, *GPS-Free Positioning in Mobile Ad-Hoc Networks*, Hawaii International Conference on System Sciences, HICCSS-34 Jan 2001

[4] L. Doherty, L. El Ghaoui and K. S. J. Pister, *Convex Position Estimation in Wireless Sensor Networks*, Proceedings of Infocom 2001, Anchorage, AK, April 2001

[5] J. Elson and D. Estrin, *Time Synchronization for Wireless Sensor Networks*, IPDPS Workshop on Parallel and Distributed Computing Issues in Wireless Networks and Mobile Computing, San Fransisco, April 22, 2001

[6] W. Garber, A. Savvides, R. Moses and M. B. Srivastava *An Analysis of Error Inducing Parameters in Multihop Sensor Node Localization*, under submission to JSAC special Issue on the Fundamentals of Sensor Networks

[7] L. Girod, V. Bychkovskiy, J. Elson, and D. Estrin *Locating tiny sensors in time and space: A case study* In Proceedings of the International Conference on Computer Design (ICCD 2002), Freiburg, Germany. September 16-18 2002. Invited paper.

[8] K. Langendoen and N. Reijers, *Distributed Localization in Wireless Sensor Networks*, Computer Networks (Elsevier), special issue on Wireless Sensor Networks, August 2003

[9] R. L. Moses and R. M. Patterson, *Self-calibration of sensor networks*, in Unattended Ground Sensors Technologies and Applications IV (Proc. SPIE Vol. 4743) (E.M. Carapezza, ed.), pp. 108-119, April 1-4, 2002

[10] A. Nasipuri and K. Li, *A Directionality Based Location Scheme for Wireless Sensor Networks*, Proceedings of First ACM International Workshop on Wireless Sensor Networks and Applications, pp. 105-111, September 28, Atlanta, Georgia 2002

[11] D. Nicolescu and B. Nath, *Ad-Hoc Positioning System*, Proceedings of IEEE GlobeCom, November 2001

[12] D. Nicolescu and B. Nath, *Ad-hoc Positioning System using AoA*, In Proceedings of the IEEE/INFOCOM 2003, San Francisco, CA, April 2003.

[13] S. Park, I. Locher, A. Savvides, M. B. Srivastava, A. Chen, R. Muntz and S. Yuen, *Design of a Wearable Sensor Badge for Smart Kindergarten*, Proceedings of the International Symposium on Wearable Computing, October 2002

[14] N. Priyantha, A. Chakraborty and H. Balakrishnan, *The Cricket Location Support System*, Proceedings of the ACM SIGMOBILE 7th Annual International Conference on Mobile Computing and Networking, Boston, USA, August 2000

[15] C. Savarese, J. Rabay and K. Langendoen, *Robust Positioning Algorithms for Distributed Ad-Hoc Wireless Sensor Networks*, USENIX Technical Annual Conference, June 2002

[16] A. Savvides, H. Park and M. B. Srivastava, *The n-hop Multilateration Primitive for Node Localization Problems*, Proceedings of Mobile Networks and Applications 8, 443-451, 2003, Kluwer Academic Publishers, Netherlands.

[17] Y. Shang, W. Ruml, Y. Zhang and M. Fromherz, *Localization from Mere Connectivity*, In proceedings of the Fourth ACM Symposium on Mobile Ad-Hoc Networking and Computing (MobiHoc), Annapolis, MD June 2003

[18] M. Srivastava, R. Muntz and M. Potkonjak, *Samrt Kindergarten: Sensor-based Wireless Networks for Smart Developmental Problem-solving Environments*, Proceedings of the ACM SIGMOBILE 7th Annual International Conference on Mobile Computing and Networking, Rome, Italy, July 2001

[19] K. Yao and R. Hudson and C. Reed and D. Chen and F. Lorenzelli, *Blind Beamforming on a Randomly Distributed Sensor Array System*, Proceedings of IEEE JSAC, Vol 18, No. 8, October 1998

[20] H. Wang and L. Yip and D. Maniezzo and J.C. Chen and R.E. Hudson and J. Elson and and K. Yao, *A Wireless Time-Synchronized COTS Sensor Platform, Part II: Applications to Beamforming*, In Proceedings of the IEEE CAS Workshop on Wireless Communications and Networking, Pasadena, California, September 5-6 2002.

[21] N. Priyantha and A. Miu and H. Balakrishnan and S. Teller, *The Cricket Compass for Context-Aware Mobile Applications*, Proceedings of the ACM SIGMOBILE 8th Annual International Conference on Mobile Computing and Networking, pp. 32-43, August 2001

Chapter 16

SENSOR MANAGEMENT

Mark Perillo
University of Rochester
Rochester, NY 14627
perillo@ece.rochester.edu

Wendi Heinzelman
University of Rochester
Rochester, NY 14627
wheinzel@ece.rochester.edu

Abstract   Sensors are deployed in a sensor network for the purpose of providing data about environmental phenomena to the sink node(s). As not all sensors may be able to transmit their data directly to the sink(s), sensors must also route other sensors' data. Therefore, sensors must assume roles of both data provider and router. However, there are often many more sensors in the network than are needed at a given time to accomplish these tasks. Sensor management is needed to assign roles to each sensor, so that nodes that are not needed at a given time can enter a sleep state to save energy until they are needed, thereby extending network lifetime. The area of sensor management, as defined here, includes topology control—choosing which nodes should be routers to ensure a connected network—and sensing mode selection—choosing which nodes should sense data to meet application requirements (application QoS). It is possible to maximize network lifetime by finding optimal schedules of when each node should perform each function. However, these optimizations are computationally intensive, require global knowledge, and are not robust to changes in network topology. Therefore, there is a need for distributed, robust, and computationally efficient sensor management protocols that extend network lifetime while ensuring that application goals are met. In this chapter, protocols for both topology control and sensing mode selection are described and a qualitative comparison of these protocols is given.

Keywords: Sensor management, topology control, sensing mode selection, application QoS, network lifetime

16.1 INTRODUCTION

Recent advances in micro-fabrication technology have spurred a great deal of interest in the use of large-scale wireless sensor networks. The sizes of these networks are expected to grow to large numbers of nodes (thousands) in the next several years as the cost of manufacturing the sensors continues to drop significantly. The goal of these large-scale sensor networks is to gather enough data to monitor the environment with an acceptable fidelity, or application quality of service (QoS). At the same time, these networks are expected to last for a long time (months or even years) without recharging the small batteries providing energy to the individual sensors. Sensor management, including topology control and sensing mode selection, is essential to ensure that application QoS is met while extending network lifetime.

A sensor network is essentially a distributed network of data sources that provide information about environmental phenomena to an end user or multiple end users. Typically, data from the individual sensors are routed via the other sensors to certain sink points in the network (base stations), through which the user accesses the data. Two of the essential services provided by each sensor are sensing the environment and routing other sensors' data. As sensor nodes are provided solely to support the sensor network application, each node should only be used for sensing or routing if this is the best role for that sensor to play to support the end goal.

Quite often, there are so many nodes deployed that not all of the nodes need to provide or route data. Sensor management protocols determine which sensors are needed to provide data (sensing mode selection) and which are needed to ensure a connected topology so data can reach the sink points (topology control). The goal of most sensor management protocols, both for topology control and for sensing mode selection, is to ensure energy efficiency to maximize network lifetime. If sensors are not needed at a given time to provide data or route other sensors' data, they can save energy by shutting down (i.e., going into "sleep" mode) or halting traffic generation until they are needed at a later time. Both topology control and sensing mode selection play critical roles in ensuring that necessary sensor data reach the sink(s) while removing any unnecessary redundancy in the network, resulting in network-wide energy efficiency and long network lifetime.

Topology Control. Topology control is used when sensors are deployed with density high enough that not all sensors are needed to route data to the sink(s). The goal of a topology control protocol is to ensure that enough nodes are activated to provide a connected network so all sensors that have data to send can get their data to the base station while turning off any unnecessary sensors to save energy. Oftentimes, topology control protocols aim to rotate active sensor nodes so that the energy drain of performing routing functions is distributed evenly among all the nodes in the network. In addition to achieving energy efficiency, topology control protocols should be fault tolerant so that the loss of one or a small number of sensors does not disconnect the network.

Sensing Mode Selection. The need for sensing mode selection arises when sensors are deployed with density high enough that activating every sensor in the network provides little more quality of service (QoS) to the sensor network application than what could have been provided with many fewer sensors (i.e., the marginal quality provided by many of the sensors is minimal). In fact, activating all of the sensors can be detrimental to the overall task of the sensor network if there is so much traffic on the network that congestion is noticeable [28]. In this case, network throughput can degrade significantly and important data may be dropped as packet queues at the sensor nodes overflow. Among the data packets that do reach the data sink(s), high packet delays may be introduced, rendering the data useless. The goal of sensing mode selection is to have only certain sensors gather data so that there is no unnecessary redundancy and the network can operate at a point where the cumulative sensor data quality is sufficient to meet the application's goals, network congestion is minimal, and energy-efficiency is achieved.

Optimizing Sensor Management. If an application is able to perform at an acceptable level using data from a number of different sensor sets, it is possible to schedule the sets so as to maximize the sum of the time that all sensor sets are used. The lifetime of an individual sensor is influenced by the amount of data that it routes as well as the amount that it generates. Thus, network lifetime can be greatly extended when sensor mode selection and routing are solved jointly.

This problem of when to schedule sensors to sense data and when to schedule them to route data can be solved optimally so that network lifetime is maximized for a particular application QoS [1, 22]. The constraints of the optimization problem include the sensor battery levels, which dictate the total amount of time any node can route other nodes' data and the total amount of time any node can be an active sensor, the fact that all data sent by a scheduled sensor must be forwarded through other sensors to reach the sink (conservation of data flow), and the minimum acceptable level of quality of service. The objective of the problem is to maximize the total time the network is operational, which is the sum of the times that each sensor set is operational, given these constraints. This problem can be formalized as a generalized maximum flow graph problem and solved via a linear program [22].

In some sensor network applications, data can be aggregated to reduce the total amount of data sent to the base station. Similar network flow optimization problems can be designed to choose not only sensing and routing roles but also data aggregator roles for each sensor such that application QoS is met while lifetime is maximized [1, 15]. The Maximum Lifetime Data Aggregation

(MLDA) problem can be solved to find an optimal schedule of when sensors should transmit data and where data should be aggregated [15]. Similar work in [1] can be used to find the upper bound of attainable lifetime using sensor management algorithms that determine sensing, routing and data fusing roles for each sensor in the network.

The above optimization problems provide upper bounds on achievable lifetime, but they are typically not feasible in real sensor networks for three reasons. First, each of these optimization problems require global knowledge of every sensor's location and battery level. Second, the amount of computation required to solve these optimization problems becomes prohibitive for large-scale networks. Finally, the solution to these optimization problems is not robust to node failures or changes in network topology (e.g., addition of new nodes, node mobility, etc.). If, for example, a critical router node fails, data may not reach the sink(s) for the entire duration of that sensor's scheduled time. Therefore, there is a need for more feasible sensor management solutions, whose properties generally include:

- significant network lifetime extension, achieved by choosing sensor roles (sensing, routing) to reduce energy dissipation at each sensor whenever possible while meeting application QoS,
- scalability for large numbers of sensors,
- distributed control and decision-making,
- robustness to individual node failures, and
- low overhead and computational feasibility.

In the remainder of this chapter, currently proposed solutions for topology control and sensing mode selection that aim to achieve these important sensor management goals are described.

16.2 TOPOLOGY CONTROL ALGORITHMS

The topic of topology control in general ad hoc networks has been studied extensively. The purpose of traditional topology control protocols has been to balance two contradictory goals—reducing energy consumption and maintaining high connectivity. Most early topology control protocols adjusted radio settings (e.g., transmission power [2, 23, 24, 29], beamforming patterns [16]) to maintain connectivity with an optimal set of neighbors. Because it is often more power-efficient to relay packets over several short hops than a single long hop, reducing transmission power is an effective means for reducing overall energy consumption. Reducing transmission power also allows the network to benefit from spatial reuse, possibly resulting in reduced congestion, higher throughput, and a reduction in the number of costly data packets that are unnecessarily overheard.

These methods may be very effective in sensor networks where energy consumption is dominated by the energy consumed in transmitting data packets. However, typical power models considered for sensor networks show that receive power and idle power are comparable to transmit power [26]. Based on this observation, further savings can surely be achieved by not only reducing transmission power, but also setting the sensors' radios into a sleep state whenever possible. Below, several topology control protocols that achieve energy efficiency through these means are described. While some of these protocols were originally designed for use in general ad hoc networks, the fact that nodes are often allowed to turn their radios off nevertheless makes them suitable protocols for sensor networks as well.

16.2.1 GAF: GEOGRAPHIC ADAPTIVE FIDELITY

The GAF protocol [31] takes advantage of the fact that neighboring nodes are often nearly identical from the perspective of data routing. In GAF, a virtual grid is formed throughout the deployed network, and each node is assigned to the virtual grid cell in which it resides. Only a single node from a cell in the virtual grid is chosen to be active at any given time (see Figure 16.1). Nodes implementing GAF initially enter a discovery state, where they listen for messages from other nodes within their cell. If the node determines that a more suitable node can handle the routing responsibilities for its cell, it falls into a sleep state, from which it periodically reenters the discovery state; otherwise, it enters the active state and participates in data routing. After a predetermined active period, active nodes fall back into the discovery state. As the density of a network implementing GAF increases, the number of activated nodes per grid cell remains constant while the number of nodes per cell increases proportionally. Thus, GAF can extend lifetime approximately linearly as a function of node density.

16.2.2 SPAN

Span [4] is a topology control protocol that allows nodes that are not involved in a routing backbone to sleep for extended periods of time. In Span, certain nodes assign themselves the position of "coordinator." These coordinator nodes are chosen to form a backbone of the network, so that the capacity of the backbone approaches the potential capacity of the complete network. Periodically, nodes that have not assigned themselves the coordinator role initiate a procedure to decide if they should become a coordinator. The criteria for this transition is if the minimum distance between any two of the node's neighbors exceeds three hops. To avoid the situation where many nodes simultaneously

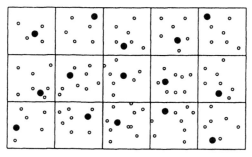

• Active Router ○ Inactive Router

Figure 16.1. Example of a GAF virtual grid [31]. Only one node per cell is activated as a router.

decide to become coordinators, backoff delays are added to nodes' coordinator announcement messages. The backoff delays are chosen such that nodes with higher remaining energy and those potentially providing more connectivity in their neighborhood are more likely to become a coordinator. To ensure a balance in energy consumption among the nodes in the network, coordinator nodes may fall back from their coordinator role if neighboring nodes can make up for the lost connectivity in the region.

16.2.3 ASCENT: ADAPTIVE SELF-CONFIGURING SENSOR NETWORKS TOPOLOGIES

ASCENT [3] is similar to Span in that certain nodes are chosen to remain active as routers while others are allowed to conserve energy in a sleep state. In ASCENT, the decision to become an active router is based not only on neighborhood connectivity, but also on observed data loss rates, providing the network with the ability to trade energy consumption for communication reliability. Nodes running the ASCENT protocol initially enter a test state where they actively participate in data routing, probe the channel to discover neighboring sensors and learn about data loss rates, and send their own "Neighborhood Announcement" messages. If, based on the current number of neighbors and current data loss rates, the sensor decides that its activation would be beneficial to the network, it becomes active and remains so permanently. If the sensor decides not to become active, it falls into a passive state, where it gathers the same information as it does in the test state (as well as any "Help" messages from neighboring sensors experiencing poor communication links), but it does not actively participate in data routing. From this state, the node may reenter the test state if the information gathered indicates poor neighborhood communication quality, or enter the sleep state, turning its radio off and saving energy.

The node periodically leaves the sleep state to listen to the channel from the passive state.

One procedure for node activation is shown in Figure 16.2. Figure 16.2a illustrates a situation in which the channel quality between a source and its sink becomes poor, prompting the sink to broadcast "Help" messages to its neighbors. A limited number of intermediate nodes that receive these messages enter the test state and broadcast "Neighborhood Announcement" messages (Figure 16.2b) to inform their neighbors that they intend to become active. Alternatively, this transition to the test state can be triggered by a node observing that it has a low number of active neighbors or high data loss rate. The network eventually stabilizes at a point in which the communication between the source and sink becomes adequately reliable, as shown in Figure 16.2c.

16.2.4 STEM: SPARSE TOPOLOGY AND ENERGY MANAGEMENT

In the case of many sensor network applications, it is expected that nodes will continuously sense the environment but transmit data to a base station very infrequently or only when an event of interest has occurred. While GAF, Span, and ASCENT save energy by reducing the number of sensors used for routing, the energy consumption of the selected routers can be further reduced by exploiting the low traffic generation rates of the sensors. STEM [25] takes advantage of this by leaving all sensors in a sleep state while monitoring the environment but not sending data, which is assumed to be the majority of time. STEM is quite different from the rest of the topology control protocols described here in that it activates nodes reactively rather than proactively. When data packets are generated, the sensor generating the traffic uses a paging channel (separate from the data channel) to awaken its downstream neighbors. Two versions of STEM have been proposed—STEM-T, which uses a tone to wake neighboring nodes, and STEM-B, in which the traffic generating node sends beacons on the paging channel and sleeping nodes turn on their radios with a low duty cycle to receive the messages. A benefit of STEM is that it can be combined with any of the aforementioned protocols, with only the currently activated nodes required to listen on the paging channel.

16.3 SENSING MODE SELECTION

The previous section outlined some topology control protocols that limit energy consumption by reducing the number of nodes actively participating in data routing. To further exploit potential energy savings in a wireless sensor network, careful consideration should be taken in the selection of the sensing modes of the nodes in the network. This can be as simple as determining which sensors should be activated or deactivated or as complex as determining cer-

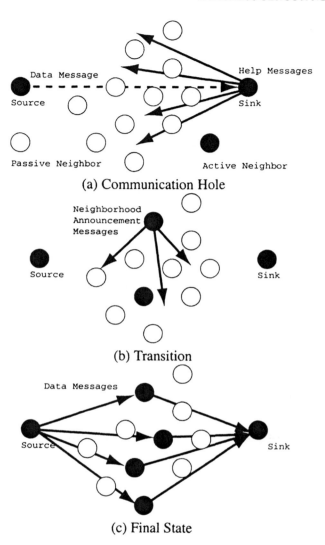

Figure 16.2. One procedure for the activation of inactive routers in ASCENT [3]. When the channel quality is poor, a node will broadcast "Help" messages to its neighbors (a), some of whom will become active and broadcast "Neighborhood Announcement" messages (b). After these nodes are activated, the communication quality improves (c).

tain features of the sensing, such as sensing frequency and data resolution. By doing so, the designer is not only influencing how the data are routed, but also what traffic is generated in the first place. This aspect of sensor management has several implications. First, it may be absolutely necessary. In simple coverage applications, if the sensing range is significantly larger than the transmission range, some of the previously described topology control protocols may be suitable alone. However, if this is not the case, additional sensors besides the selected active routers will need to be activated. Also, with the use of an energy-efficient MAC protocol [8, 9, 34], the average energy consumption of a node when it is not transmitting or receiving data may not approach the levels of the transmit and receive power, as was the assumption for the topology control protocols. In this case, the amount of data generated on the network may influence network energy consumption (and consequently, network lifetime) in a more drastic manner. Furthermore, deactivating sensors can reduce any congestion that may arise in low bandwidth networks [28].

In this section, several protocols for sensing mode selection in wireless sensor networks are described. In most cases, sensing mode selection is as simple as deciding which sensors should send data to a base station or other nodes. The process of determining the subset of sensors chosen for activation should be influenced by QoS (or fidelity) requirements of the application. Given the many proposed applications for wireless sensor networks, these requirements can be quite wide-ranging. In this chapter, the focus is primarily on applications that require coverage of the entirety or a portion of a region where the sensors are deployed [19]. Examples of sensing mode selection in other applications include

- edge detection [18], where only a certain subset of the nodes is expected to change state when the edge moves,

- oversampling of bandlimited phenomena [12], where nodes can adjust the resolution of their data in dense regions without significant information loss,

- target localization/tracking [20, 35], where selection of only the subset of sensors in the current vicinity of the target is beneficial, and

- general target classification [5, 6].

The reader is referred to the cited literature for more detailed descriptions of this work.

The primary goal of the protocols described in this section is coverage preservation rather than network connectivity (as in Section 16.2). Coverage preserving protocols and algorithms have many potential applications, including intruder detection, biological/chemical agent detection, and fire detection. Also, these protocols and algorithms can be used in the initial stages of many target tracking architectures, where a more detailed description or location estimate of a phenomenon is required only when a "tripwire" threshold is crossed in the measurements of some of the active sensors [11].

16.3.1 PEAS: PROBING ENVIRONMENT AND ADAPTIVE SLEEPING

PEAS [33] is a protocol that was developed to provide consistent environmental coverage and robustness to unexpected node failures. Nodes begin in a sleeping state, from which they periodically enter a probing state. In the probing state, a sensor transmits a probe packet, to which its neighbors will reply after a random backoff time if they are within the desired probing range. If no replies are received by the probing node, the probing sensor will become active; otherwise, it will return to the sleep state. The probing range is chosen to meet the more stringent of the density requirements imposed by the sensing radius and the transmission radius. The probing rate of PEAS is adaptive and is adjusted to meet a balance between energy savings and robustness. Specifically, a low probing rate may incur long delays before the network recovers following an unexpected node failure. On the other hand, a high probing rate may lead to expensive energy waste. Basically, the probing rate of individual nodes should increase as more node failures arise, so that a consistent expected recovery time is maintained.

16.3.2 NSSS: NODE SELF-SCHEDULING SCHEME

A node self-scheduling scheme for coverage preservation in sensor networks is presented in [27]. In NSSS, a node measures its neighborhood redundancy as the union of the sectors/central angles covered by neighboring sensors within the node's sensing range. At decision time, if the union of a node's "sponsored" sectors covers the full 360° (see Figure 16.3), the node will decide to power off. It should be noted that additional redundancy may exist between sensors and that the redundancy model is simplified at a cost of not being able to exploit this redundancy. In NSSS, at the beginning of each round, there is a short self-scheduling phase where nodes first exchange location information and then decide whether or not to turn off after some backoff time. NSSS avoids scenarios of unattended areas due to the simultaneous deactivation of nodes by requiring nodes to double check their eligibility to turn off after making the decision.

16.3.3 GUR GAME

In [13], the problem of optimal sensor selection for a given resolution was modeled as a Gur game. It is assumed that a base station wishes to receive packets from a predetermined number of sensors. Each sensor operates as a single chain finite state machine, where one side of the chain represents states in which the sensor sends a packet to the base station and the other side represents states in which the sensor does not (see Figure 16.4). After each round,

Sensor Management

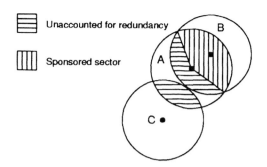

Figure 16.3. A sponsored sector, as defined by NSSS [27]. Sensor A admits the redundant coverage of sensor B in the vertically shaded regions. The additional redundancy of sensors B and C shown in the horizontally shaded regions is not accounted for.

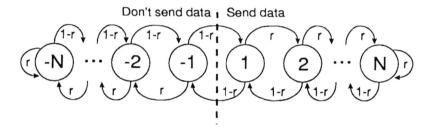

Figure 16.4. State machine for the Gur game at an individual node [13].

the base station calculates a reward probability r from a function of the number of packets received. The reward function has its maximum value at the desired resolution (number of packets received). The base station broadcasts the value of this reward function to the sensors, which move toward the edges of their state machine with probability r and toward the center with probability $(1 - r)$. The network settles at the desired resolution and is robust to sensor failures and additional sensor deployment. This work was further developed into a hierarchical model, providing better scalability, in [14].

16.3.4 REFERENCE TIME-BASED SCHEDULING SCHEME

In the reference time-based scheduling scheme presented in [32], the environment is divided into a grid and coverage is maintained continuously at every grid point while minimizing the number of active sensors. During an initialization process, each node broadcasts a randomly chosen reference time (uniformly distributed on $[0, T)$, where T is the round length) to all neighbor-

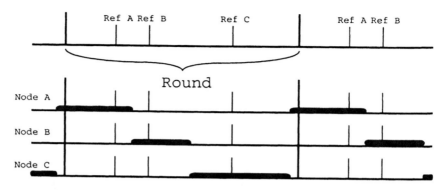

Figure 16.5. Schedule calculation for a single grid point, as proposed in the reference time-based scheduling scheme [32]. After all grid point schedules are calculated, the schedules are merged and a sensor's overall schedule is the union of all of its grid point schedules.

ing sensors within twice its sensing radius. For each location in the grid that the sensor is capable of monitoring, a sensor sorts the reference times of all sensors capable of monitoring that grid point. For a given grid point, the sensor schedules itself to be active beginning halfway between its reference time and the reference time of the sensor immediately preceding it in the sorted list. Similarly, its scheduled slot for the grid point ends halfway between its reference time and the reference time of the sensor immediately after it in the sorted list (see Figure 16.5). The sensor remains active during the union of the scheduled slots calculated for each grid point within its sensing range. This algorithm is also enhanced to guarantee coverage by multiple sensors in selected areas as well as provide robustness to node failures.

16.3.5 CCP: COVERAGE CONFIGURATION PROTOCOL

In CCP [30], an eligibility rule is proposed to maintain a certain degree of coverage (coverage at every location by a given number of sensors). First, each node finds all intersection points between the borders of its neighbors' sensing radii and any edges in the desired coverage area. The CCP rule assigns a node as eligible for deactivation if each of these intersection points is K-covered, where K is the desired sensing degree. The CCP scheme assumes a Span-like protocol and state machine that can use the Span rule for network connectivity or the proposed CCP rule for K-coverage, depending on the application requirements and the relative values of the communication radius and sensing radius. An example of how the CCP rule is applied is given in Figure 16.6. In Figure 16.6a, node D, whose sensing range is represented by the bold circle, must decide whether it should become active in order to meet a coverage con-

Sensor Management

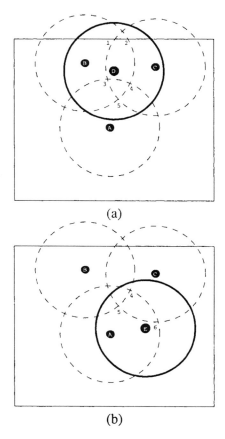

(a)

(b)

Figure 16.6. Illustration of the CCP activation rule for K-coverage, K = 1 [30]. Nodes D and E decide whether or not to become active in (a) and (b), respectively, knowing that neighbors A, B, and C, are already active. Node D may remain inactive since all of its intersection points are K-covered, but the CCP rule dictates that E must become active since intersection point 6 is not covered by any of E's neighbors.

straint of $K = 1$. It is assumed that D knows that A, B, and C, whose sensing ranges are represented by the dashed circles, are currently active. The intersection points within D's sensing range are found and enumerated 1-5 in the figure. Since B covers points 1 and 3, C covers points 2 and 4, and A covers point 5, D deduces that the coverage requirements have already been met and remains inactive. In the case of node E, illustrated in Figure 16.6b, there is an intersection point (labeled 6 in the figure) that is not covered by any of E's neighbors. Thus, E must become active and sense the environment.

16.4 INTEGRATED SENSING MODE SELECTION AND ROUTING PROTOCOLS

Sensing mode selection can potentially have implications on topology control and data routing. Accordingly, some sensor management protocols integrate these functions [7, 21]. To see why this is beneficial, consider that topology control protocols that do not use knowledge of traffic patterns may activate many more nodes than the necessary amount to accommodate the sensed data. For example, in the network shown in Figure 16.7, four nodes are required to send raw data to the base station. Without knowledge of traffic patterns and using the GAF topology control protocol discussed previously, the nodes selected as routers might be the set shown in Figure 16.7(a). Given that traffic will only be generated from the small subset of selected sensor nodes, many of these nodes should actually be allowed to sleep, and only the nodes used in the optimally calculated routes, shown in Figure 16.7(b), need to remain active.

16.4.1 CONNECTED SENSOR COVER

The Connected Sensor Cover algorithm presented in [7] provides a joint sensing mode selection and topology control solution. The problem addressed in this work is to find a minimum set of sensors and additional routing nodes necessary in order to efficiently process a query over a given geographical region. In the centralized version of the algorithm, an initial sensor within the query region is randomly chosen, following which additional sensors are added by means of a greedy algorithm. At each step in this algorithm, all sensors that redundantly cover some area that is already covered by the current active subset are considered candidate sensors and calculate the shortest path to one of the sensors already included in the current active subset. For each of these candidate sensors, a heuristic is calculated based on the number of unique sections in the query region that the sensor and its routers would potentially add and the number of sensors on its calculated path. The sensor with the most desirable heuristic value and those along its path are selected for inclusion in the sensor set. This process continues until the query region is entirely covered. The algorithm has been extended to account for node weighting, so that low energy nodes can be avoided, and to be implemented through distributed means, with little loss in solution optimality compared with the centralized version.

16.4.2 DAPR: DISTRIBUTED ACTIVATION WITH PRE-DETERMINED ROUTES

The integration of sensing mode selection and data routing/topology control can also be accomplished by choosing routes so as to avoid sensors that are

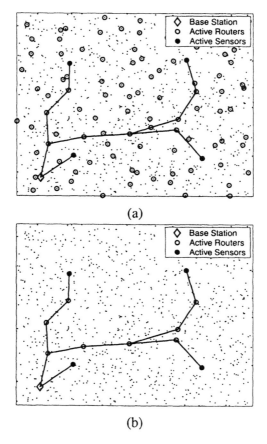

Figure 16.7. Example of a situation where knowledge of traffic patterns can potentially benefit topology control. In (a), many nodes have unnecessarily been activated as routers. In (b), only the necessary routers for the given traffic conditions are activated.

critical for maintaining the desired coverage. Sensors that are more important to the sensing application and those whose potential active lifetime is shortest should not be chosen as routers over those who are less important to the application and those with potentially long active lifetimes (see Figure 16.8).

DAPR [21] is an algorithm for integrated sensing mode selection and data routing that follows this intuition. In DAPR, nodes assign themselves an application cost—a measure of their value to the application. The application cost is calculated based on a node's remaining energy and the remaining energy of the node's neighbors that provide redundant coverage. The application cost was chosen to consider a node's neighbors' energy as well as its own because neighbors covering the same region redundantly are equivalent from the

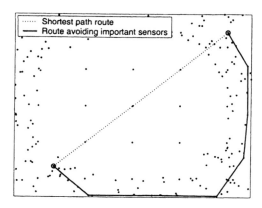

Figure 16.8. Calculated routes using shortest path and an approach that considers the importance of the sensors to the application.

application's perspective. This is because DAPR is less concerned with prolonging individual sensors' lifetime as it is with prolonging the time before a spot in the network is left uncovered. A round in DAPR begins with a route discovery phase, in which the base station broadcasts a query packet across the network and nodes discover their optimal routes based on smallest cumulative application cost along the path. Following this phase is a role discovery phase, consisting of an Opt In and an Opt Out portion. In the Opt In portion, nodes attempt to activate themselves after a backoff delay proportional to their cumulative route cost. If a sensor discovers that its neighborhood has already been covered by the time it wishes to broadcast its activation message, it withholds this message. In the Opt Out phase, nodes check to see if their neighborhood has been completely redundantly covered by other nodes, some of which may have been activated after their own activation message was broadcast. Nodes are given the opportunity to opt out in the reverse order in which they opted in, so that nodes with high cumulative route cost are still favored for deselection. After this set-up phase, activated sensors send data along the pre-calculated routes to the base station.

16.5 DISCUSSION

The properties of the protocols and schemes described in this chapter are summarized in Table 16.1. It should be noted that the classifications and descriptions in this table are not straightforward and that simple modifications can often eliminate any undesirable properties of a protocol. For example, some topology control protocols could be used as sensing mode selection protocols with some simple modifications, and vice versa. Also, some of the protocols that require location information can be modified to operate without location information, but this usually comes at the cost of reduced performance (e.g., higher energy dissipation, no application QoS guarantees, etc.). For example, DAPR can be modified so that coverage quality is measured as the number of active neighbors that a node overhears rather than the actual calculation of area covered by the active neighbors. This approach removes the need for location information at the cost of not guaranteeing complete coverage. In general, the protocols are summarized in Table 16.1 as they were presented in their original literature.

This chapter has described methods for the efficient management of the scarce resources typically available in wireless sensor networks. Specifically, some proposed methods for choosing which sensors should act as routers (through topology control protocols) and which should actively sense the environment and send traffic to any data sinks in the network (through sensing mode selection) have been presented. The choice of which protocol to use in a specific network is subject to factors such as

- the application requirements,
- the choice of MAC protocol,
- the bandwidth resources, and
- the availability of certain network services (e.g., sensor node localization, time synchronization).

For example, in applications where data need to be sent only after an event of interest occurs, a reactive solution such as STEM [25] might be preferable. However, in applications that require constant streaming traffic and have little tolerance for delays, protocols such as GAF [31] or Span [4] that proactively maintain a connected network might be desirable. Application requirements also dictate what level of service is needed. For example, an application that requires an absolute guarantee that the entirety of a region is covered by at least one sensor may want to use an approach such as the node self-scheduling scheme in [27] or the reference time-based scheduling scheme in [32], whereas if the application requires only that a given number of sensors send data, an approach such as the Gur game modeling presented in [13] may be acceptable and require less communication overhead, enabling the sensor nodes to save even more energy and extend network lifetime.

When using a MAC protocol that consumes significant idle power, topology control protocols by themselves may be suitable since the power consumption is dominated by the idle power consumption of the nodes chosen as active routers and the traffic induced on the network may be of little consequence from an energy consumption standpoint. With a MAC protocol that allows nodes to sleep during periods of inactivity, however, sensing mode selection protocols can be much more effective in reducing power consumption and extending network lifetime. Also, sensing mode selection protocols may relieve network congestion in low bandwidth networks.

Finally, a cross-layer approach that selects routes in conjunction with choosing sensors to sense the environment may be the best approach in many applications with static sensor nodes and data sinks. However, in some situations, such as battlefield scenarios in which mobile soldiers and vehicles may act as sensors or data sinks, the network may consist of a dynamic topology. With constant route updates becoming necessary, a layered approach that handles sensing mode selection and topology control/routing separately may be necessary.

In addition to the important roles of sensing and routing, careful consideration should be taken when determining which nodes should assume other network roles such as cluster heads and data fusion points [10, 17]. While this chapter has not addressed this, some research has been devoted to this area.

Sensor Management

Protocol	TC/MS	Loc?	D/C	Energy balancing	Robustness	Comp.	Overhead
GAF [31]	TC	Y	D	Activity rotation	Good	Low	Neighbor state exchange
Span [4]	TC	N	D	Activity rotation, energy weighting	Good	Low	Neighbor state exchange
ASCENT [3]	TC	N	D	No	Good	Low	Neighbor state exchange
STEM [25]	TC	N	D	N/A	Good	Low	Paging channel
PEAS [33]	TC,MS	Y	D	No	Good, dep. on probing frequency	Low	Neighbor state exchange
NSSS [27]	MS	Y	D	No	Dep. on round ln.	Low	Neighbor state exchange
Gur game [13]	MS	N	D	Dep. on N, reward function	Coverage not guaranteed	Low	BS queries
Ref. time-based scheduling [32]	MS	Y	D	Activity rotation	K-degree coverage	Low	Reference time exchange
CCP [30]	TC,MS	Y	D	Activity rotation	K-degree coverage	Low	Neighbor state exchange
Connected Sensor Cover [7]	TC,MS	Y	C/D	Energy weighting possible	Dep. on query frequency	Low	BS queries, neighbor state exchange (D)
DAPR [21]	MS	Y	D	Activity rotation, energy weighting	Dep. on round ln.	Low	BS queries, neighbor state exchange
Optimization	MS	Y	C	Yes	Poor	High	N/A

Table 16.1. Summary of protocols/schemes described.

TC/MS: Topology Control/Mode Selection   Loc?: Location information required?   D/C: Distributed/Centralized

REFERENCES

[1] Bhardwaj, M. and A. Chandrakasan: 2002, 'Bounding the Lifetime of Sensor Networks Via Optimal Role Assignments'. In: *Proceedings of the Twenty First International Annual Joint Conference of the IEEE Computer and Communications Societies (INFOCOM)*.

[2] Blough, D., M. Leoncini, G. Resta, and P. Santi: 2003, 'The K-Neigh Protocol for Symmetric Topology Control in Ad Hoc Networks'. In: *Proceedings of MobiHOC*.

[3] Cerpa, A. and D. Estrin: 2002, 'ASCENT: Adaptive Self-Configuring sEnsor Networks Topologies'. In: *Proceedings of the Twenty First International Annual Joint Conference of the IEEE Computer and Communications Societies (INFOCOM)*.

[4] Chen, B., K. Jamieson, H. Balakrishnan, and R. Morris: 2000, 'Span: An Energy-Efficient Coordination Algorithm for Topology Maintenance in Ad Hoc Wireless Networks'. In: *Proceedings of the Sixth Annual International Conference on Mobile Computing and Networking*.

[5] Chu, M., H. Haussecker, and F. Zhao: 2002, 'Scalable Information-Driven Sensor Querying and Routing for Ad Hoc Heterogeneous Sensor Networks'. *International Journal of High Performance Computing Applications* 16(3).

[6] Ertin, E., J. W. F. III, and L. C. Potter: 2003, 'Maximum Mutual Information Principle for Dynamic Sensor Query Problems'. In: *Proceedings of the Second International Workshop on Information Processing in Sensor Networks*.

[7] Gupta, H., S. Das, and Q. Gu: 2003, 'Connected Sensor Cover: Self-Organization of Sensor Networks for Efficient Query Execution'. In: *Proceedings of the Fourth ACM International Symposium on Mobile Ad Hoc Networking and Computing (MobiHoc)*.

[8] Gutierrez, J. A., M. Naeve, E. Callaway, M. Bourgeois, V. Mitter, and B. Heile: 2001, 'IEEE 802.15.4: A Developing Standard for Low-Power, LowCost Wireless Personal Area Networks'. *IEEE Network* 15(5), 12–19.

[9] Haartsen, J., M. Naghshineh, J. Inouye, O. Joeressen, and W. Allen: 1998, 'Bluetooth: Vision, Goals, and Architecture'. *Mobile Computing and Communications Review* 2(4), 38–45.

[10] Heinzelman, W., A. Chandrakasan, and H. Balakrishnan: 2002, 'An Application-Specific Protocol Architecture for Wireless Microsensor Networks'. *IEEE Transactions on Wireless Communications* 1(4), 660–670.

[11] Hui, J., Z. Ren, and B. H. Krogh: 2003, 'Sentry-Based Power Management in Wireless Sensor Networks'. In: *Proceedings of the Second International Workshop on Information Processing in Sensor Networks*.

[12] Ishwar, P., A. Kumar, and K. Ramchandran: 2003, 'Distributed Sampling for Dense Sensor Networks: A Bit-Conservation Principle'. In: *Proceedings of the Second International Workshop on Information Processing in Sensor Networks*.

[13] Iyer, R. and L. Kleinrock: 2003a, 'QoS Control For Sensor Networks'. In: *Proceedings of the IEEE International Conference on Communications*.

[14] Iyer, R. and L. Kleinrock: 2003b, 'Scalable Sensor Network Resolution'. In: *Proceedings of Sensys*.

[15] Kalpakis, K., K. Dasgupta, and P. Namjoshi: 2002, 'Maximum Lifetime Data Gathering and Aggregation in Wireless Sensor Networks'. In: *Proceedings of the 2002 IEEE International Conference on Networking*.

[16] Li, Q. and D. Rus: 2000, 'Sending Messages to Mobile Users in Disconnected Ad-Hoc Wireless Networks'. In: *Proceedings of the Sixth Annual International Conference on Mobile Computing and Networking.*

[17] Lindsey, S., C. Raghavendra, and K. Sivalingam: 2002, 'Data Gathering Algorithms in Sensor Networks Using Energy Metrics'. *IEEE Transactions on Parallel and Distributed Systems* 13(9), 924–935.

[18] Liu, J., P. Cheung, L. Guibas, and F. Zhao: 2002, 'A Dual-Space Approach to Tracking and Sensor Management in Wireless Sensor Networks'. In: *Proceedings of the ACM International Workshop on Wireless Sensor Networks and Applications.*

[19] Meguerdichian, S., F. Koushinfar, M. Potkonjak, and M. Srivastava: 2001, 'Coverage Problems in Wireless Ad-hoc Sensor Networks'. In: *Proceedings of the Twentieth International Annual Joint Conference of the IEEE Computer and Communications Societies (INFOCOM).*

[20] Pattem, S., S. Poduri, and B. Krishnamachari: 2003, 'Energy-Quality Tradeoffs for Target Tracking in Wireless Sensor Networks'. In: *Proceedings of the Second International Workshop on Information Processing in Sensor Networks.*

[21] Perillo, M. and W. Heinzelman: 2003a, 'Sensor Management and Routing Protocols For Prolonging Network Lifetime (Technical Report)'. In: *University of Rochester Technical Report.*

[22] Perillo, M. and W. Heinzelman: 2003b, 'Simple Approaches for Providing Application QoS Trhough Intelligent Sensor Management'. *Elsevier Ad Hoc Networks Journal* 1(2-3), 235–246.

[23] Ramanathan, R. and R. Hain: 2000, 'Topology Control of Multihop Wireless Networks Using Transmit Power Adjustment'. In: *Proceedings of the Nineteenth International Annual Joint Conference of the IEEE Computer and Communications Societies (INFOCOM).*

[24] Rodoplu, V. and T. Meng: 1998, 'Minimum Energy Mobile Wireless Networks'. In: *Proceedings of the IEEE International Conference on Communications.*

[25] Schurgers, C., V. Tsiatsis, S. Ganeriwal, and M. Srivastava: 2002, 'Optimizing Sensor Networks in the Energy-Latency-Density Design Space'. *IEEE Transactions on Mobile Computing* 1(1), 70–80.

[26] Stemm, M. and R. H. Katz: 1997, 'Measuring and reducing energy consumption of network interfaces in hand-held devices'. *IEICE Transactions on Communications* E80-B(8), 1125–31.

[27] Tian, D. and N. Georganas: 2003, 'A Node Scheduling Scheme for Energy Conservation in Large Wireless Sensor Networks'. *Wireless Communications and Mobile Computing Journal* 3(2), 271–290.

[28] Tilak, S., N. Abu-Ghazaleh, and W. Heinzelman: 2002, 'Infrastructure Tradeoffs for Sensor Networks'. In: *Proceedings of the First ACM International Workshop on Wireless Sensor Networks and Applications.*

[29] Tseng, Y., Y. Chang, and P. Tseng: 2002, 'Energy-Efficient Topology Control for Wireless Ad Hoc Sensor Networks'. In: *Proceedings of the International Computer Symposium.*

[30] Wang, X., G. Xing, Y. Zhang, C. Lu, R. Pless, , and C. Gill: 2003, 'Integrated Coverage and Connectivity Configuration in Wireless Sensor Networks'. In: *Proceedings of Sensys.*

[31] Xu, Y., J. Heidemann, and D. Estrin: 2001, 'Geography-informed Energy Conservation for Ad Hoc Routing'. In: *Proceedings of the ACM/IEEE International Conference on Mobile Computing and Networking.*

[32] Yan, T., T. He, and J. A. Stankovic: 2003, 'Differentiated Surveillance for Sensor Networks'. In: *Proceedings of Sensys*.

[33] Ye, F., G. Zhong, J. Cheng, S. Lu, and L. Zhang: 2003, 'PEAS: A Robust Energy Conserving Protocol for Long-lived Sensor Networks'. In: *Proceedings of the Twenty Third International Conference on Distributed Computing Systems*.

[34] Ye, W., J. Heidemann, and D. Estrin: 2002, 'An Energy-Efficient MAC Protocol for Wireless Sensor Networks'. In: *Proceedings of the Twenty First International Annual Joint Conference of the IEEE Computer and Communications Societies (INFOCOM)*.

[35] Zou, Y. and K. Chakrabarty: 2003, 'Target Localization Based on Energy Considerations in Distributed Sensor Networks'. In: *Proceedings of the First IEEE International Workshop on Sensor Network Protocols and Applications*.

VI

APPLICATIONS

Chapter 17

DETECTING UNAUTHORIZED ACTIVITIES USING A SENSOR NETWORK

Thomas Clouqueur, Parameswaran Ramanathan and Kewal K.Saluja
Department of Electrical and Computer Engineering, University of Wisconsin-Madison
1415 Engineering Drive, Madison, WI 53706, USA
{ clouqueu,parmesh,saluja } @ece.wisc.edu

Abstract Sensing devices can be deployed to form a network for monitoring a region of interest. This chapter investigates the detection of a target in the region being monitored by using collaborative target detection algorithms among the sensors. The objective is to develop a low cost sensor deployment strategy to meet a performance criteria. A path exposure metric is proposed to measure the goodness of deployment. Exposure can be defined and efficiently computed for various target activities, targets traveling at variable speed and in the presence of obstacles in the region.
Using exposure to evaluate the detection performance, the problem of random sensor deployment is formulated. The problem defines cost functions that take into account the cost of single sensors and the cost of deployment. A sequential sensor deployment approach is then developed. The chapter illustrates that the overall cost of deployment can be minimized to achieve a desired detection performance by appropriately choosing the number of sensors deployed in each step of the sequential deployment strategy.

Keywords: Collaborative target detection, Deployment, Exposure, Sensor Networks, Value fusion.

17.1 INTRODUCTION

Recent advances in computing hardware and software are responsible for the emergence of sensor networks capable of observing the environment, processing the data and making decisions based on the observations. Such a network can be used to monitor the environment, detect, classify and locate specific events, and track targets over a specific region. Examples of such systems are in surveillance, monitoring of pollution, traffic, agriculture or civil infrastructures [15]. The deployment of sensor networks varies with the application considered. It can be predetermined when the environment is sufficiently known and under control, in which case the sensors can be strategically hand placed. In some other applications when the environment is unknown or hostile, the deployment cannot be a priori determined, for example if the sensors are air-dropped from an aircraft or deployed by other means, generally resulting in a random placement.

This chapter investigates the deployment of sensor networks performing target detection over a region of interest. In order to detect a target moving in the region, sensors make local observations of the environment and collaborate to produce a global decision that reflects the status of the region covered [3]. In general, this collaboration requires local processing of the observations, communication between different nodes, and information fusion [16]. Since the local observations made by the sensors depend on their position, the performance of the detection algorithm is a function of the deployment, i.e. the number of sensors and their location. Recently, several studies have addressed the problem of deployment with the goal of determining the performance of a given deployment of sensors or determining an optimal deployment to achieve a desired coverage of the monitored region. The analysis of the sensor network performance relies on the model chosen for the detection capacity of the sensors. This chapter presents several models that differ in their level of abstraction. Then, it focuses on a formulation that models the energy measured by the sensors as a function of the distance from the target to the sensors. Detection algorithms using these energy values for decision making are developed and their performance is analyzed in terms of false alarm and detection probability. A probabilistic measure of the goodness of deployment for target detection, called *path exposure*, is then defined. It is a measure of the likelihood of detecting a target traveling in the region using a given path. The higher the exposure of a path, the better the deployment is to detect that path. A set of possible paths can be associated to a given target activity, for example the traversal of the region from one end to another is one such activity. The deployment performance for detecting an activity is measured by the minimum exposure over all possible paths. In this chapter, methods to efficiently determine the least exposed path are derived for three different target activities, namely unauthorized traversing, reaching and idling. The evaluation of exposure in the presence of obstacles and for variable speed target is also considered. Finally, this chapter presents a deployment strategy to achieve a given detection performance. The deployment is assumed to be random which corresponds to many practical applications where the region to be monitored is not accessible for precise placement of sensors. The tradeoffs of deployment lie between the network performance, the cost of the sensors deployed, and the cost of deploying the sensors.

This chapter is organized as follows. In Section 17.2, models for target detection and associated problems are presented. In Section 17.3, two algorithms for distributed detection, namely value and decision fusion, are proposed and analyzed. In Section 17.4, a definition for path exposure is proposed and a method to evaluate the exposure of a given path is developed for various applications. In Section 17.5, the problem of random deployment is formulated and a solution using exposure is presented. The chapter concludes with Section 17.6.

17.2 VARIOUS MODELS FOR DETECTION AND ASSOCIATED PROBLEMS

In this section, a literature survey is conducted and various detection sensor network models are presented. These models capture the ease for sensors to detect a target, and they differ in their level of abstraction. Both sensor models and target models have been considered. Sensor models directly consider the sensors individual detection performance using sensing range or detection probability. Target models considers the signal emitted by the target and measured by the sensors.

17.2.1 MODELS ON INDIVIDUAL SENSOR PERFORMANCE

A property common to most studies on deployment is that sensing ability decreases with distance between the sensor and the phenomenon observed. Relying merely on this assumption, the study presented in [12] relates the detectability of a path P to the distance from P to the sensors. Two paths of interest are identified. The maximal breach path is defined as a path from which the closest distance to any of the sensors is as large as possible, hence it is the path hardest to detect. The maximal support path is defined as a path for which the farthest distance from the closest sensor is as small as possible, hence it is the path easiest to detect.

In [9], authors propose a probabilistic model for the detection capability of individual sensors. They assume that the probability of a sensor detecting a target decreases exponentially with the distance between the target and the sensor, and this probability can also be altered by the possible presence of obstacle between the target and the sensor. Instead of considering the detection of target following paths, the study considers that the sensors have to fully cover the region in which they are deployed. The problem is to reach a desired average coverage as well as to maximize the coverage of the most vulnerable points of the region. Under these assumptions, they propose algorithms for placing sensors on grid points to maximize the average or minimum detection probability of a target that could be located at any grid point. An extension of that study developed similar algorithm considering that part of the sensors can fail [8]. Also, using the same model but assuming that the sensors can move, an algorithm for finding the optimal motion of the sensors to cover the region after being randomly deployed is developed in [17].

In [4], a model is developed that assumes that each sensor has a sensing range in which it is guaranteed to detect any target. This is similar to the probabilistic model developed in [9] assuming that the detection probability of a target is 1 (resp. 0) whenever the distance between the target and the sensor is less (resp. higher) than the sensing range. Again, the goal of deployment is to fully cover the region of interest. The study develops an algorithm based on integer linear programming that minimizes the cost of sensors when requiring complete coverage of the sensor field.

17.2.2 TARGET SIGNAL MODEL

In [13], the authors model the signal intensity and assume it decays as a power of the distance from the target to the sensor. The goal of the deployment is to prevent targets from traversing a monitored region without being detected. The measure of *exposure* of a path P is defined as the total signal intensity emitted by the target moving through the field following P and measured by the closest or all the sensors. The smaller this signal intensity, the lesser the likelihood of detecting the target, thus the coverage provided by a deployment is measured by the minimum exposure when considering all possible paths across the region.

17.2.3 MODEL USED IN THIS CHAPTER

Now, the model used for the remaining of this chapter is presented. Sensor nodes with possibly different sensing modalities are deployed over a region R to perform target detection. Sensors measure signals at a given sampling rate to produce time series that are processed by the nodes. The nodes fuse the information obtained from every sensor accordingly to the sensor type and location to provide an answer to a detection query. The nodes are assumed to have the ability to communicate to each other. However this work is not concerned with communication issues and therefore the peer to peer communication among nodes is assumed to be reliable through the use of appropriate communication techniques [1, 14].

In this chapter, the sensor nodes are assumed to obtain a target energy measurement every T seconds that accounts for the behavior of the target during time T. The detection algorithm consists of exchanging and fusing the energy values produced by the sensor nodes to obtain a detection query answer.

A target at location u emits a signal which is measured by the sensors deployed at locations s_i, $i = 1, \ldots, N$. The strength of the signal S emitted by the target decays as a power k of the distance. Energy measurements at a sensor are usually corrupted by noise. If N_i denotes the noise energy at sensor i during a particular measurement, then the total energy measured by sensor i at location s_i when the target is at location u is given by

$$E_i(u) = S_i(u) + N_i = \frac{K}{||u - s_i||^k} + N_i. \qquad (17.1)$$

where K is the energy emitted by the target and $||u - s_i||$ is the geometric distance between the target and the sensor. Depending on the environment the value k typically ranges from 2.0 to 5.0 [11].

Although any noise distribution can be chosen in this model, simulations performed in this chapter makes the conservative assumption that the signals measured by the sensors are corrupted by Gaussian noise. Therefore, the noise energy has a Gaussian square distribution or chi-square distribution of degree 1.

17.3 DETECTION ALGORITHMS

Using the energy model described in the previous section, the detection probability of targets can be derived for a given detection algorithm used by the sensors. This section develops two detection algorithms, namely value and decision fusion, and evaluates their performance.

17.3.1 ALGORITHMS

The algorithms considered are non parametric detection algorithms that let the nodes share their information and use a fusion rule to arrive at a decision. Different fusion algorithms can be derived by varying the size of the information shared between sensor nodes. Two cases are explored: 1) *value fusion* where the nodes exchange their raw energy measurements and 2) *decision fusion* where the nodes exchange local detection decisions based on their energy measurement [6]. Both algorithms are described in Figure 17.1.

17.3.2 ANALYTICAL EVALUATION OF THE ALGORITHMS PERFORMANCE

False alarm and detection probability are now derived for value and decision fusion. The false alarm probability is the probability to conclude that a target is present given that the target is absent. The detection probability is the probability to conclude that a target is present given that a target is actually present. False alarm and detection probability are determined by the threshold

Value Fusion	Decision Fusion
at each node: 1. obtain energy from every node 2. compute average of values 3. compare average to threshold for final decision	*at each node:* 1. obtain local decision from every node 2. compute average of local decisions 3. compare average to threshold for final decision

Figure 17.1. Value and decision fusion algorithms

defined in the algorithms, the noise level, the target position and the sensor deployment.

Let N be the total number of sensors, η_v and η_d be the thresholds for value and decision fusion, α be the second threshold for decision fusion, and $FA_{v,d}$ and $D_{v,d}$ denote the false alarm and detection probabilities of value and decision fusion. $f_X(x)$ is the probability density function (pdf) of noise energy that is assume to be χ_1^2 (chi-square with one degree of freedom) [10]. $F_X(x)$ is the cumulative distribution function (cdf) of noise ($F_X(x) = P[X \le x]$). Furthermore, the noise at different sensors is assumed to be independent.

Value fusion. False alarms occur when the average of the N values measured by the sensors is above the threshold η_v in the absence of target. The measured values contain only noise, and the false alarm probability is given by:

$$FA_v = P\left[\frac{1}{N}\sum_{i=1}^{N} N_i > \eta_v\right] \quad (17.2)$$

$\sum_{i=1}^{N} N_i$ is chi-square noise (χ_N^2) with N degrees of freedom with a cumulative distribution function $F_{\chi_N^2}(x)$. Therefore equation 17.2 becomes:

$$FA_v = 1 - F_{\chi_N^2}(N\eta_v) \quad (17.3)$$

Detections occur when the average of the N values measured by the sensors is above the threshold η_v in the presence of a target. The values measured consist of energy (function of the distance from the target to the sensor) plus noise, and the detection probability for a given position of target u is given by:

$$D_v(u) = P\left[\frac{1}{N}\sum_{i=1}^{N}(E_i(u) + N_i) > \eta_v\right] \quad (17.4)$$

For varying position of target, the average detection probability is given by:

$$D_v = \left\langle 1 - F_{\chi_N^2}\left(N\eta_v - \sum_{i=1}^{N} E_i(u)\right)\right\rangle \quad (17.5)$$

where $\langle f(u) \rangle$ denotes the average of f over different positions u of the target in the region considered.

Decision fusion. For decision fusion, false alarms occur when more than αN sensors have a value above the threshold η_d in the absence of target, where α is the threshold used in step 3 of the decision fusion algorithm. The probability that i sensors have a value above η_d is $(1 - F_X(\eta_d))^i$ and the probability that the remaining $N - i$ sensors have a value below η_d is $F_X(\eta_d)^{N-i}$. Since there are $\binom{N}{i}$ ways of choosing the i sensors among N sensors, and i can vary from $\lceil \alpha N \rceil$ to N for a false alarm to occur, the false alarm probability is given by the following equation:

$$FA_d = \sum_{i=\lceil \alpha N \rceil}^{N} \binom{N}{i} F_X(\eta_d)^{N-i}(1 - F_X(\eta_d))^i \quad (17.6)$$

Detections occur when $i \geq \lceil \alpha N \rceil$ sensors have a value above the threshold η_d in the presence of a target. For a given set of detecting sensors defined by the permutation h (such that the set $\{h(j), 1 \leq j \leq i\}$ are the indices of detecting sensors), the probability of detection is $\prod_{j=1}^{i}(1 - F_X(\eta_d - E_{h(j)}(u))) \prod_{j=i+1}^{N} F_X(\eta_d - E_{h(j)}(u))$ (see [5] for details). The detection probability for a given position of target is the sum of these terms for different combinations h and different number of detecting sensors (from $\lceil \alpha N \rceil$ to N). The detection probability is the average of this expression over different position u of the target in the region:

$$D_d = \left\langle \sum_{i=\lceil \alpha N \rceil}^{N} \sum_{h \in C_{i,N}} \left[\prod_{j=1}^{i}(1 - F_X(\eta_d - E_{h(j)}(u))) \prod_{j=i+1}^{N} F_X(\eta_d - E_{h(j)}(u))\right]\right\rangle$$

(17.7)

17.4 EXPOSURE

This section proposes a probabilistic definition of exposure to evaluate the quality of sensor deployments. A general formulation is first developed. Then, several problems are formulated and solved to derive the minimum exposure of sensor deployments for the various target activities, variable speed targets, and sensor fields with obstacles.

17.4.1 DEFINITION

The notion of path exposure introduced in the Section 17.2 measures the ease of detecting a target traveling along a given path, which is referred as "detecting the path". Therefore, the minimum exposure when considering all possible paths in the region is a measure of the quality of a deployment. The probabilistic exposure is now defined to directly measure the likelihood of detecting a path.

Let P denote a path in the sensor field. A target that follows P is detected if and only if it is detected at any time while it is on that path. The *exposure* of path P is defined as the net probability of detecting a target that travels along the path P. First, only the west-east unauthorized traversal problem is addressed. The target is assumed to be crossing the region from west to east following any path through the sensor field. The problem is to find the path P with the least exposure.

17.4.2 SOLUTION TO THE WEST-EAST UNAUTHORIZED TRAVERSING PROBLEM

Let P denote a path from the west to the east periphery through the sensor field. Since detection attempts by the sensor network occur at a fixed frequency, each detection attempt can be easily associated with a point $u \in P$ when assuming that the target traverses the field at a constant speed. The detection attempts are based on energy measured over a period of time T during which the target is moving. Therefore, the detection probability associated with each point u reflects the measurements performed during time T. Considering the path, the net probability of not detecting a target traversing the field using P is the product of the probabilities of not detecting it at any point $u \in P$. These "point detection" probabilities depend on the algorithm used for distributed detection. Algorithms presented in Section 17.3, namely value or decision fusion can be used, and the corresponding detection probabilities when a target is at position u are denoted as $D_v(u)$ and $D_d(u)$ respectively. Closed form expressions for $D_v(u)$ and $D_d(u)$ can be found in Section 17.3. To account for the movement of the target, the energy used to compute $D_v(u(i))$ and $D_d(u(i))$ is the average energy measured as the target travels from $u(i-1)$ to $u(i)$. Note that other algorithms for collaborative target detection can be used by the sensors. Finding the exposure of a path requires knowing the detection probability $D(u)$ when the sensors collaborate to detect a target located at a given point u in the region. If $G(P)$ denotes the net probability of not detecting a target as it traverses through path P, then,

$$\log G(P) = \sum_{u \in P} \log(1 - D(u)),$$

Since the exposure of P is $(1 - G(P))$, the problem is to find the path which minimizes $(1 - G(P))$ or equivalently the path that minimizes $|\log G(P)|$ (Note that, $G(P)$ lies between 0 and 1 and thus $\log G(P)$ is negative).

In general, the path P that minimizes $|\log G(P)|$ can be of fairly arbitrary form. The proposed solution does not exactly compute this path. Instead, it relies on the following approximation: the sensor field is divided into a fine grid and the target is assumed to only move along this grid. The problem then is to find the path P on this grid that minimizes $|\log G(P)|$. Clearly, the finer the grid the closer the approximation.

For the target not to be detected at any point $u \in P$, it must not be detected at any point u between two adjacent grid points of P. Therefore, path P is divided as a chain of grid segments. Let v_1 and v_2 be two adjacent points on the grid. Let l denote the line segment between v_1 and v_2. Also, let m_l denote the probability of not detecting a target traveling between v_1 and v_2 on the line segment l. Then, from the discussion above,

$$\log m_l = \sum_{u \in l} \log(1 - D(u)), \qquad (17.8)$$

where $D(u)$ is either $D_v(u)$ or $D_d(u)$ depending on whether the sensors are using value or decision fusion. The probability m_l can be evaluated by finding the detection probability $D(u)$ at each point $u \in l$. Note that, m_l lies between 0 and 1 and, therefore, $\log m_l$ is negative.

To find the least exposed path, a non-negative weight equal to $|\log m_l|$ is assigned to each segment l on this grid. Also, a fictitious point a is created and a line segment is added from a to each grid point on the west periphery of the sensor field. A weight of 0 is assigned to each of these line segments. Similarly, a fictitious point b is created and a line segment is added from b to each grid point on the east periphery of the sensor field and a weight of 0 is assigned to each of these line segments.

The problem of finding the least exposed path from west periphery to east periphery is then equivalent to the problem of finding the least weight path from a to b on this grid. Such a path can be efficiently determined using the Dijkstra's shortest path algorithm [2]. A pseudo-code of the overall algorithm is shown in Figure 17.2.

Example: Figure 17.3 shows a sensor field with eight sensors at locations marked by dark circles. Assume the noise process at each sensor is Additive White Gaussian with mean 0 and variance 1. Further assume that the sensors use value fusion to arrive at a consensus decision. Then, from Equation 17.3, a threshold $\eta = 3.0$ is chosen to achieve a false alarm probability of 0.187%. The field has been divided into a 10×10 grid. The target emits an energy $K = 12$ and the energy decay factor is $k = 2$. The figure shows the weight assigned to each line segment in the grid as described above. The least exposure path

384   WIRELESS SENSOR NETWORKS

1. **Generate** a suitably fine rectangular grid.
2. **For each** line segment l between adjacent grid points
3.   **Compute** $|\log m_l|$ using Equation 17.8
4.   **Assign** l a weight equal to $|\log m_l|$
5. **Endfor**
6. **Add** a link from virtual point a to each grid point on the west
7. **Add** a link from virtual point b to each grid point on the east
8. **Assign** a weight of 0 to all the line segments from a and b
9. **Find** the least weight path from a to b using Dijkstra's algorithm
10. **Let** w equal the total weight of P.
11. **Return** P as the least exposure path with an exposure of 10^{-w}.

Figure 17.2. Pseudo-code of the proposed solution for the UT problem.

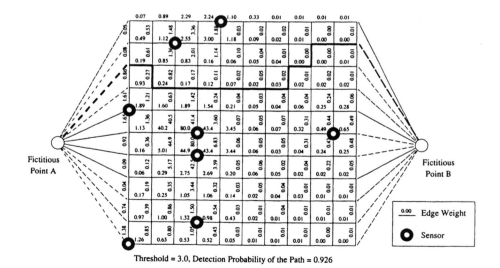

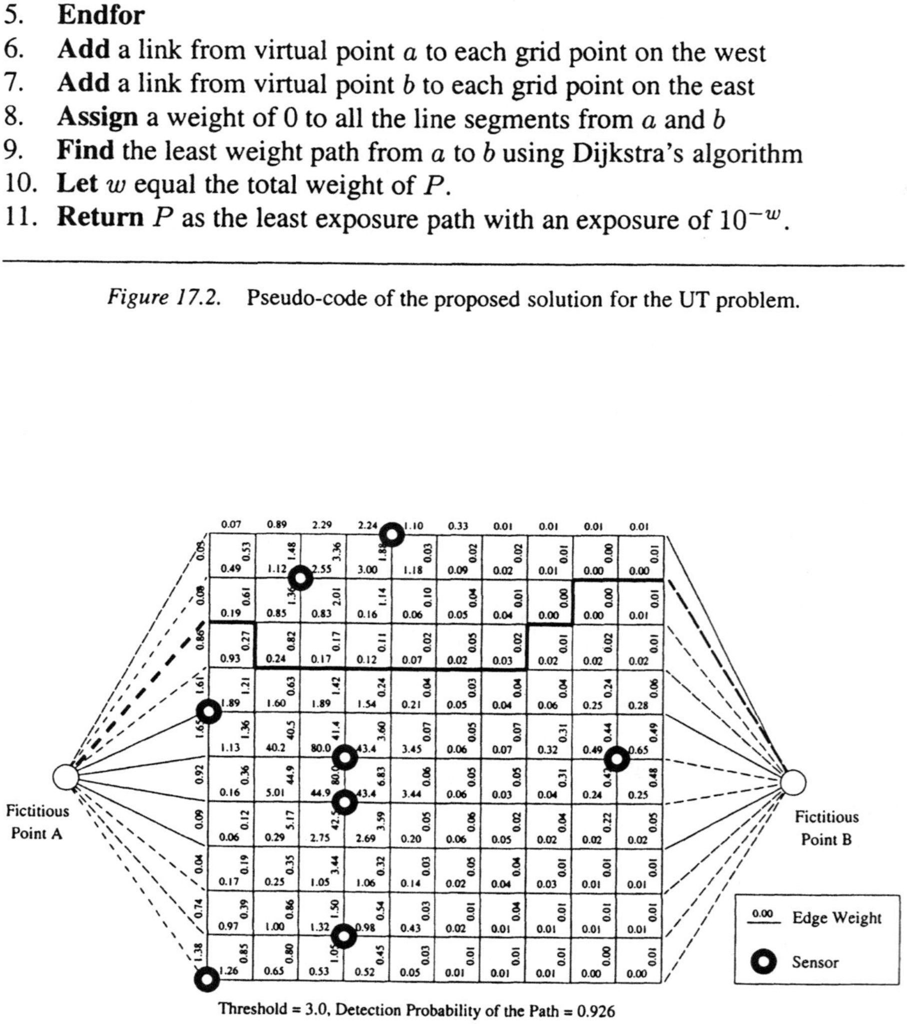

Figure 17.3. Illustration of the proposed solution for an example UT problem.

found by the Dijkstra's algorithm for this weighted grid is highlighted. The probability of detecting the target traversing the field using the highlighted path is 0.926.

Comparison of intensity and probabilistic exposure. The probabilistic exposure defined here is similar to the intensity exposure defined in [13]. In general, the higher the energy measured by the sensors the higher the probability of detecting the target. Therefore, the two metrics measure the same effect of energy decreasing with distance from the target. However, a given total target signal energy measured by the sensors as the target travels along a path can correspond to different probabilities of detecting the path depending on the energy profile along the path. If one can derive the probability of detection of targets, the probabilistic exposure has the advantage to directly measure the performance of the sensor deployment.

Nevertheless, according to the method for finding least exposed paths developed in this section, probabilistic exposure remains limited in its accuracy. Indeed, the method presented in this section relies on the assumption that paths lie on a square grid and the grid size has to be a multiple of the distance covered by a target in T seconds. The method for finding intensity exposure developed in [13] proposes to use higher order grids instead of the rectangular grid used here to improve the accuracy of the measure. However such grids cannot be used to derive probabilistic exposure because the grid size is not constant and it is in general not a multiple of the distance covered by a target in T seconds assuming targets travels at a constant speed. So far, the efficient method for determining minimum exposure across a region developed in this section can only consider coarse approximations of the possible paths because of the incompatibility of finer grids.

17.4.3 EXPOSURE OF DIFFERENT TARGET ACTIVITIES

In the previous subsection, the problem of finding the minimum exposure of targets traversing the region from west to east is solved. Here, other target activities are considered.

Unauthorized traversing problem. The problem is to find the minimum exposure of targets that are expected to traverse the sensor field from one side to the opposite side. Two cases need to be considered, namely west to east traversing, which is equivalent to traversing from east to west, and north to south traversing, which is equivalent to traversing from south to north.

Solution. A method to solve the west-east unauthorized traversal problem was proposed in the previous subsection. A similar method can be applied to find the least exposed path from north periphery to the south periphery of the sensor field by connecting fictitious points to the north and south peripheries. Also, if either of these activities is expected, the minimum exposure is determined by finding the minimum of the two exposures.

Unauthorized reaching problem. The problem is to find the minimum exposure of targets that are expected to enter the sensor field through any point of the periphery, reach a specific point of interest in the region and then leave the sensor field through any point of the periphery.

Solution. The least exposed path to reach the point of interest B from the periphery can be found by connecting a fictitious point A to all the grid points of the periphery with a segment weight of 0 and use Dijkstra's shortest path algorithm to find the least weighted path from A to B. That path only accounts for half of the target activity since the target has to leave the region. However, the least exposed path for leaving the region is the same as the least exposed path to enter the region. If the exposure of the path to reach the point of interest is e_{in}, then the total path exposure is given by $1 - (1 - e_{in})^2$.

Unauthorized idling problem. For unauthorized idling activity, the target is expected to remain at a fixed point u in the sensor field and observe a point of interest for a given duration of $T_i * T$ so that it is detectable for T_i attempts by the sensors. Assuming a sensing range model, the target is assumed to be able to observe the point of interest only if it lies within a given distance d_i to the point of interest.

Solution. The minimum exposure in this scenario is the minimum of $1 - (1 - D(u))^{T_i}$ when considering all the grid point positions u that are less than d_i away from the point of interest.

17.4.4 ALGORITHM COMPLEXITY

The actual implementation of the algorithm of Figure 17.2 can be tailored to reduce the complexity of finding the least exposed path. Considering computation cost, computing the weight of each line segment and finding the least exposed path using Dijkstra's algorithm are the most expensive operations of the algorithm. Computing the weight for an edge in the graph requires finding the distance from the edge to every sensor. Actually, the energy associated with several points lying on the edge is calculated and averaged to estimate the energy measured by the sensors as a target travels along the edge. To account for the presence of obstacles, it must be checked for every point if an obstacle lies between the point and the sensor considered (as will be explained in Section 17.4.6). Dijkstra's algorithm is an efficient minimum weight path search algorithm with complexity $O(m \log n)$ where m is the number of edges and n the number of vertices in the graph.

When specifying a source vertex v, Dijkstra's algorithm finds the minimum distance from every other vertex in the graph. When searching for the least exposed path traversing a region from west to east, Dijkstra's algorithm can be stopped as soon as a vertex on the east side of the region is reached. Therefore, not all the vertices need to be visited by the algorithm to find the minimum weight path in general. A consequence of this observation is that Dijkstra's algorithm does not need to know all the segment weights initially to find the shortest path between two points. Therefore, the computation cost can be reduced by computing the segment weights dynamically, as needed by the Dijkstra's algorithm. Simulations show that 50% to 80% of the weights need to be computed when finding the least exposed path for traversing activity, while 20% to 45% of the weights need to be computed for reaching activity. These results vary with the number of sensors deployed, the energy level of the signal emitted by the target and the size of the region.

17.4.5 EXPOSURE OF VARIABLE SPEED TARGETS

The method proposed above to solve the unauthorized traversing problem assumed that targets cross the region at a constant speed. That assumption is not valid in many applications, for example when the targets are vehicles driving on a road. Below, a method for determining the minimum exposure of variable speed targets is developed.

Problem formulation. Assume that the target can travel with variable speed v, $v_{min} \leq v \leq v_{max}$. Traveling at a faster speed reduces the amount of time the target spends crossing the region and therefore reduces the number of times the sensors have to detect the target. If the energy emitted by the target is independent of the velocity of the target, it can be argued that the least exposed path is obtained when the target travels at maximum speed. However, the energy emitted by the target is expected to be an increasing function of its speed. Thus, as the energy emitted by the target increases, the probability that the sensors detect the target increases and the exposure of the target increases. Therefore, there is a tradeoff between detectability and speed when considering the speed of the target. This tradeoff depends on the energy model as a function of the speed and the sensors topology. In general, the speed along the least exposed path is not expected to be constant.

Multiple speed solution. In order to find the least exposed path across the region, the variable speed problem is transformed into a multiple speed problem. The target is assumed to be traveling at discrete speeds v_i, $i = 1, \cdots, I$. Note that this approximation can be made arbitrarily fine by increasing the number of speeds, I, to achieve desired accuracy. For speed v_i, the region can be divided into a grid with grid size $g_i = v_i.T$ so that each segment corresponds to a detection attempt by the sensors. The solution consists in constructing a graph with all vertices on a fine grid and the edges connecting vertices that are $g_i, 1 \leq i \leq I$, apart from each other. Let the speeds considered be $v_i = p_i.v_0$ where p_i are integers and v_0 is a unit speed. Note that this is possible in general by making v_0 small and v_0 defines the granularity of speeds for all practical purposes. Let p_0 be the greatest common divisor of $p_i, 1 \leq i \leq I$ and let the grid size be $g = p_i.v_0 * T$, so that all the points of the grid can be reached (except certain points close to the region boundaries). Each grid point is connected to other grid points that are p_i away and each edge is given a weight of value $|\log(1 - D(u))|$ where the detection probability $D(u)$ depends on the energy emitted by the target on the segment, and therefore on the target speed. Note that it is assumed that, during a period T, the target travels in one direction only (east, west, north or south). This assumption simplifies the problem by reducing the number of edges in the graph. Once the graph is constructed, the least exposed path is found using the same approach as in the constant speed problem. The higher the number of speeds, I, considered, the higher the complexity of building the graph and finding the shortest path.

Demonstration. A simulator implementing the algorithm described in the previous section was developed to solve the problem for variable number of sensors, speeds and target energies.

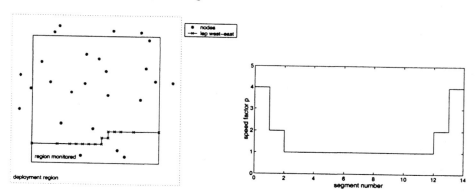

Figure 17.4. Shape and speed profile of the least exposed path (lep) through a region with 25 sensors and for speeds $p_1 = 1, p_2 = 2$ and $p_3 = 4$.

The simulator assumes that value fusion is used for collaboration among the sensors and the threshold of value fusion can be chosen to obtain a desired false alarm probability for each detection attempt [7]. For the simulation performed, it was assumed that the detection attempts occur every $T = 2$ seconds and the false alarm probability is chosen so that the expected number of false alarms is one per hour. The energy emitted by the target, i.e. coefficient K in Equation 17.1, was assumed proportional to the square root of the speed. This assumption used in the experiment can be easily replaced if one knows the characteristics of the target that is expected to cross the region. An example of least exposed path through a region of size 20×20 with 25 sensors deployed is shown in Figure 17.4. Both the shape and the speed profile of the least exposed path are presented. In that example, the target is assumed to have three different possible speeds, $v_1 = 1.v_0$, $v_2 = 2.v_0$ and $v_3 = 4.v_0$, and the energy parameters are set to $K = 20$ (maximum energy) and $k = 2$ (decay coefficient). The minimum exposure was found to be 36%. The example in Figure 17.4 exhibits a general trend of least exposed paths that was observed throughout simulations: the least exposed path corresponds to high speeds when the target crosses parts of the region with low sensor density and low speed when it crosses parts of the region with high sensor density. The proximity of sensors to the path make high speed targets (i.e. high energy) easy to detect.

To compare the multi-speed results with constant speed results, the minimum exposure was found for the same sensor deployment as in Figure 17.4 but when assuming the target can only travel at a constant speed $v_1 = v_0$, or $v_2 = 2.v_0$ or $v_3 = 4.v_0$. The minimum exposures found are respectively 38%, 49% and 64%. It is expected that the minimum exposure for constant speed targets is greater than for multiple speeds because constant speed paths are a special case of the multiple speed problem. Another observation made during the simulation is that the least exposed path can be substantially different in shape when considering variable or constant speed. Figure 17.5 shows the least exposed path for a target traveling at constant speed $v_2 = 2.v_0$ which is different from the least exposed path for variable speed shown in Figure 17.4.

17.4.6 MINIMUM EXPOSURE IN THE PRESENCE OF OBSTACLES

Problem formulation in the presence of obstacles. The presence of obstacles in the sensor field impacts the minimum exposure in two ways. First, the obstacles modify the propagation of the signal emitted by the target and this is modeled by a coefficient α_i in the equation for the energy measured by each node ($S_i(u) = \frac{K*\alpha_i(u)}{||u-s_i||^k}$). Second, the obstacles are assumed to completely obstruct the motion of the target, i.e., paths intersecting the obstacles need not be considered. The coefficients α_i are dependent on the nature, size and position of the obstacles in the sensor field. It is assumed that an array is given that specifies the values of α_i for each sensor and for every possible position of the target. It is also assumed that the position and size of the obstacles is given so that only paths not intersecting the obstacles are considered.

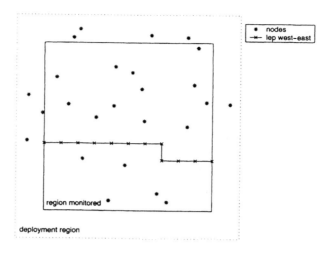

Figure 17.5. Least exposed path (lep) for constant speed $p = 2$.

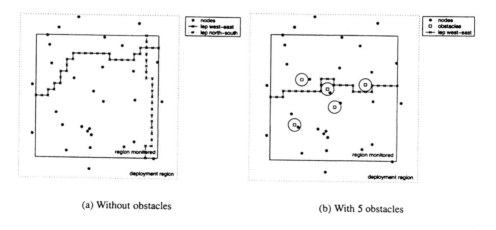

(a) Without obstacles (b) With 5 obstacles

Figure 17.6. Least exposed paths (lep) for traversing a region with 30 nodes. With no obstacles east-west minimum exposure=.87, north-south minimum exposure = 0.66. With obstacles east-west minimum exposure=0.84.

The solution consists of modifying the energy measured by the sensor with the obstacle coefficient α_i and ignoring the paths that cross over obstacles. The general methods developed to find the least exposed path for each target activity remain unchanged.

Figure 17.6(a) shows a region with 30 nodes and with no obstacles. The least exposed paths for east-west and north-south target traversals are shown.

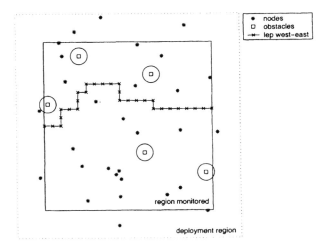

Figure 17.7. Least exposed paths (lep) for traversing the region horizontally (minimum exposure=.93) with 30 nodes deployed and 5 obstacles.

Figure 17.6(b) shows the results for the same deployment but in the presence of 5 obstacles. Obstacles are assumed to have a circular shape of radius 1 and the obstacle coefficients α_i associated with a node at position s_i when a target is present at grid point u is set as follows: $\alpha_i = 0$ whenever the line (s_i, u) intersects with the obstacle, $\alpha_i = 1$ whenever the distance between the line (s_i, u) and the obstacle is more than twice the radius of the obstacle, and $0 < \alpha_i < 1$ and grows linearly with the distance between the line (s_i, u) and the obstacle. Comparing Figures 17.6(a) and 17.6(b) shows that the least exposed path for east-west traversal changes when obstacles are present in the region. In this example, the minimum exposure decreases from .87 to .84 due to the fact that some path segments are hidden from some sensors. It is also possible that the minimum exposure increases when obstacles are added in the region as shown in Figure 17.7. This is due to extra constraints added by the presence of obstacles that prevent targets from using some paths across the region since targets are assumed not to be able to cross over the obstacles. In the example of Figure 17.7, the least exposed path for east-west traversal identified in Figure 17.6(a) is not a valid path in the presence of an obstacle on that path. As a result, the exposure of the least exposed path increases from .87 to .93 for the same node deployment.

17.5 DEPLOYMENT

In this section, the problem of sensor deployment for unauthorized traversal detection is formulated and solutions are identified.

17.5.1 PROBLEM FORMULATION

Consider a region to be monitored for unauthorized traversal using a sensor network. The energy level K emitted by a target of interest and the noise statistics in the region are known. The sensors are to be deployed over the region in a random fashion where the sensors locations in the region cannot be determined a priori and only the number or density of sensors can be chosen. The problem is to find a deployment strategy that results in a desired performance level in unauthorized traversal monitoring of the region.

The performance is measured by the false alarm probability and the path exposure defined in Section 17.4. The false alarm probability is assumed to be fixed in this study so that the problem consists of maximizing the exposure at constant false alarm rate. Since targets can traverse the region through any path, the goal of deployment is to maximize the exposure of the least exposed path in the region.

In most cases, the minimum exposure in the region increases (if false alarm rate is kept constant) as more sensors are deployed in the region [5]. However, since the deployment is random, there are no guarantees that the desired exposure level is achieved with a given number of sensors. A study of the statistical distribution of exposure for varying sensor placement for a given number of sensors can provide a confidence level that the desired detection level is achieved. In practical situations, only a limited number of sensors are available for deployment and only a limited detection level with associated confidence level is achievable for a fixed false alarm rate.

17.5.2 SOLUTION

Based on the above discussion, a solution method is developed to solve the deployment problem when a maximum of M sensors can be used. Deploying the M sensors results in the maximum achievable detection level but this is not optimal when considering the cost of sensors. To reduce the number of sensors deployed, only part of the available sensors should be deployed first and the sensors can then report their position. The random sensor placement obtained can be analyzed to determine if it satisfies the desired performance level. If it does not, additional sensors can be deployed until the desired exposure level is reached or until all M available sensors are deployed.

The number of sensors used in this strategy can be minimized by deploying one sensor at a time. However, a cost is usually associated with each deployment of sensors and deploying one sensor at a time may not be most cost effective if the cost of deployment is sufficiently large with respect to the cost of single sensors. By assigning distinct costs to both single sensors and deployment, the optimal number of sensors to be deployed at first and thereafter can be determined. Below, analytical expressions are developed for finding the optimal solution. In general, the optimal cost solution is neither deploying one sensor at a time nor deploying all the sensors at once.

17.5.3 ANALYTICAL SOLUTION

Let e_d be the desired minimum exposure for the sensor network to be deployed when a maximum of M sensors are available for deployment. The position of sensors are random in the region of interest R and for a given number of sensors n, the least exposure e is a random variable. Let $F_n(x)$ denote the cumulative distribution function of e, i.e. the probability that e is less than x, when n sensors are deployed.

As the minimum exposure e is a random variable, the cost of deploying the sensors in steps until the desired exposure is reached is also a random variable C. The expression for the expected value of C is now derived. To evaluate the probability that the exposure e_d is reached after additional sensor deployment, the following approximation is made: the distribution of exposure for n sensors is independent of the exposure corresponding to k of these n sensors, $1 \leq k \leq n-1$. Let n_i be the total number of sensors deployed after step i. Let S be the maximum number of steps so that $n_S = M$. Note that $n_i - n_{i-1}$ is the number of sensors deployed at step i. Also let C_d be the cost of deploying the sensors at each step and C_s be the cost of each sensor. If the desired exposure is obtained after the first step, the total cost of deployment is $C_d + n_1 C_s$, and this event happens with probability $1 - F_{n_1}(e_d)$. Considering all the possible events, the expected cost is given by

$$E\{C\} = \sum_{i=1}^{S-1} (i.C_d + n_i.C_s) \left(\prod_{j=1}^{i-1} F_{n_j}(e_d) \right) (1 - F_{n_i}(e_d)) + (S.C_d + M.C_s) \prod_{j=1}^{S-1} F_{n_j}(e_d)$$
(17.9)

Note that a different expression is needed for the cost of step S since no additional sensors are deployed after this step even when the desired exposure is not achieved.

17.5.4 SIMULATION

This section presents results of simulations that were performed to collect the exposure distribution function of the number of sensors deployed.

Method. The exposure distribution is obtained by collecting statistics on the exposure when deploying sensors randomly in a predefined region. The region monitored is of size 20×20 and the random deployment is assumed to be Gaussian, centered at the center of the region monitored and with standard de-

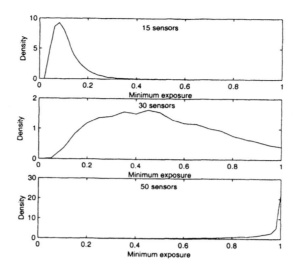

Figure 17.8. Probability density function for the distribution of minimum exposure for deployments of 15, 30 and 50 sensors.

viation 10. In the simulation conducted, the deployment region has dimensions 33% larger than the monitored region, i.e. size 26.6 × 26.6.

For every deployment, the minimum exposure is found using a simulator implementing the algorithm presented in Section 17.4. A decay factor of $k = 2$ and maximum energy of $K = 30$ are chosen to model the energy emitted by targets (cf Equation 17.1). The sensors use value fusion to collaborate when making a common decision on the presence of a target in the region. The noise $\sum_{i=1}^{N} N_i^2$ in the sum of energies from N sensors, computed in value fusion and appearing in Equation 17.4, is from a Chi-square distribution with N degrees of freedom. The threshold for detection is chosen as a function of the number of sensors to give a constant false alarm probability. The false alarm probability for each detection attempt is chosen so that the expected number of false alarms is one per hour, assuming that detection attempts occur every 2 seconds.

Distribution of minimum exposure. The distribution of minimum exposure were found for the number of sensor deployed varying from 1 to 100. To illustrate the results, the probability density functions for 15, 30 and 50 sensors are shown in Figure 17.8.

For 15 sensors deployed, the minimum exposure has zero density for values less than the false alarm probability of .02. The highest density is obtained for values around .08 and then drops exponentially towards zero for higher values of exposure. For deployment of 30 sensors, the minimum exposure has zero density for values below .02, then increases and has the shape of a bell curve

Detecting Unauthorized Activities Using A Sensor Network 395

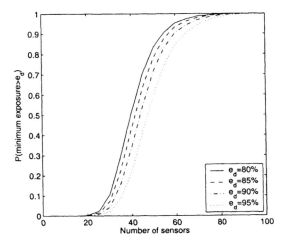

Figure 17.9. Probability that the minimum exposure is above e_d for varying number of sensors and e_d=80%,85%,90% and 95%.

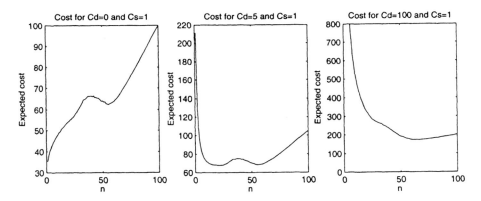

Figure 17.10. Expected cost of achieving minimum exposure of 90% as function of the number of sensors for three different cost assignments.

centered around .4. For deployment of 50 sensors, densities start at zero for small values and remain very small for most values of minimum exposure. The density slowly increases and has a large peak for minimum exposure of 1.

As expected, the minimum exposure increases on average as the number of sensors deployed increases. When randomly deploying 15 sensors, it is very unlikely to obtain a placement providing a desirable minimum exposure. When deploying 30 sensors, most of the exposure levels are equally likely and only poor confidence is given to obtain a desirable exposure level. When deploying 50 sensors, it is very likely that the sensor placement will give good exposure and this likelihood keeps increasing with the number of sensors deployed.

The cumulative distribution function obtained from the statistics collected was used to evaluate the likelihood that the desired level of exposure e_d is obtained for varying number of sensors. The graph of Figure 17.9 shows the probability that the minimum exposure is above e_d as a function of the number of sensors deployed for $e_d = 80\%, 85\%, 90\%$ and 95%. These values can be used to evaluate the cost expressed in Equation 17.9. The graph shows that the confidence level to obtain a given minimum exposure level e_d increases with the number of sensors deployed. The confidence for e_d when deploying 100 sensors is above .999, which is sufficient for most applications, and therefore the distribution of exposure was not evaluated when deploying more than 100 sensors.

17.5.5 RESULTS

The expected cost of deploying sensors is now evaluated using the simulation results and assuming that the number of sensors deployed at every step is constant so that $n_i - n_{i-1} = n$ for all $1 \leq i \leq S$. The expected cost as a function of n was evaluated for three different cost assignments with a desired exposure of $e_d = 90\%$. The three corresponding graphs are shown in Figure 17.10. The first cost assignment is $(C_d = 0, C_s = 1)$ so that the expected cost is the expected number of sensors to be used to achieve an exposure of 90%. Since $C_d = 0$, the number of steps used to deploy the sensors doesn't affect the cost and it is therefore optimal to deploy one sensor at a time until the minimum exposure e_d is reached, as shown on the graph. Overall, the expected number of sensor to be deployed increases with n but there is a local minimum for $n = 55$ that can be explained by the following analysis. The expected number of sensors is a weighted sum of $i.n, 1 \leq i \leq S$ that are the different number of sensors than can be deployed at a time when deploying n sensors at each step. For n around 50, the probability that the minimum exposure is above e_d varies a lot as shown in Figure 17.9 and the weight associated with the first term of the sum (n) increases rapidly while the weights associated with higher number of sensors decrease. This is the cause of the local minimum and the cost starts to increase again when the increase in n compensates for the decrease in weights. In other words, the probability to achieve the desired exposure is much higher when deploying 55 sensors randomly than when deploying 40 sensors randomly. Therefore, it is less costly to deploy 55 sensors at every step since one deployment is likely to be sufficient whereas two or more deployments, and thus a total of 80 or more sensors, are most likely to be needed when deploying 40 sensors at every step.

The second cost assignment is $(C_d = 5, C_s = 1)$ so that the cost of a deployment is equal to the cost of five sensors (note that only the relative cost of C_d/C_s determines the shape of the graphs). In this case, deploying one sensor at a time is prohibited by the cost of deployment and the optimal number of sensors to deploy at every step is 22. Again, the curve presents a local minimum for $n = 55$ that is due to the variations in weights. The last cost assignment is $(C_d = 100, C_s = 1)$ and the minimum cost is achieved when deploying 65 sensors at every step. These results are specific to the region and the parameters characterizing the signal emitted by the target that were chosen for the simulation. Similar results can be derived for other parameters, most of the effort residing in finding the exposure distributions through simulation.

17.6 CONCLUSION

This chapter addressed the problem of sensor deployment in a region to be monitored for target intrusion. Several models for target detection were presented with corresponding problem formulations. For a model considering target energy, algorithms for sensor collaboration to perform target detection were proposed. False alarm and detection probability of these algorithms were derived. The chapter then proposed a probabilistic definition of path exposure that measures the likelihood for a target to be detected as it travels along a path. The minimum exposure is used as a measure of the quality of deployment to detect a given target activity, the goal being to maximize the exposure of the least exposed path in the region.

A Method was developed to evaluate the exposure of paths in the region. Also, algorithms for determining the least exposed path in a region are proposed for different target activities, targets with variable speed and when obstacles are present in the monitored region. In the case where sensors are randomly placed in a region to be monitored, a mechanism for sequential deployment in steps is developed. The strategy consists of deploying a limited number of sensors at a time until the desired minimum exposure is achieved. The cost function used in this study depends on the number of sensors deployed in each step and the cost of each deployment. Through simulation, the distribution of minimum exposure obtained by random deployment was evaluated for varying number of sensors deployed. These results were used to evaluate the cost of deployment for varying number of sensors deployed in each step.

REFERENCES

[1] I. Akyildiz, W. Su, Y. Sankarasubramaniam, and E. Cayirci, "A survey on sensor networks," *IEEE Communications Magazine*, vol. 40, no. 8, pp. 102–114, August 2002.

[2] S. Baase and A. V. Gelder, *Computer algorithms: Introduction to design and analysis*, Addison-Wesley, 2000.

[3] R. R. Brooks and S. S. Iyengar, *Multi-Sensor Fusion: Fundamentals and Applications with Software*, Prentice Hall, 1998.

[4] K. Chakrabarty, S. S. Iyengar, H. Qi, and E. Cho, "Grid coverage for surveillance and target location in distributed sensor networks," *IEEE Transactions on Computers*, vol. 51, no. 12,

December 2002.

[5] T. Clouqueur, "Deployment of fault-tolerant sensor networks for target detection," *PhD thesis, Department of Electrical and Computer Engineering, University of Wisconsin-Madison*, 2003.

[6] T. Clouqueur, P. Ramanathan, K. K. Saluja, and K.-C. Wang, "Value-Fusion versus Decision-Fusion for Fault-tolerance in Collaborative Target Detection in Sensor Networks," in *Proceedings of the 4th Ann. Conf. on Information Fusion*, pp. TuC2/25–TuC2/30, August 2001.

[7] T. Clouqueur, K. K. Saluja, and P. Ramanathan, "Fault tolerance in collaborative sensor networks for target detection," *IEEE Transaction on Computers*, 2003.

[8] S. S. Dhillon and K. Chakrabarty, "A fault-tolerant approach to sensor deployment in distributed sensor networks," in *Proc. Army Science Conference*, 2002.

[9] S. S. Dhillon, K. Chakrabarty, and S. S. Iyengar, "Sensor placement for effective grid coverage and surveillance," in *Workshop on Signal Processing, Communications, Chaos and Systems, Newport, RI*, 2002.

[10] G. Grimmett and D. Stirzaker, *Probability and Random Processes*, Oxford Science Publications, 1992.

[11] M. Hata, "Empirical formula for propagation loss in land mobile radio services," *IEEE Transactions on Vehicular Technology*, vol. 29, pp. 317–325, August 1980.

[12] S. Meguerdichian, F. Koushanfar, M. Potkonjak, and M. Srivastava, "Coverage problems in wireless ad-hoc sensor networks," in *Proceedings of INFOCOM*, pp. 1380–1387, April 2001.

[13] S. Meguerdichian, F. Koushanfar, G. Qu, and M. Potkonjak, "Exposure in wireless ad-hoc sensor networks," in *Proceedings of MOBICOM*, pp. 139–150, July 2001.

[14] P. Rentala, R. Musunnuri, S. Gandham, and U. Saxena, "Survey on sensor networks," website http://citeseer.nj.nec.com/479874.html.

[15] Sensor Information Technology Website, http://www.darpa.mil/ito/research/sensit/index.html.

[16] P. Varshney, *Distributed Detection and Data Fusion*, Springer-Verlag New-York, 1996.

[17] Y. Zou and K. Chakrabarty, "A practical sensor deployment strategy based on virtual forces," in *Proceedings of INFOCOM*, volume 2, April 2003.

Chapter 18

ANALYSIS OF WIRELESS SENSOR NETWORKS FOR HABITAT MONITORING

Joseph Polastre[1], Robert Szewczyk[1], Alan Mainwaring[2], David Culler[1,2] and John Anderson[3]

[1] *University of California at Berkeley*
Computer Science Department
Berkeley, CA 94720
{ polastre,szewczyk,culler } @cs.berkeley.edu

[2] *Intel Research Lab, Berkeley*
2150 Shattuck Ave. Suite 1300
Berkeley, CA 94704
{ amm,dculler } @intel-research.net

[3] *College of the Atlantic*
105 Eden Street
Bar Harbor, ME 04609
jga@ecology.coa.edu

Abstract We provide an in-depth study of applying wireless sensor networks (WSNs) to real-world habitat monitoring. A set of system design requirements were developed that cover the hardware design of the nodes, the sensor network software, protective enclosures, and system architecture to meet the requirements of biologists. In the summer of 2002, 43 nodes were deployed on a small island off the coast of Maine streaming useful live data onto the web. Although researchers anticipate some challenges arising in real-world deployments of WSNs, many problems can only be discovered through experience. We present a set of experiences from a four month long deployment on a remote island. We analyze the environmental and node health data to evaluate system performance. The close integration of WSNs with their environment provides environmental data at densities previously impossible. We show that the sensor data is also useful for predicting system operation and network failures. Based on over one million data readings, we analyze the node and network design and develop network reliability profiles and failure models.

Keywords: Wireless Sensor Networks, Habitat Monitoring, Microclimate Monitoring, Network Architecture, Long-Lived Systems

18.1 Introduction

The emergence of wireless sensor networks has enabled new classes of applications that benefit a large number of fields. Wireless sensor networks have been used for fine-grain distributed control [27], inventory and supply-chain management [25], and environmental and habitat monitoring [22].

Habitat and environmental monitoring represent a class of sensor network applications with enormous potential benefits for scientific communities. Instrumenting the environment with numerous networked sensors can enable long-term data collection at scales and resolutions that are difficult, if not impossible, to obtain otherwise. A sensor's intimate connection with its immediate physical environment allows each sensor to provide localized measurements and detailed information that is hard to obtain through traditional instrumentation. The integration of local processing and storage allows sensor nodes to perform complex filtering and triggering functions, as well as to apply application-specific or sensor-specific aggregation, filtering, and compression algorithms. The ability to communicate not only allows sensor data and control information to be communicated across the network of nodes, but nodes to cooperate in performing more complex tasks. Many sensor network services are useful for habitat monitoring: localization [4], tracking [7, 18, 20], data aggregation [13, 19, 21], and, of course, energy-efficient multihop routing [9, 17, 32]. Ultimately the data collected needs to be meaningful to disciplinary scientists, so sensor design [24] and in-the-field calibration systems are crucial [5, 31]. Since such applications need to run unattended, diagnostic and monitoring tools are essential [33].

In order to deploy dense wireless sensor networks capable of recording, storing, and transmitting microhabitat data, a complete system composed of communication protocols, sampling mechanisms, and power management must be developed. We let the application drive the system design agenda. Taking this approach separates actual problems from potential ones, and relevant issues from irrelevant ones from a biological perspective. The application-driven context helps to differentiate problems with simple, concrete solutions from open research areas.

Our goal is to develop an effective sensor network architecture for the domain of monitoring applications, not just one particular instance. Collaboration with scientists in other fields helps to define the broader application space, as well as specific application requirements, allows field testing of experimental systems, and offers objective evaluations of sensor network technologies. The impact of sensor networks for habitat and environmental monitoring will be measured by their ability to enable new applications and produce new, otherwise unattainable, results.

Few studies have been performed using wireless sensor networks in long-term field applications. During the summer of 2002, we deployed an outdoor habitat monitoring application that ran unattended for four months. Outdoor applications present an additional set of challenges not seen in indoor experiments. While we made many simplifying assumptions and engineered out the need for many complex services, we were able to collect a large set of environmental and node diagnostic data. Even though the collected data was not high enough quality to make scientific conclusions, the fidelity of the sensor data yields important observations about sensor network behavior. The data analysis discussed in this paper yields many insights applicable to most wireless sensor deployments. We examine traditional quality of service metrics such as packet loss; however, the sensor data combined with network metrics provide a deeper understanding of failure modes including those caused by the sensor node's close integration with its environment. We anticipate that with system evolution comes higher fidelity sensor readings that will give researchers an even better better understanding of sensor network behavior.

In the following sections, we explain the need for wireless sensor networks for habitat monitoring in Section 18.2. The network architecture for data flow in a habitat monitoring deployment is presented in Section 18.3. We describe the WSN application in Section 18.4 and analyze the network behaviors deduced from sensor data on a network and per-node level in Section 18.5. Section 18.6 contains related work and Section 18.7 concludes.

18.2 Habitat Monitoring

Many research groups have proposed using WSNs for habitat and microclimate monitoring. Although there are many interesting research problems in sensor networks, computer scientists must work closely with biologists to create a system that produces useful data while leveraging sensor network research for robustness and predictable operation. In this section, we examine the biological need for sensor networks and the requirements that sensor networks must meet to collect useful data for life scientists.

18.2.1 The Case For Wireless Sensor Networks

Life scientists are interested in attaining data about an environment with high fidelity. They typically use sensors on probes and instrument as much of the area of interest as possible; however, densely instrumenting any area is expensive and involves a maze of cables. Examples of areas life scientists currently monitor are redwood canopies in forests, vineyard microclimates, climate and occupancy patterns of seabirds, and animal tracking. With these applications in mind, we examine the current modes of sensing and introduce wireless sensor networks as a new method for obtaining environmental and habitat data at scales and resolutions that were previously impractical.

Traditional data loggers for habitat monitoring are typically large in size and expensive. They require that intrusive probes be placed in the area of interest and the corresponding recording and analysis equipment immediately adjacent. Life scientists typically use these data loggers since they are commercially available, supported, and provide a variety of sensors. Probes included with data loggers may create a "shadow effect"–a situation that occurs when an organism alters its behavioral patterns due to an interference in their space or lifestyle [23]. Instead, biologists argue for the miniaturization of devices that may be deployed on the surface, in burrows, or in trees. Since interference is such a large concern, the sensors must be inconspicuous. They should not disrupt the natural processes or behaviors under study [6]. One such data logger is the Hobo Data Logger [24] from Onset Corporation. Due to size, price, and organism disturbance, using these systems for fine-grained habitat monitoring is inappropriate.

Other habitat monitoring studies install one or a few sophisticated weather stations an "insignificant distance" (as much as tens of meters) from the area of interest. A major concern with this method is that biologists cannot gauge whether the weather station actually monitors a different microclimate due to its distance from the organism under study [12]. Using these readings, biologists make generalizations through coarse measurements and sparsely deployed weather stations. A revolution for biologists would be the ability to monitor the environment on the scale of the organism, not on the scale of the biologist [28].

Life scientists are increasingly concerned about the potential impacts of direct human interaction in monitoring plants and animals in field studies. Disturbance effects are of particular concern in a small island situation where it may be physically impossible for researchers to avoid impacting an entire population. Islands often serve as refugia for species that cannot adapt to the presence of terrestrial mammals. In Maine, biologists have shown that even a 15 minute visit to a seabird colony can result in up to 20% mortality among eggs and chicks in a given breeding year [2]. If the disturbance is repeated, the entire colony may abandon their breeding site. On Kent Island, Nova Scotia, researchers found that nesting Leach's Storm Petrels are likely to abandon their burrows if disturbed during their first 2 weeks of incubation. Additionally, the hatching success of petrel eggs was reduced by 56% due to investigator disturbance compared to a control group that was not disturbed for the duration of their breeding cycle [3].

Sensor networks represent a significant advance over traditional, invasive methods of monitoring. Small nodes can be deployed prior to the sensitive period (*e.g.*, breeding season for animals, plant dormancy period, or when the ground is frozen for botanical studies). WSNs may be deployed on small islets where it would be unsafe or unwise to repeatedly attempt field studies. A key difference between wireless sensor networks and traditional probes or data loggers is that WSNs permit real-time data access without repeated visits to sensitive habitats. Probes provide real-time data, but require the researcher to be present on-site, while data in data loggers is not accessible until the logger is collected at a point in the future.

Deploying sensor networks is a substantially more economical method for conducting long-term studies than traditional, personnel-rich methods. It is also more economical than installing many large data loggers. Currently, field studies require substantial maintenance in terms of logistics and infrastructure. Since sensors can be deployed and left, the logistics are reduced to initial placement and occasional servicing. Wireless sensor network may organize themselves, store data that may be later retrieved, and notify operates that the network needs servicing. Sensor networks may greatly increase access to a wider array of study sites that are often limited by concerns about disturbance or lack easy access for researchers.

18.2.2 Great Duck Island

Great Duck Island (GDI), located at (44.09N, 68.15W), is a 237 acre island located 15 km south of Mount Desert Island, Maine. The Nature Conservancy, the State of Maine, and the College of the Atlantic (COA) hold much of the island in joint tenancy. Great Duck contains approximately 5000 pairs of Leach's Storm Petrels, nesting in discrete "patches" within the three major habitat types (spruce forest, meadow, and mixed forest edge) found on the island [1]. COA has ongoing field research programs on several remote islands with well established on-site infrastructure and logistical support. Seabird researchers at COA study the Leach's Storm Petrel on GDI. They are interested in four major questions [26]:

1 What is the usage pattern of nesting burrows over the 24-72 hour cycle when one or both members of a breeding pair may alternate incubation duties with feeding at sea?

2 What environmental changes occur inside the burrow and on the surface during the course of the seven month breeding season (April-October)?

3 What is the variation across petrel breeding sites? Which of these conditions yield an optimal microclimate for breeding, incubation, and hatching?

4 What are the differences in the micro-environments between areas that contain large numbers of nesting petrels and those areas that do not?

Petrels nest in underground burrows that provide relatively constant conditions in terms of temperature and humidity. Burrows are usually within 2–6 cm of the surface and range from 40 cm to over one meter in length with an internal diameter of approximately 6 cm. One sensor node per burrow is sufficient for data sampling but it must be small enough in size such that the sensor and petrel can coexist without interfering with the petrel's activities and does not obstruct the passage. Burrows occur in discrete "patches" around the island that may be hundreds of meters from locations that can support network and power infrastructure. Each patch may contain over 50 burrows; a large number of these burrows should be populated with sensors. Some should be left unpopulated to evaluate if there are disturbance effects caused by wireless sensors. Sensors should cover as many petrel burrows as possible.

Above ground, the environmental conditions vary widely, depending on vegetation type, density, exposure, and location. Humidity readings at a given point in time will vary with vegetation type; a forested area will have higher humidity due to moisture retained by trees and an open meadow will have lower humidity due to direct sunlight and evaporation. Monitoring the environment above each burrow, biologists can examine differences between the above-ground and in-burrow microclimates. Variations in local microclimates may provide clues to nest site selection and overall fitness.

Overall, the petrel cycle lasts approximately 5 months [16]. The deployed system must efficiently manage its power consumption through low duty cycle operation in order to operate for an entire field season. In order to adequately monitor the habitat, it must be monitored on the spatial scale of the organism at frequencies that match environmental changes and organism behavior. By increasing the size of the area monitored and the number of sampling locations, we can obtain data at resolutions and densities not possible using traditional methods. Temporally, sensors should collect data at a rate equal or greater to changing environmental conditions that the organism experiences (on the order of 5-10 times per hour). Traditional data collection systems calculate the average, minimum, and maximum over 24 hour periods as well as time series data. This methodology runs the risk of missing significant but short-term variations in conditions and presents an overly uniform view of micro-environmental conditions. Data analysis must be able to capture duration of events in addition to environmental changes.

It is unlikely that any one parameter or sensor reading could determine why petrels choose a specific nesting site. Predictive models will require multiple measurements of many variables or sensors. These models can then be used to determine which conditions seabirds prefer. To link organism behavior with environmental conditions, sensors may monitor micro-environmental conditions (temperature, humidity, pressure, and light levels) and burrow occupancy (by detecting infrared radiation) [22].

Finally, sensor networks should be run alongside traditional methods to validate and build confidence in the data. The sensors should operate reliably and predictably.

18.3   Network Architecture

In order to deploy a network that satisfies the requirements of Section 18.2, we developed a system architecture for habitat monitoring applications. Here, we describe the architecture, the functionality of individual components, and the interoperability between components.

The system architecture for habitat monitoring applications is a tiered architecture. Samples originate at the lowest level that consists of *sensor nodes*. These nodes perform general purpose computing and networking in addition to application-specific sensing. Sensor nodes will typically be deployed in dense *sensor patches* that are widely separated. Each patch encompasses a particular geographic region of interest. The sensor nodes transmit their data through the sensor patch to the sensor network *gateway*. The gateway is responsible for transmitting sensor data from the *sensor patch* through a local transit network to the remote *base station* that provides WAN connectivity and data logging. The base station connects to database replicas across the Internet. Finally, the data is displayed to scientists through any number of user interfaces. Mobile devices may interact with any of the networks–whether it is used in the field or across the world connected to a database replica. The full architecture is depicted in Figure 18.1.

Sensor nodes are small, battery-powered devices are placed in areas of interest. Each sensor node collects environmental data about its immediate surroundings. The sensor node computational module is a programmable unit that provides computation, storage, and bidirectional communication with other nodes in the system. It interfaces with analog and digital sensors on the sensor module, performs basic signal processing (*e.g.,* simple translations based on calibration data or threshold filters), and dispatches the data according to the application's needs. Compared with traditional data logging systems, it offers two major advantages: it can *communicate* with the rest of the system in real time and can be *retasked* in the field. WSNs may coordinate to deliver data and be reprogrammed with new functionality.

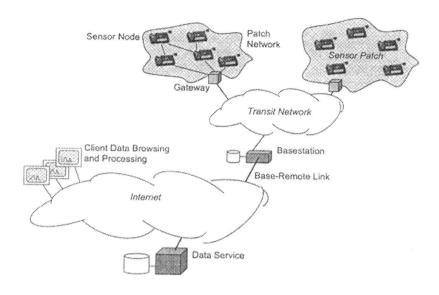

Figure 18.1. System architecture for habitat monitoring

Individual sensor nodes communicate and coordinate with one another in the same geographic region. These nodes make up a *sensor patch*. The sensor patches are typically small in size (tens of meters in diameter); in our application they correspond to petrel nesting patches.

Using a multi-tiered network is particularly advantageous since each habitat involves monitoring several particularly interesting areas, each with its own dedicated sensor patch. Each sensor patch is equipped with a *gateway*. The gateway provides a bridge that connects the sensor network to the base station through a transit network. Since each gateway may include more infrastructure (*e.g.,* solar panels, energy harvesting, and large capacity batteries), it enables deployment of small devices with low capacity batteries. By relying on the gateway, sensor nodes may extend their lifetime through extremely low duty cycles. In addition to providing connectivity to the base station, the gateway may coordinate the activity within the sensor patch or provide additional computation and storage. In our application, a single repeater node served as the transit network gateway. It retransmitted messages to the base station using a high gain Yagi antenna over a 350 meter link. The repeater node ran at a 100% duty cycle powered by a solar cell and rechargeable battery.

Ultimately, data from each sensor needs to be propagated to the Internet. The propagated data may be raw, filtered, or processed. Bringing direct wide area connectivity to each sensor patch is not feasible–the equipment is too costly, it requires too much power and the installation of all required equipment is quite intrusive to the habitat. Instead, the wide area connectivity is brought to a *base station*, where adequate power and housing for the equipment is provided. The base station communicates with the sensor patches using the transit network. To provide data to remote end-users, the *base station* includes wide area network (WAN) connectivity and persistent data storage for the collection of sensor patches. Since many habitats of interest are quite remote, we chose to use a two-way satellite connection to connect to the Internet.

Data reporting in our architecture may occur both spatially and temporally. In order to meet the network lifetime requirements, nodes may operate in a phased manner. Nodes primarily sleep; periodically, they wake, sample, perform necessary calculations, and send readings through the network Data may travel spatially through various routes in the sensor patch, transit network, or wide area network; it is then routed over long distances to the wide area infrastructure.

Users interact with the sensor network data in two ways. Remote users access the replica of the base station database (in the degenerate case they interact with the database directly). This approach allows for easy integration with data analysis and mining tools, while masking the potential wide area disconnections with the base stations. Remote control of the network is also provided through the database interface. Although this control interface is is sufficient for remote users, on-site users may often require a more direct interaction with the network. Small, PDA-sized devices enables such interaction. The data store replicates content across the WAN and its efficiency is integral to live data streams and large analyses.

18.4 Application Implementation

In the summer of 2002, we deployed a 43 node sensor network for habitat monitoring on Great Duck Island. Implementation and deployment of an experimental wireless sensor network platform requires engineering the application software, hardware, and electromechanical design. We anticipated contingencies and feasible remedies for the electromechanical, system, and networking issues in the design of the application that are discussed in this section.

18.4.1 Application Software

Our approach was to simplify the system design wherever possible, to minimize engineering and development efforts, to leverage existing sensor network

Figure 18.2 Mica mote (left) with Mica Weather Board sensor board for habitat monitoring includes sensors for light, temperature, humidity, pressure, and infrared radiation.

platforms and components, and to use off-the-shelf products where appropriate to focus attention upon the sensor network itself. We chose to use the Mica mote developed by UC Berkeley [14] running the TinyOS operating system [15].

In order to evaluate a long term deployment of a wireless sensor network, we installed each node with a simple periodic application that meets the biologists requirements defined in Section 18.2. Every 70 seconds, each node sampled each of its sensors and transmitted the data in a single 36 byte data packet. Packets were timestamped with 32-bit sequence numbers kept in flash memory. All motes with sensor boards were transmit-only devices that periodically sampled their sensors, transmitted their readings, and entered their lowest-power state for 70 seconds. We relied on the underlying carrier sense MAC layer protocol in TinyOS to prevent against packet collisions.

18.4.2 Sensor board design

To monitor petrel burrows below ground and the microclimate above the burrow, we designed a specialized sensor board called the Mica Weather Board. Environmental conditions are measured with a photoresistive sensor, digital temperature sensor, capacitive humidity sensor, and digital pressure sensor. To monitor burrow occupancy, we chose a passive infrared detector (thermopile) because of its low power requirements. Since it is passive, it does not interfere with the burrow environment. Although a variety of surface mount and probe-based sensors were available, we decided to use surface mount components because they were smaller and operated at lower voltages. Although a probe-based approach has the potential to allow precise co-location of a sensor with its phenomenon, the probes typically operated at higher voltages and currents than equivalent surface mount parts. Probes are more invasive since they puncture the walls of a burrow. We designed and deployed a sensor board with all of the sensors integrated into a single package. A single package permitted miniaturization of the node to fit in the size-constrained petrel burrow.

Even if this initial generation of devices were larger than a highly engineered, application-specific platform would be, we wanted to push in the direction of miniaturized wireless sensors. All sensors resided on a single sensor board, one per mote. This preserved the form factor of the underlying mote platform and limited the circuit board design and simplified manufacturing. The board includes a separate 12-bit ADC to maximize sensor resolution isolate analog noise, and allow concurrent sensor processing and node operation. One consequence of a single integrated design is the amount of shared fate between sensors; a failure of one sensor is likely affects all other sensors. The design did not consider fault isolation among independent sensors or controlling the effects of malfunctioning sensors on shared hardware resources.

18.4.3 Packaging strategy

The environmental conditions on offshore islands are diverse. In-situ instrumentation experiences rain, often with pH readings of less than 3, dew, dense fog, and flooding. They could experience direct sunlight and extremes of warm and cold temperatures. Waterproofing was a primary packaging concern.

Sealing electronics from the environment could be done with conformal coating, packaging, or combinations of the two. Since our sensors were surface mounted and needed to be exposed to the environment, we sealed the entire mote with a parylene sealant. Through successful tests in our lab, we concluded the mote's electronics could be protected from the environment with sealant. A case provides mechanical protection but would not be required for waterproofing. Our survey of off-the-shelf enclosures found many that were slightly too small for the mote or too large for tunnels. Custom enclosures were too costly. Above ground motes were placed in ventilated acrylic enclosures. In burrows, motes were deployed without enclosures.

Of primary concern for the packaging was the effect it has on RF propagation. We decided to use board-mounted miniature whip antennas. There were significant questions about RF propagation from motes inside burrows, above ground on the surface, within inches of granite rocks, tree roots and low, dense vegetation. When we deployed the motes we noted the ground substrate, distance into the burrow, and geographic location of each mote to assist in the analysis of the RF propagation for each mote.

18.4.4 Experiment goals

Since our deployment was the first long term use of the mote platform, we were interested in how the system would perform. Specifically, this deployment served to prove the feasibility of using a miniature low-power wireless sensor network for long term deployments where robustness and predictable operation are essential. We set out to evaluate the efficacy of the sealant, the

Figure 18.3. Acrylic enclosures used at different outdoor applications.

radio performance in and out of burrows, the usefulness of the data for biologists including the occupancy detector, and the system and network longevity. Since each hardware and software component was relatively simple, our goal was to draw significant conclusions about the behavior of wireless sensor networks from the resulting data.

After 123 days of the experiment, we logged over 1.1 million readings. During this period, we noticed abnormal operation among the node population. Some nodes produced sensor readings out of their operating range, others had erratic packet delivery, and some failed. We sought to understand why these events had occurred. By evaluating these abnormalities, future applications may be designed to isolate problems and provide notifications or perform self-healing. The next section analyzes node operation and identifies the causes of abnormal behavior.

18.5 System Analysis

Before the disciplinary scientists perform the analysis of the sensor data, we need convincing evidence that the sensor network is functioning correctly. We look at the performance of the network as a whole in Section 18.5.1 as well as the failures experienced by individual nodes in Section 18.5.2.

In order to look more closely at the network and node operation, we would like to introduce you to the node community that operated on Great Duck Island. Shown in Table 18.1, each of the nodes is presented with their node ID and lifetime in days. Some of the nodes had their batteries replaced and ran for a second "life". Of importance is that some of the nodes fell victim to raw humidity readings of zero or significant clock skew. The number of days after the first sign of either abnormality is referred to as the amount of time on "death row". We discuss individual nodes highlighted in Table 18.1 throughout our analysis in this section and explain their behavior.

Table 18.1. The node population and their properties. RH indicates whether the node experienced raw relative humidity readings of zero during its lifetime (if RH=1). CS indicates that the node experienced excessive clock skew (if CS=1). After the first sign of abnormal humidity readings or clock skew, the node's remaining lifetime (in days) is given in the "death row" (DR) column. The lifetime (in days) of each node on their first and second set of batteries is listed. The total RH and CS counts how many nodes exhibited those properties. The total DR and lifetime is the average time on "death row" and average lifetime over the entire population. Shaded nodes appear in our analysis and graphs.

Node	RH	CS	DR	Life 1	Life 2	Node	RH	CS	DR	Life 1	Life 2
2	1	1	4	14	-	39	0	0	41	44	-
3	1	1	12	14	-	40	1	0	6	6	-
4	1	1	2	2	-	41	1	1	60	67	-
5	1	0	1	13	-	42	0	1	1	6	-
9	1	1	12	12	-	43	1	1	11	12	-
10	1	0	1	1	-	44	1	1	1	1	-
12	0	0	0	25	-	45	0	0	11	13	-
13	0	0	0	31	40	46	1	1	7	67	-
15	0	0	0	31	40	47	0	0	0	16	-
16	1	0	1	1	-	48	1	1	12	16	-
17	0	1	1	27	-	49	1	1	1	1	-
18	0	1	6	44	-	50	1	1	8	8	-
19	1	1	6	2	-	51	1	1	2	2	-
22	1	0	1	1	-	52	1	1	5	6	-
24	1	0	1	14	35	53	0	0	2	8	-
25	1	1	1	1	-	54	1	1	2	4	-
26	0	0	6	6	-	55	0	0	1	54	-
29	0	1	0	56	66	57	0	1	0	67	-
30	0	1	0	51	28	58	1	1	6	6	-
32	1	1	1	44	-	59	1	1	2	2	-
35	0	0	0	54	33	90	1	1	1	1	-
38	0	0	0	35	-	Total	26	26	5.5	20.7	-

18.5.1 Network Analysis

We need evaluate the behavior of the sensor network to establish convincing evidence that the system is operating correctly. Our application was implemented as a single hop network, however the behavior in a single hop is equivalent to what occurs in any WSN radio cell. We begin by examining WSN operation and its performance over time in order to evaluate network cell characteristics.

Packet loss. A primary metric of network performance is packet loss in the network over time. Packet loss is a quality-of-service metric that indicates the effective end-to-end application throughput [8]. The average daily packet delivery is shown in Figure 18.4. Two features of the packet delivery plot de-

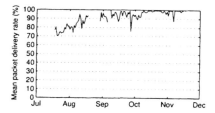 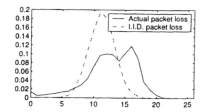

Figure 18.4. Average daily packet delivery in the network throughout the deployment. The gap in the second part of August corresponds to a database crash.

Figure 18.5. Distribution of packet losses in a time slot. Statistically, the losses are not independently distributed.

mand explanation: (1) why was the initial loss rate high and (2) why does the network improve with time? Note that the size of the sensor network is declining over time due to node failures. Either motes with poor packet reception die quicker or the radio channel experiences less contention and packet collisions as the number of nodes decreases. To identify the cause, we examine whether a packet loss at a particular node is dependent on losses from other nodes.

The periodic nature of the application allows us to assign virtual time slots to each data packet corresponding with a particular sequence number from each node. After splitting the data into time slices, we can analyze patterns of loss within each time slot. Figure 18.6 shows packet loss patterns within the network during the first week of August 2002. A black line in a slot indicates that a packet expected to arrive was lost, a white line means a packet was successfully received. If all packet loss was distributed independently, the graph would contain a random placement of black and white bars appearing as a gray square. We note that 23 nodes do not start to transmit until the morning of August 6; that reflects the additional mote deployment that day. Visual inspection reveals patterns of loss: several black horizontal lines emerge, spanning almost all nodes, *e.g.*, midday on August 6, 7, and 8. Looking closer at the packet loss on August 7, we note it is the only time in the sample window when motes 45 and 49 transmit packets successfully; however, heavy packet loss occurs at most other nodes. Sequence numbers received from these sensors reveal they transmitted data during every sample period since they were deployed even though those packets were not received.

More systematically, Figure 18.5 compares the empirical distribution of packet loss in a slot to an independent distribution. The hypothesis that the two distributions are the same is rejected by both parametric (χ^2 test yields 10^8) and non-parametric techniques (rank test rejects it with 99% confidence). The empirical distribution appears a superposition of two Gaussian functions: this is not particularly surprising, since we record packet loss at the end of the

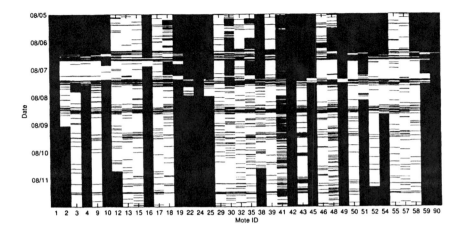

Figure 18.6. Packet loss patterns within the deployed network during a week in August. Y-axis represents time divided into virtual packet slots (note: time increases downwards). A black line in the slot indicates that a packet expected to arrive in this time slot was missed, a white line means that a packet was successfully received.

path (recall network architecture, Section 18.4). This loss is a combination of potential losses along two hops in the network. Additionally, packets share the channel that varies with the environmental conditions, and sensor nodes are likely to have similar battery levels. Finally, there is a possibility of packet collisions at the relay nodes.

Network dynamics. Given that the expected network utilization is very low (less than 5%) we would not expect collisions to play a significant role. Conversely, the behavior of motes 45 and 49 implies otherwise: their packets are only received when most packets from other nodes are lost. Such behavior is possible in a periodic application: in the absence of any backoff, the nodes will collide repeatedly. In our application, the backoff was provided by the CSMA MAC layer. If the MAC worked as expected, each node would backoff until it found a clear slot; at that point, we would expect the channel to be clear. Clock skew and channel variations might force a slot reallocation, but such behavior should be infrequent.

Looking at the timestamps of the received packets, we can compute the phase of each node, relative to the 70 second sampling period. Figure 18.7 plots the phase of selected nodes from Figure 18.6. The slope of the phase corresponds to a drift as a percentage of the 70-second cycle. In the absence of clock drift and MAC delays, each mote would always occupy the same time

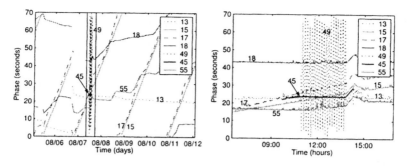

Figure 18.7. Packet phase as a function of time; the right figure shows the detail of the region between the lines in the left figure.

slot cycle and would appear as a horizontal line in the graph. A 5 ppm oscillator drift would result in gently sloped lines, advancing or retreating by 1 second every 2.3 days. In this representation, the potential for collisions exists only at the intersections of the lines.

Several nodes display the expected characteristics: motes 13, 18, and 55 hold their phase fairly constant for different periods, ranging from a few hours to a few days. Other nodes, *e.g.*, 15 and 17 appear to delay the phase, losing 70 seconds every 2 days. The delay can come only from the MAC layer; on average they lose 28 msec, which corresponds to a single packet MAC backoff. We hypothesize that this is a result of the RF automatic gain control circuits: in the RF silence of the island, the node may adjust the gain such that it detects radio noise and interprets it as a packet. Correcting this problem may be done by incorporating a signal strength meter into the MAC that uses a combination of digital radio output and analog signal strength. This additional backoff seems to capture otherwise stable nodes: *e.g.*, mote 55 on August 9 transmits in a fixed phase until it comes close to the phase of 15 and 17. At that point, mote 55 starts backing off before every transmission. This may be caused by implicit synchronization between nodes caused by the transit network.

We note that potential for collisions does exist: the phases of different nodes do cross on several occasions. When the phases collide, the nodes back off as expected, *e.g.*, mote 55 on August 9 backs off to allow 17 to transmit. Next we turn to motes 45 and 49 from Figure 18.6. Mote 45 can collide with motes 13 and 15; collisions with other nodes, on the other hand, seem impossible. In contrast, mote 49, does not display any potential for collisions; instead it shows a very rapid phase change. Such behavior can be explained either though a clock drift, or through the misinterpretation of the carrier sense (*e.g.*, a mote determines it needs to wait a few seconds to acquire a channel). We associate such behavior with faulty nodes, and return to it in Section 18.5.2.

18.5.2 Node Analysis

Nodes in outdoor WSNs are exposed to closely monitor and sense their environment. Their performance and reliability depend on a number of environmental factors. Fortunately, the nodes have a local knowledge of these factors, and they may exploit that knowledge to adjust their operation. Appropriate notifications from the system would allow the end user to pro-actively fix the WSN. Ideally, the network could request proactive maintenance, or self-heal. We examine the link between sensor and node performance. Although the particular analysis is specific to this deployment, we believe that other systems will be benefit from similar analyses: identifying outliers or loss of expected sensing patterns, across time, space or sensing modality. Additionally, since battery state is an important part of a node's self-monitoring capability [33], we also examine battery voltage readings to analyze the performance of our power management implementation.

Sensor analysis. The suite of sensors on each node provided analog light, humidity, digital temperature, pressure, and passive infrared readings. The sensor board used a separate 12-bit ADC to maximize the resolution and minimize analog noise. We examine the readings from each sensor.

Light readings. The light sensor used for this application was a photoresistor that we had significant experience with in the past. It served as a confidence building tool and ADC test. In an outdoor setting during the day, the light value saturated at the maximum ADC value, and at night the values were zero. Knowing the saturation characteristics, not much work was invested in characterizing its response to known intensities of light. The simplicity of this sensor combined with an *a priori* knowledge of the expected response provided a valuable baseline for establishing the proper functioning of the sensor board. As expected, the sensors deployed above ground showed periodic patterns of day and night and burrows showed near to total darkness. Figure 18.8 shows light and temperature readings and average light and temperature readings during the experiment.

The light sensor operated most reliably of the sensors. The only behavior identifiable as failure was disappearance of diurnal patterns replaced by high value readings. Such behavior is observed in 7 nodes out of 43, and in 6 cases it is accompanied by anomalous readings from other sensors, such as a $0°C$ temperature or analog humidity values of zero.

Temperature readings. A Maxim 6633 digital temperature sensor provided the temperature measurements While the sensor's resolution is $0.0625°C$, in our deployment it only provided a $2°C$ resolution: the hardware always supplied readings with the low-order bits zeroed out. The enclosure was IR

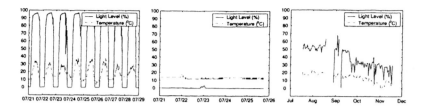

Figure 18.8. Light and temperature time series from the network. From left: outside, inside, and daily average outside burrows.

transparent to assist the thermopile sensor; consequently, the IR radiation from direct sunlight would enter the enclosure and heat up the mote. As a result, temperatures measured inside the enclosures were significantly higher than the ambient temperatures measured by traditional weather stations. On cloudy days the temperature readings corresponded closely with the data from nearby weather buoys operated by NOAA.

Even though motes were coated with parylene, sensor elements were left exposed to the environment to preserve their sensing ability. In the case of the temperature sensor, a layer of parylene was permissible. Nevertheless the sensor failed when it came in direct contact with water. The failure manifested itself in a persistent reading of $0^{\circ}C$. Of 43 nodes, 22 recorded a faulty temperature reading and 14 of those recorded their first bad reading during storms on August 6. The failure of temperature sensor is highly correlated with the failure of the humidity sensor: of 22 failure events, in two cases the humidity sensor failed first and in two cases the temperature sensor failed first. In remaining 18 cases, the two sensors failed simultaneously. In all but two cases, the sensor did not recover.

Humidity readings. The relative humidity sensor was a capacitive sensor: its capacitance was proportional to the humidity. In the packaging process, the sensor needed to be exposed; it was masked out during the parylene sealing process, and we relied on the enclosure to provide adequate air circulation while keeping the sensor dry. Our measurements have shown up to 15% error in the interchangeability of this sensor across sensor boards. Tests in a controlled environment have shown the sensor produces readings with 5% variation due to analog noise. Prior to deployment, we did not perform individual calibration; instead we applied the reference conversion function to convert the readings into SI units.

In the field, the protection afforded by our enclosure proved to be inadequate. When wet, the sensor would create a low-resistance path between the power supply terminals. Such behavior would manifest itself in either abnor-

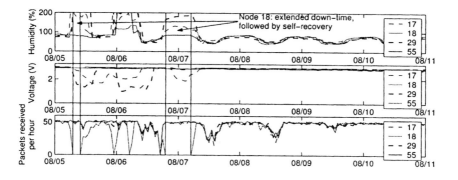

Figure 18.9. Sensor behavior during the rain. Nodes 17 and 29 experience substantial drop in voltage, while node 55 crashes. When the humidity sensor recovers, the nodes recover.

mally large (more than 150%) or very small humidity readings (raw readings of 0V). Figure 18.9 shows the humidity and voltage readings as well as the packet reception rates of selected nodes during both rainy and dry days in early August. Nodes 17 and 29 experienced a large drop in voltage while recording an abnormally high humidity readings on Aug 5 and 6. We attribute the voltage drop to excessive load on the batteries caused by the wet sensor. Node 18 shows an more severe effect of rain: on Aug 5, it crashes just as the other sensors register a rise in the humidity readings. Node 18, on the other hand, seems to be well protected: it registers high humidity readings on Aug 6, and its voltage and packet delivery rates are not correlated with the humidity readings. Nodes that experienced the high humidity readings typically recover when they dried up; nodes with the unusually low readings would fail quickly. While we do not have a definite explanation for such behavior, we evaluate that characteristics of the sensor board as a failure indicator below.

Thermopile readings. The data from the thermopile sensor proved difficult to analyze. The sensor measures two quantities: the ambient temperature and the infrared radiation incident on the element. The sum of thermopile and thermistor readings yields the object surface temperature, *e.g.*, a bird. We would expect that the temperature readings from the thermistor and from the infrared temperature sensor would closely track each other most of the time. By analyzing spikes in the IR readings, we should be able to deduce the bird activity.

The readings from the thermistor do, in fact, track closely with the temperature readings. Figure 18.10 compares the analog thermistor with the digital maxim temperature sensor. The readings are closely correlated although dif-

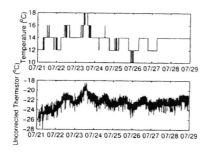
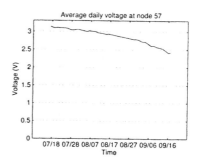

Figure 18.10. The digital temperature sensor (top) and analog thermistor (bottom), though very different on the absolute scale, are closely correlated: a linear fit yields a mean error of less than 0.8°C.

Figure 18.11. Voltage readings from node 57. Node 57 operates until the voltage falls below 2.3V; at this point the alkaline cells can not supply enough current to the boost converter.

ferent on an absolute scale. A best linear fit of the temperature data to the thermistor readings on a per sensor per day basis yields a mean error of less than 0.9°C, within the half step resolution of the digital sensor. The best fit coefficient varies substantially across the nodes.

Assigning biological significance to the infrared data is a difficult task. The absolute readings often do not fall in the expected range. The data exhibits a lack of any periodic daily patterns (assuming that burrow occupancy would exhibit them), and the sensor output appears to settle quickly in one of the two extreme readings. In the absence of any ground truth information, *e.g.,* infrared camera images corresponding to the changes in the IR reading, the data is inconclusive.

Power Management. As mentioned in Section 18.4, one of the main challenges was sensor node power management. We evaluate the power management in the context of the few nodes that did not exhibit other failures. Motes do not have a direct way of measuring the energy they consumed, instead we use battery voltage as an indirect measure. The analysis of the aggregate population is somewhat complicated by in-the-field battery replacements, failed voltage indicators, failed sensors and gaps in the data caused by the database crashes. Only 5 nodes out of 43 have clearly exhausted their original battery supply. This limited sample makes it difficult to perform a thorough statistical analysis. Instead we examine the battery voltage of a single node without other failures. Figure 18.11 shows the battery voltage of a node as a function of time. The batteries are unable to supply enough current to power the node once the voltage drops below 2.30V. The boost converter on the Mica mote is able to extract only 15% more energy from the battery once the voltage drops below 2.5V (the lowest operating voltage for the platform without the voltage regulation). This fell far short of our expectations of being able to drain the batteries down to 1.6V, which represents an extra 40% of energy stored in a cell [10]. The periodic, constant power load presented to the batteries is ill suited to extract the maximum capacity. For this class of devices, a better solution would use batteries with stable voltage, *e.g.,* some of the lithium-based chemistries. We advocate future platforms eliminate the use of a boost converter.

Node failure indicators. In the course of data analysis we have identified a number of anomalous behaviors: erroneous sensor readings and application phase skew. The humidity sensor seemed to be a good indicator of node health. It exhibited 2 kinds of erroneous behaviors: very high and very low readings. The high humidity spikes, even though they drained the mote's batteries, correlated with recoverable mote crashes. The humidity readings corresponding to a raw voltage of 0V correlated with permanent mote outage: 55% of the nodes with excessively low humidity readings failed within two days. In the course of packet phase analysis we noted some motes with slower than usual clocks. This behavior also correlates well with the node failure: 52% of nodes with such behavior fail within two days.

These behaviors have a very low false positive detection rate: only a single node exhibiting the low humidity and two nodes exhibiting clock skew (out of 43) exhausted their battery supply instead of failing prematurely. Figure 18.12 compares the longevity of motes that have exhibited either the clock skew or a faulty humidity sensor against the survival curve of mote population as a whole. We note that 50% of motes with these behaviors become inoperable within 4 days.

18.6 Related Work

As described in Section 18.2, traditional data loggers are typically large and expensive or use intrusive probes. Other habitat monitoring studies install weather stations an "insignificant distance" from the area of interest and make coarse generalizations about the environment. Instead, biologists argue for the miniaturization of devices that may be deployed on the surface, in burrows, or in trees.

Habitat monitoring for WSNs has been studied by a variety of other research groups. Cerpa et. al. [7] propose a multi-tiered architecture for habitat monitoring. The architecture focuses primarily on wildlife tracking instead of habitat monitoring. A PC104 hardware platform was used for the implementation with future work involving porting the software to motes. Experimentation using a hybrid PC104 and mote network has been done to analyze acoustic signals [30], but no long term results or reliability data has been published. Wang

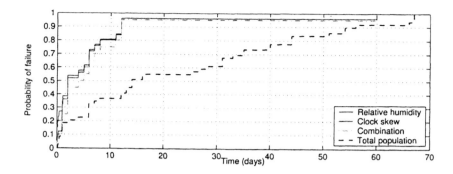

Figure 18.12. Cumulative probability of node failure in the presence of clock skew and anomalous humidity readings compared with the entire population of nodes.

et. al. [29] implement a method to acoustically identify animals using a hybrid iPaq and mote network.

ZebraNet [18] is a wireless sensor network design for monitoring and tracking wildlife. ZebraNet uses nodes significantly larger and heavier than motes. The architecture is designed for an always mobile, multi-hop wireless network. In many respects, this design does not fit with monitoring the Leach's Storm Petrel at static positions (burrows). ZebraNet, at the time of this writing, has not yet had a full long-term deployment so there is currently no thorough analysis of the reliability of their sensor network algorithms and design.

The number of deployed wireless sensor network systems is extremely low. There is very little data about long term behavior of sensor networks, let alone wireless networks used for habitat monitoring. The Center for Embedded Network Sensing (CENS) has deployed their Extensible Sensing System [11] at the James Mountain Reserve in California. Their architecture is similar to ours with a variety of sensor patches connected via a transit network that is tiered. Intel Research has recently deployed a network to monitor Redwood canopies in Northern California and a second network to monitor vineyards in Oregon. Additionally, we have deployed a second generation multihop habitat monitoring network on Great Duck Island, ME. As of this writing, these systems are still in their infancy and data is not yet available for analysis.

18.7 Conclusion

We have presented the need for wireless sensor networks for habitat monitoring, the network architecture for realizing the application, and the sensor network application implementation. We have shown that much care must be taken when deploying a wireless sensor network for prolonged outdoor operation keeping in mind the sensors, packaging, network infrastructure, application software. We have analyzed environmental data from one of the first outdoor deployments of WSNs. While the deployment exhibited very high node failure rates and failed to produce meaningful data for the disciplinary sciences, it yielded valuable insight into WSN operation that could not have been obtained in simulation or in an indoor deployment. We have identified sensor features that predict a 50% node failure within 4 days. We analyzed the application-level data to show complex behaviors in low levels of the system, such as MAC-layer synchronization of nodes.

Sensor networks do not exist in isolation from their environment; they are embedded within it and greatly affected by it. This work shows that the anomalies in sensor readings can be used to predict node failures with high confidence. Prediction enables pro-active maintenance and node self-maintenance. This insight will be very important in the development of self-organizing and self-healing WSNs.

Notes

Data from the wireless sensor network deployment on Great Duck Island can be view graphically at http://www.greatduckisland.net. Our website also includes the raw data for researchers in both computer science and the biological sciences to download and analyze.

This work was supported by the Intel Research Laboratory at Berkeley, DARPA grant F33615-01-C1895 (Network Embedded Systems Technology "NEST"), the National Science Foundation, and the Center for Information Technology Research in the Interest of Society (CITRIS).

References

[1] Julia Ambagis. Census and monitoring techniques for Leach's Storm Petrel (Oceanodroma leucorhoa). Master's thesis, College of the Atlantic, Bar Harbor, ME, USA, 2002.

[2] John G. T. Anderson. Pilot survey of mid-coast maine seabird colonies: An evaluation of techniques. In *Report to the State of Maine Department of Inland Fisheries and Wildlife*, Bangor, ME, USA, 1995.

[3] Alexis L. Blackmer, Joshua T. Ackerman, and Gabrielle A. Nevitta. Effects of investigator disturbance on hatching success and nest-site fidelity in a long-lived seabird, Leach's Storm-Petrel. *Biological Conservation*, 2003.

[4] Nirupama Bulusu, Vladimir Bychkovskiy, Deborah Estrin, and John Heidemann. Scalable, ad hoc deployable, RF-based localization. In *Proceedings of the Grace Hopper Conference on Celebration of Women in Computing*, Vancouver, Canada, October 2002.

[5] Vladimir Bychkovskiy, Seapahn Megerian, Deborah Estrin, and Miodrag Potkonjak. Colibration: A collaborative approach to in-place sensor calibration. In *Proceedings of the 2nd International Workshop on Information Processing in Sensor Networks (IPSN'03)*, Palo Alto, CA, USA, April 2003.

[6] Karen Carney and William Sydeman. A review of human disturbance effects on nesting colonial waterbirds. *Waterbirds*, 22:68–79, 1999.

[7] Alberto Cerpa, Jeremy Elson, Deborah Estrin, Lewis Girod, Michael Hamilton, and Jerry Zhao. Habitat monitoring: Application driver for wireless communications technology. In *2001 ACM SIGCOMM Workshop on Data Communications in Latin America and the Caribbean*, San Jose, Costa Rica, April 2001.

[8] Dan Chalmers and Morris Sloman. A survey of Quality of Service in mobile computing environments. *IEEE Communications Surveys*, 2(2), 1992.

[9] Benjie Chen, Kyle Jamieson, Hari Balakrishnan, and Robert Morris. Span: An energy-efficient coordination algorithm for topology maintenance in ad hoc wireless networks. In *Proceedings of the 7th Annual International Conference on Mobile Computing and Networking*, pages 85–96, Rome, Italy, July 2001. ACM Press.

[10] Eveready Battery Company. Energizer no. x91 datasheet. http://data.energizer.com/datasheets/library/primary/alkaline/energizer_e2/x91.pdf.

[11] Michael Hamilton, Michael Allen, Deborah Estrin, John Rottenberry, Phil Rundel, Mani Srivastava, and Stefan Soatto. Extensible sensing system: An advanced network design for microclimate sensing. http://www.cens.ucla.edu, June 2003.

[12] David C.D. Happold. The subalpine climate at smiggin holes, Kosciusko National Park, Australia, and its influence on the biology of small mammals. *Arctic & Alpine Research*, 30:241–251, 1998.

[13] Tian He, Brian Blum, John Stankovic, and Tarek Abdelzaher. AIDA: Adaptive Application Independant Data Aggregation in Wireless Sensor Networks. *ACM Transactions in Embedded Computing Systems (TECS), Special issue on Dynamically Adaptable Embedded Systems*, 2003.

[14] Jason Hill and David Culler. Mica: a wireless platform for deeply embedded networks. *IEEE Micro*, 22(6):12–24, November/December 2002.

[15] Jason Hill, Robert Szewczyk, Alec Woo, Seth Hollar, David Culler, and Kristofer Pister. System architecture directions for networked sensors. In *Proceedings of the 9th International Conference on Architectural Support for Programming Languages and Operating Systems (ASPLOS-IX)*, pages 93–104, Cambridge, MA, USA, November 2000. ACM Press.

[16] Chuck Huntington, Ron Butler, and Robert Mauck. *Leach's Storm Petrel (Oceanodroma leucorhoa)*, volume 233 of *Birds of North America*. The Academy of Natural Sciences, Philadelphia and the American Orinthologist's Union, Washington D.C., 1996.

[17] Chalermek Intanagonwiwat, Ramesh Govindan, and Deborah Estrin. Directed diffusion: a scalable and robust communication paradigm for sensor networks. In *Proceedings of the 6th Annual International Conference on Mobile Computing and Networking*, pages 56–67, Boston, MA, USA, August 2000. ACM Press.

[18] Philo Juang, Hidekazu Oki, Yong Wang, Margaret Martonosi, Li-Shiuan Peh, and Daniel Rubenstein. Energy-efficient computing for wildlife tracking: Design tradeoffs and early experiences with ZebraNet. In *Proceedings of the 10th International Conference on Architectural Support for Programming Languages and Operating Systems (ASPLOS-X)*, pages 96–107, San Jose, CA, USA, October 2002. ACM Press.

[19] Bhaskar Krishanamachari, Deborah Estrin, and Stephen Wicker. The impact of data aggregation in wireless sensor networks. In *Proceedings of International Workshop of Distributed Event Based Systems (DEBS)*, Vienna, Austria, July 2002.

[20] Jie Liu, Patrick Cheung, Leonidas Guibas, and Feng Zhao. A dual-space approach to tracking and sensor management in wireless sensor networks. In *Proceedings of the 1st ACM International Workshop on Wireless Sensor Networks and Applications*, pages 131–139, Atlanta, GA, USA, September 2002. ACM Press.

[21] Samuel Madden, Michael Franklin, Joseph Hellerstein, and Wei Hong. TAG: a Tiny AGgregation service for ad-hoc sensor networks. In *Proceedings of the 5th USENIX Symposium on Operating Systems Design and Implementation (OSDI '02)*, Boston, MA, USA, December 2002.

[22] Alan Mainwaring, Joseph Polastre, Robert Szewczyk, David Culler, and John Anderson. Wireless sensor networks for habitat monitoring. In *Proceedings of the 1st ACM International Workshop on Wireless Sensor Networks and Applications*, pages 88–97, Atlanta, GA, USA, September 2002. ACM Press.

[23] Ian Nisbet. Disturbance, habituation, and management of waterbird colonies. *Waterbirds*, 23:312–332, 2000.

[24] Onset Computer Corporation. HOBO weather station. http://www.onsetcomp.com.

[25] Kris Pister, Barbara Hohlt, Jaein Jeong, Lance Doherty, and J.P. Vainio. Ivy: A sensor network infrastructure for the University of California, Berkeley College of Engineering. http://www-bsac.eecs.berkeley.edu/projects/ivy/, March 2003.

[26] Joseph Polastre. Design and implementation of wireless sensor networks for habitat monitoring. Master's thesis, University of California at Berkeley, Berkeley, CA, USA, 2003.

[27] Bruno Sinopoli, Cory Sharp, Luca Schenato, Shawn Schaffert, and Shankar Sastry. Distributed control applications within sensor networks. *Proceedings of the IEEE*, 91(8):1235–1246, August 2003.

[28] Marco Toapanta, Joe Funderburk, and Dan Chellemi. Development of Frankliniella species (Thysanoptera: Thripidae) in relation to microclimatic temperatures in vetch. *Journal of Entomological Science*, 36:426–437, 2001.

[29] Hanbiao Wang, Jeremy Elson, Lewis Girod, Deborah Estrin, and Kung Yao. Target classification and localization in habitat monitoring. In *Proceedings of IEEE International Conference on Acoustics, Speech, and Signal Processing (ICASSP 2003)*, Hong Kong, China, April 2003.

[30] Hanbiao Wang, Deborah Estrin, and Lewis Girod. Preprocessing in a tiered sensor network for habitat monitoring. *EURASIP JASP Special Issue on Sensor Networks*, 2003(4):392–401, March 2003.

[31] Kamin Whitehouse and David Culler. Calibration as parameter estimation in sensor networks. In *Proceedings of the 1st ACM International Workshop on Wireless Sensor Networks and Applications*, pages 59–67, Atlanta, GA, USA, September 2002. ACM Press.

[32] Ya Xu, John Heidemann, and Deborah Estrin. Geography-informed energy conservation for ad hoc routing. In *Proceedings of the 7th Annual International Conference on Mobile Computing and Networking*, pages 70–84, Rome, Italy, July 2001. ACM Press.

[33] Jerry Zhao, Ramesh Govindan, and Deborah Estrin. Computing aggregates for monitoring wireless sensor networks. In *Proceedings of the 1st IEEE International Workshop on Sensor Network Protocols and Applications*, Anchorage, AK, USA, May 2003.

Index

Actuated boundary detection, 16
Ad hoc positioning, 143
Ambient conditions, 24
Application areas, 23
APS, 143
ASCENT, 356
Attribute-based naming, 33
Automatic localization, 13
Automatic Repeat reQuest, 40
Beamforming, 334
Congestion Control, 28
Connected sensor cover, 364
Continuous functions, 129
Cost-field based dissemination, 112, 117
Coverage Configuration Protocol, 362
Cramer Rao bound, 340
CSMA-Based Medium Access, 39
Data aggregation, 33, 353
Data gathering, 234
Data-centric routing, 186
Data-centric storage, 188
Data-centric, 33
Datalink Layer, 37
Directed Diffusion, 12
Discrete wavelet transform, 209
DISCUS, 210
Distributed Activation with Pre-determined Routes, 365
Distributed sensing, 8
Distributed source coding, 211
Energy aware, 32
Energy efficient route, 33
Environmental monitoring, 400
Error Control, 39
ESPIHT, 207
ESRT, 30
Event-to-sink reliability, 29
Failure indicators, 419
Fidelity, 352
Flow optimization, 234
Generic Security Model for WSN, 267
Geographic Adaptive Fidelity protocol, 355
Geographic Hash Tables (GHTs), 191
GPSR, 192

Great Duck Island Project, 403
Gur Game, 360
Habitat Monitoring, 400
Hardware design, 408
Hybrid TDMA/FDMA Based Medium Access, 39
In-network processing, 112, 124
Infrared communication, 42
Interest dissemination, 26, 33
Jiggle switch, 332
Key distribution, 277
Key establishment, 277
Key management, 269
Least square estimation, 219
Linear Constant Rate Model, 239
Link reliability, 40
Load balancing, 120, 125
LPS, 144
M-ary modulation scheme, 43
Medium access control, 37, 73, 93
MICA motes, 23
Mobility and security, 307
Modulation scheme, 43
Multi-dimensional range queries, 201
Multi-tiered networks, 406
Multihop communication, 32
Network architecture, 405
Network layer schemes, 36
Network lifetime, 236
Network Time Protocol, 45
Non-linear Adaptive Rate Models, 237
NSSS scheduling, 360
Packaging design, 409
Path exposure, 376
PEAS protocol, 360
Position centric routing, 129
Power management, 418
PSFQ, 31
QoS, 352
Rate-adaptation, 236
Real-time delivery, 125
Received signal strength, 335
Reference time-based scheduling, 361
Reference-Broadcast Synchronization, 45
Reliable Transport, 28

Reverse-path forwarding, 112, 114
Robustness, 118–119
Security models for WSNs, 268
Sensing mode selection, 352, 357
Sensor management, 352
Sensor network characteristics, 22
Sensor query and tasking language (SQTL), 26
Sensor reading calibration, 211
Sensor synchronization, 211
Sift MAC Protocol, 95
Smart Kindergarten, 334, 347
Span protocol, 355
Spatio-temporal correlation, 207
Spatio-temporal queries, 197
STEM protocol, 357
Structured replication, 194

Target detection, 375
TDMA-Based Medium Access, 38
Tiered architecture, 11
Time difference of arrivals, 333
Time synchronization, 44
Time-Diffusion Synchronization Protocol, 45
Timing techniques, 45
TinyOS, 11
Topology control, 352, 354
Trajectory based forwarding, 129
Trajectory, 131
Ultra Wideband, 43, 337
UNPF protocol framework, 99
Virtual hierarchy, 121
Vulnerabilities and threats, 255
Watermarking technique, 306, 310
Wavelet packet decomposition, 211

Printed in the United States
131037LV00002B/166-171/A